DPH MATHEMATICS SERIES

TEXT BOOK OF
3-D
(CO-ORDINATE SYSTEMS AND STRAIGHT LINES)

By
A.K. Sharma

DISCOVERY PUBLISHING HOUSE
NEW DELHI-110002

First Published-2005

ISBN 81-7141-976-3

Published by

DISCOVERY PUBLISHING HOUSE
4831/24, Ansari Road, Prahlad Street,
Darya Ganj, New Delhi-110002 (India)
Phone: 23279245 • Fax: 91-11-23253475
E-mail:dphtemp@indiatimes.com

Printed at:
Arora Offset Press
Laxmi Nagar, Delhi 110 092.

Preface

This book Text Book of Straight Line is intended as an introduction to analytical solid geometry and covers as much of the subject as is generally expected of students going up for the B.A., B.Sc., pass and Honours examination of all Indian Universities.

I have endeavoured to develop the subject in a systematic and logical manner. To help the beginner, elementary parts of the subject have been presented in as simple and lucid a manner as possible and fairly large number of solved examples to illustrate various types have been introduced. The book already existing in the market cover a rather extensive ground and consequently comparative lesser attention is paid to the introductory portion than is necessary for a beginner.

The book contains numerous exercise of varied types in a graded form. Some of these have been selected from various examination papers and standards works to whose publishers and authors, I offer my best thanks.

I shall be very grateful for any suggestions for improvements for correction of text or examples.

A.K. Sharma

Contents

1

Systems of Co-ordinates

INTRODUCTION

In a plane the position of a point is determined by an ordered pair (x, y) of real numbers, obtained with reference to two straight lines in the plane generally at right angles. The position of a point in space is however determined by an ordered triad (x, y, z) of real numbers, we now proceed as to how thus is done.

RECTANGULAR CARTESIAN CO-ORDINATES

This system of co-ordinates is most commonly used in three dimensional analytical geometry.

Draw two mutually perpendicular lines Y'OY and ZOZ' in the plane of the paper intersecting each other at O.

Through O imagine a third line X'OX perpendicular to both the above lines, as shown in the adjoining figure.

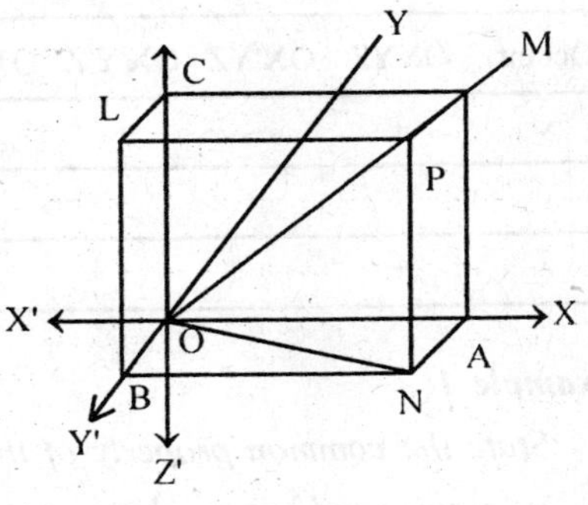

O is called, the origin and the set of these three mutually perpendicular lines X'OX, Y'OY and ZOZ' are called the co-ordinate axes (rectangular). The co-ordinate axes XOX', YOY' and ZOZ' are called x-axis, y-axis and z-axis repectively.

These three axes taken in pairs give us three planes YOZ, ZOX and XOY and they are called yz, zx and xy-planes respectively. These planes are called *co-ordinate planes.*

CO-ORDINATES OF A POINT IN SPACE

Let P be any point in space. Through P pass planes parallel to the three co-ordinate planes and cutting x, y and z axes in A,B and C respectively as shown.

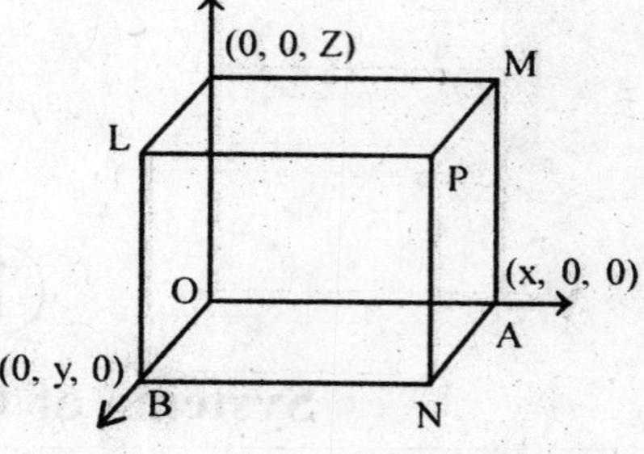

The position of P relative to the co-ordinate system is given by its perpendicular distances from the co-ordinate planes and these distances are given by lengths OA, OB and OC.

Let OA = a, OB = b and OC = c. Then a, b, c are called x-co-ordinate, y-co-ordinate, and z-co-ordinate respectively of the point P. The point P is referred as (a, b, c).

The co-ordinates of the origin O are (0, 0, 0) and those of A, B, C, N, K, and M (a, 0, 0); (0, b, 0); (0, 0, c); (a, b, 0); (0, b, c) and (a, 0, c) respectively.

The three co-ordinate planes divide space into eight parts and these parts are called *octants.* The sign of co-ordinates of a point determine the octant in which it lies.

The signs for the eight octants are given in the tabular form below:

Octant	OXYZ	OX'YZ	OX'Y'Z	OXY'Z	OXY'Z	OX'YZ'	OX'Y'Z'	OXY'Z'
x	+	–	–	+	+	–	–	+
y	+	+	–	–	–	+	–	–
z	+	+	+	+	–	–	–	–

Example 1:

State the common property of the co-ordinates of points lying on :

(i) z-axis, and (ii) xz-plane.

Solution:

(1) The x an y-co-ordinates of all points on the z-axis are zero i.e

$$x = 0 = y.$$

(2) The y-co-ordinates of all points on the xz-plane is zero *i.e.,* y = 0. Hence the required common property is y = 0. **Ans.**

Example 2:

What are the perpendicular distance of the point (x, y, z) from the co-ordinate axes?

Solution:

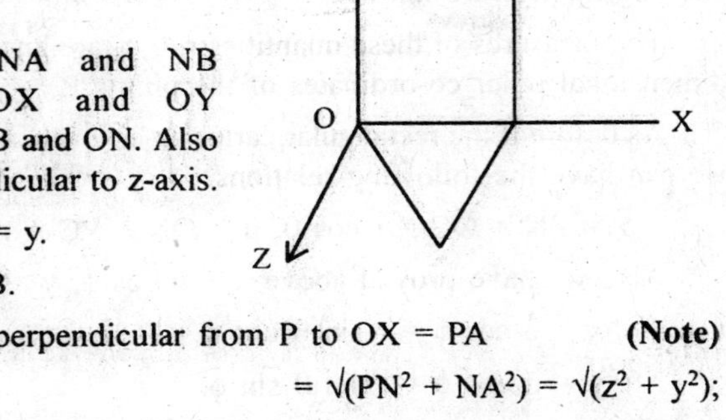

Let P be the point (x, y, z). From P draw PN perpendicular to xy-plane (*i.e.*, XOY plane).

From N draw NA and NB perpendiculars to OX and OY respectively. Join PA, PB and ON. Also from P draw PC perpendicular to z-axis.

Then PN = z, NA = y.

Then OA = x = AB.

Now the length of perpendicular from P to OX = PA **(Note)**

$$= \sqrt{(PN^2 + NA^2)} = \sqrt{(z^2 + y^2)};$$

The length of perpendicular from P to OY = PB **(Note)**

$$= \sqrt{(PN^2 + NB^2)} = \sqrt{(z^2 + x^2)};$$

The length of perpendicular from P to OZ = PC = ON **(Note)**

$$= \sqrt{(OA^2 + NA^2)} = \sqrt{(x^2 + y^2)};$$

OTHER METHODS OF DEFINING THE POSITION OF ANY POINT P IN SPACE

Cylindrical Co-ordinates

If XOX', YOY', ZOZ' are the rectangular axes and PN the perp. form P to the XOY plane the position of the point P can be determined if ON, the angle XON and NP are known. The measures of the quantities u, ϕ and z are the cylindrical co-ordinates of the point P.

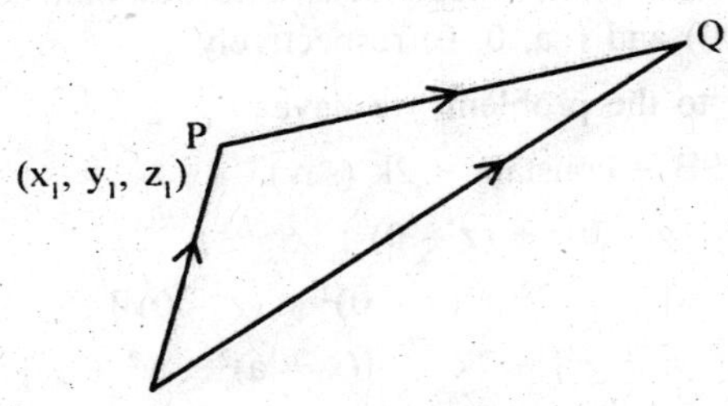

Let the rectangular cartesian co-ordinates of P are (x, y, z) then those of N are (x, y, 0) and we can easily have the following relations $x = u \cos \phi$, $y = u \sin \phi$ *and* $z = z$ hence $u^2 = x^2 + y^2$ and $\phi = \tan^{-1}(y/x)$.

SPHERICAL POLAR CO-ORDINATES

The position of the point P can also be determined when OP, angle ZOP and angle XON are known.

The measures of these quantities r, θ, ϕ are known as spherical or three dimensional polar co-ordinates of the points P.

As before if the rectangular cartesian co-ordinates of P are (x, y, z), then we can have the following relations.

$$z = PN = OC = r \cos \theta;\ u = ON = PC = r \sin \theta.$$

Also we have proved above $x = u \cos \phi$, $y = u \sin \phi$,

$$\therefore\ x = u \cos \phi = r \sin \theta \cos \phi,$$

$$y = u \cos \phi = r \sin \theta \sin \phi.$$

Thus we have $x = r \sin \theta \cos \phi$,

$$y = r \sin \theta \sin \phi$$

and $$z = r \cos \theta.$$

Also $r^2 = OP^2 = PN^2 + ON^2 = z^2 + u^2 = z^2 + (x^2 + y^2) = x^2 + y^2 + z^2$

and $$\tan \theta = \frac{u}{z} = \frac{\sqrt{(x^2 + y^2)}}{z};\ \tan \phi = \frac{y}{x}$$

Example 1:

Find the locus of a point which moves so that the sum of its distance from the points (a, 0, 0) and (–a, 0, 0) is constant.

Solution:

Let P(x, y, z) be the point whose locus is to be obtained Let A and B be the points (a, 0, 0) and (–a, 0, 0) respectively.

Then according to the problem, we have

$$PA + PB = \text{constant} = 2k \text{ (say).}$$ **(Note)**

$$\Rightarrow \sqrt{[(x - a)^2 + (y - 0)^2 + (z - 0)^2]} + \sqrt{[(x + a)^2 + (y - 0)^2 + (z - 0)^2]} = 2k$$

$$\Rightarrow \sqrt{[(x - a)^2 + y^2 + z^2]} = 2k - \sqrt{[(x + a)^2 + y^2 + z^2]}.$$

Squaring both sides, we have

$$(x - a)^2 + y^2 + z^2 = 4k^2 + [(x + a)^2 + y^2 + z^2] - 4k \sqrt{[(x + a)^2 + y^2 + z^2]}$$

$$\Rightarrow 4k\sqrt{[(x + a)^2 + y^2 + z^2]} = 4k^2 + (x + a)^2 - (x - a)^2$$

$$= 4k^2 + 4ax, \text{ on simplyfying}$$

$$\Rightarrow \quad \sqrt{[(x + a)^2 + y^2 + z^2]} = k + (ax/k)$$

Again squaring both sides we have

$$(x + a)^2 + y^2 + z^2 = k^2 + (a^2 x^2/k^2) + 2ax$$

$$\Rightarrow \quad x^2 + y^2 + z^2 + a^2 + k^2 + (a^2 x^2/k^2)$$

$$\Rightarrow \quad x^2 [1 - (a^2/k^2)] + y^2 + z^2 = k^2 - a^2,$$

which is the required locus. **Ans.**

Example 2:

Find the cylindrical co-ordinates of the points whose rectangular co-ordinates are (2, 3, 5), (1, 2, 3) and (–2, 5, 2)

Solution:

Let (u, ϕ, z) be the cylindrical coordinates of the point (2, 3, 5) *i.e.,* the point whose cartesian co-ordinates are $x = 2$, $y = 3$, $z = 5$.

We know that

$$u^2 = x^2 + y^2 = 2^2 + 3^2 = 13 \text{ or } u = \sqrt{(13)}$$

$$\text{And } \tan\phi = \frac{y}{x} = \frac{3}{5} \text{ or } \phi = \tan^{-1}\left(\frac{3}{5}\right)$$

∴ The required cylindrical co-ordinates are given as

$$[\sqrt{(13)}, \tan^{-1}(3/5), 5]$$ **Ans.**

Similarly find cylindrical co-ordinates for other points.

SECTION OF THE JOIN OF TWO POINTS

To find the coordinates of the point which divides the line joining two given points $P(x_1, y_1, z_1)$ and $P(x_2, y_2, z_2)$ in a given ratio m : n.

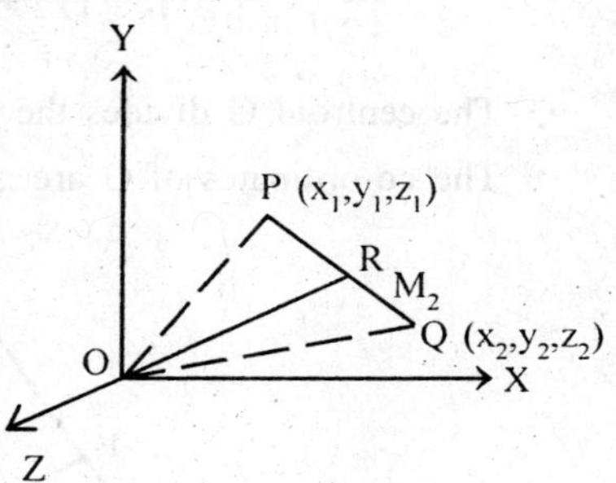

Let Q(x, y, z) be the point which divides the line joining P and Q in the ratio m : n.

Let planes PNA, QMB and RKC through P, Q and R parallel to the plane YOZ meet the axis of x at A,B and C respectively.

Then as these planes divide any two straight lines proportionately, therefore, we have

$$\frac{AC}{CB} = \frac{PR}{RQ} = \frac{m}{n} \text{ or } \frac{(OC - OA)}{(OB - OC)} = \frac{m}{n} \text{ or } \frac{x - x_1}{x_2 - x} = \frac{m}{n}$$

$$\Rightarrow \quad nx - nx_1 = mx_2 - mx \quad \Rightarrow \quad x(n + m) = mx_2 + nx_1$$

$$\Rightarrow \quad x = \frac{nx_1 + mx_2}{n + m}$$

Similarly $y = \dfrac{ny_1 + my_2}{m + n}$ and $z = \dfrac{nz_1 + mz_2}{n + m}$

Cor. 1. The mid-point of the line PQ is given by

$$\left[\frac{1}{2}(x_1 + x_2), \frac{1}{2}(y_1 + y_2), \frac{1}{2}(z_1 + z_2)\right].$$

Cor. 2. *General co-ordinates of a point on the line PQ.*

Let m : n = λ : 1, then the co-ordinates of the point R are

$$\left[\frac{x_1 + kx_2}{1 + k}, \frac{y_1 + ky_2}{1 + k}, \frac{z_1 + kz_2}{1 + k}\right].$$

Corresponding to every value of λ we shall get a point on the line PQ.

CENTROID OF A TRIANGLE

To find the co-ordinates of the centroid of a triangle whose vertices are $A(x_1, y_1, z_1)$, $B(x_2, y_2, z_2)$, and $C(x_3, y_3, z_3)$.

The coordinates of D, the mid-point of BC are given as

$$\left\{\frac{1}{2}(x_2 + x_3), \frac{1}{2}(y_2 + y_3), \frac{1}{2}(z_2 + z_3)\right\}$$

∵ The centroid G divides the median AD in the ration 2: 1.

∴ The co-ordinates of G are given as

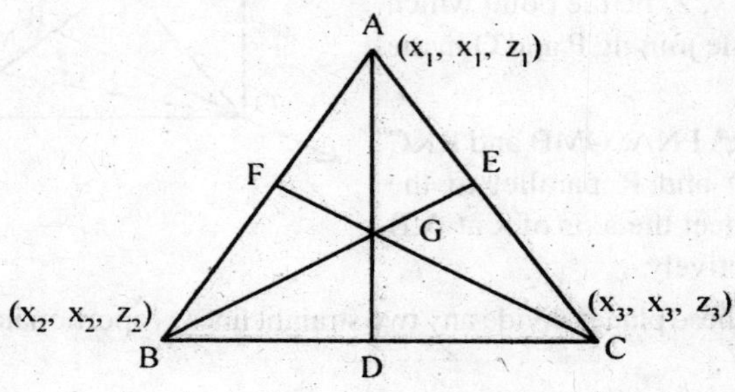

$$\left[\frac{1.x_1 + 2.\frac{1}{2}(x_2 + x_3)}{1+2}, \frac{1.y_1 + 2.\frac{1}{2}(y_2 + y_3)}{1+2}, \frac{1.z_1 + 2.\frac{1}{2}(z_2 + z_3)}{1+2}\right]$$

i.e., $$\left[\frac{1}{3}(x_1 + x_2 + x_3), \frac{1}{3}(y_1 + y_2 + y_3), \frac{1}{3}(z_1 + z_2 + z_3)\right]$$

i.e., $$\left[\frac{\text{sum of x} - \text{co} - \text{ordinates}}{3}, \frac{\text{sum of y} - \text{co} - \text{ordinates}}{3}, \frac{\text{sum of z} - \text{co} - \text{ordinates}}{3}\right]$$

CENTROID OF A TETRAHEDRON

To find the co-ordinates of the centroid of the tetrahedron whose vertices are $A(x_1, y_1, z_1)$, $B(x_2, y_2, z_2)$, $C(x_3, y_3, z_3)$ *and* $D(x_4, y_4, z_4)$.

We know that the centroid of the tetrahedron divides the line joining any vertex and the C.G. of the opposite face in the ratio 3 : 1.

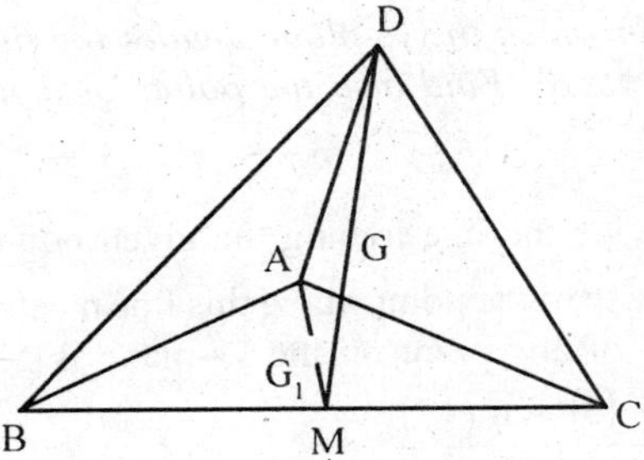

Let P be the C.G. of the triangle BCD. Join PA and divide it in the ratio 3 : 1. Let G be the point on AP such that AG: GP = 3 : 1, then G is the required centroid of the given tetrahedron.

Now the co-ordinates of P, the centroid of ΔBCD, are given by

$$\left[\frac{1}{3}(x_2 + x_3 + x_4), \frac{1}{3}(y_2 + y_3 + y_4), \frac{1}{3}(z_2 + z_3 + z_4)\right]$$

∴ If $(\bar{x}\ \ \bar{y}\ , \bar{z}\ )$ be the co-ordinates of G, which divides AP in the ratio 3 : 1, then

$$\bar{x} = \frac{1.x_1 + 3.\frac{1}{3}(x_2 + x_3 + x_4)}{1+3}$$

$$= \frac{1}{4}(x_1 + x_2 + x_3 + x_4)$$

Similarly $\bar{y} = \frac{1}{4}(y_1 + y_2 + y_3 + y_4)$

and $\bar{z} = \frac{1}{4}(z_1 + z_2 + z_3 + z_4)$.

Hence the *co-ordinates of the centroid G of the tetrahedron* are given as

$$\left[\frac{1}{4}(x_1 + x_2 + x_3 + x_4), \frac{1}{4}(y_1 + y_2 + y_3 + y_4), \frac{1}{4}(z_1 + z_2 + z_3 + z_4)\right]$$

i.e., $\left[\frac{\text{sum of x-co-ordinates}}{4}, \frac{\text{sum of y-co-ordinates}}{4}, \frac{\text{sum of z-co-ordinates}}{4}\right]$

Example 1:

Find the ratio in which the yz-plane divides the line joining the points (3, 5, – 7) and (–2, 1, 8). Find also the points of division.

Solution:

Let yz-plane divide the line joining the given points in the ratio m : n.

Then x-co-ordinate of the point where this line meets the yz-plane is zero, as the x-co-ordinate of any point on the yz-plane is zero.

$$\therefore \quad 0 = \frac{m(3) + n(-2)}{m + n}$$

or $3m - 2n = 0$

or $m : n = 2 : 3.$ **Ans.**

Also if $(0, y_1, z_1)$ be this point of division, then

$$y_1 = \frac{2(5) + 3(1)}{2 + 3} = \frac{10 + 3}{5} = \frac{13}{5}$$

and $$z_1 = \frac{2(-7) + 3(8)}{2 + 3} = \frac{-14 + 24}{5} = \frac{10}{5} = 2$$

$\therefore$ The required point is (0, 13/5, 2). **Ans.**

Example 2:

Given three collinear points A (3, 2, – 4), B (5, 4, – 6), C (9, 8, – 10), find the ratio in which B divides AC.

Solution:

Let B divide AC in the ratio m : n.

Then x-co-ordinate of $B = \frac{m.9 + n.3}{m + n}$ or $5 = \frac{9m + 3n}{m + n}$

$\Rightarrow \quad 5m + 5n = 9m + 3n$

$\Rightarrow \quad 4m = 2n$

$\Rightarrow \quad m : n = 1 : 2.$ **Ans.**

MISCELLANEOUS EXAMPLES

Example 1:

Prove that the four points whose co-ordinates are (5, – 1, 1), (7, – 4, 7), (1, – 6, 10), (–1, – 3, 4) are the vertices of a rhombus.

Solution:

Let the given points be A(5, – 1, 1) B(7, – 4, 7) C(1, – 6, 10) and D(–1, – 3, 4).

Then we have

$AB = \sqrt{[(5 - 7)^2 + (-1 + 4)^2 + (1 - 7)^2]} = \sqrt{(4 + 9 + 36)} = 7;$

$BC = \sqrt{[(7 - 1)^2 + (-4 + 6)^2 + (7 - 10)^2]} = \sqrt{(36 + 4 + 9)} = 7;$

$CD = \sqrt{[(1 + 1)^2 + (-6 + 3)^2 + (10 - 4)^2]} = \sqrt{(4 + 9 + 36)} = 7;$

and $DA = \sqrt{[(-1 - 5)^2 + (-3 + 1)^2 + (4 - 1)^2]} = \sqrt{(36 + 4 + 9)} = 7.$

Also length of diagonal AC

$= \sqrt{[(5 - 1)^2 + (-1 + 6)^2 + (1 - 10)^2]}$

$= \sqrt{(16 + 25 + 81)} = \sqrt{(122)}$

and length of diagonal BD

$= \sqrt{[(7 + 1)^2 + (-4 + 3)^2 + (7 - 4)^2]}$

$= \sqrt{(64 + 1 + 9)} = \sqrt{(74)}.$

Thus, we prove that AB = BC = CD = DA *i.e.,* four sides of the figure ABCD are equal and the diagonals AC and BD are *not equal.*

∴ The four given point are the vertices of a rhombus.

Example 2:

Find the ratios in which the sphere $x^2 + y^2 + z^2 = 504$ divides the line joining the point (12, – 4, 8) and (27, – 9, 18).

Solution:

Let the sphere meet the line through given points in (x_1, y_1, z_1).

Then $\quad x_1^2 + y_1^2 + z_1^2 = 504.$ ...(i)

Also let the point (x_1, y_1, z_1) divide the line joining the given points in the ratio m: n.

Then $\quad x_1 = \dfrac{m(27) + n(12)}{m+n},\ y_1 = \dfrac{m(-9) + n(-4)}{m+n},\ z_1 = \dfrac{m(18) + n(8)}{m+n}$

Substituting these values of x_1, y_1, z_1 in (i) we get

$$\frac{(27m + 12n)^2}{(m+n)^2} + \frac{(-9m - 4n)^2}{(m+n)^2} + \frac{(18m + 8n)^2}{(m+n)^2} = 504$$

$\Rightarrow \quad (9m + 4n)^2 + (9m + 4n)^2 + 4(9m + 4n)^2 = 504\,(m + n)^2$

$\Rightarrow \quad 14(9m + 4n)^2 = 504\,(m + n)^2,$

$\Rightarrow \quad (9m + 4n)^2 = 36\,(m + n)^2$

$\Rightarrow \quad 9m + 4n = \pm 6\,(m + n)$, taking square root of both sides.

Taking + sign we get $3m = 2n$

or $\quad m : n = 2 : 3$ **Ans.**

Taking – sign we get $15m = -10n$

or $\quad m : n = -2 : 3$ **Ans.**

Example 3:

Show that the plane $ax + by + cz + d = 0$ divides the line joining the points (x_1, y_1, z_1) and (x_2, y_2, z_2) in the ratio

$-(ax_1 + by_1 + cz_1 + d)/(ax_2 + by_2 + cz_2 + d).$

Solution:

Let the given plane meet the line joining the given points in (x_3, y_3, z_3). Then

$$ax_3 + by_3 + cz_3 + d = 0. \qquad \text{...(i)}$$

Also let the point (x_3, y_3, z_3) divide the line joining the given points in the ratio m : n.

Then $\quad x_3 = \dfrac{mx_1 + nx_2}{m+n};\ y_3 = \dfrac{my_1 + ny_2}{m+n};\ z_3 = \dfrac{mz_1 + nz_2}{m+n}$

Substituting these valuse is (i) we get

$$a\left(\frac{mx_1 + nx_2}{m+n}\right) + b\left(\frac{my_1 + ny_2}{m+n}\right) + c\left(\frac{mz_1 + nz_2}{m+n}\right) + d = 0$$

$\Rightarrow \quad a(mx_1 + nx_2) + b(my_1 + ny_2) + c(mz_1 + nz_2) + d(m+n) = 0$

$\Rightarrow \quad m(ax_1 + by_1 + cz_1 + d) + n(ax_2 + by_2 + cz_2 + d) = 0$

$\Rightarrow \quad \frac{n}{m} = -\left(\frac{ax_1 + by_1 + cz_1 + d}{ax_2 + by_2 + cz_2 + d}\right)$ **Hence porved.**

Example 4:

The point P lies on the line whose end points are A(7, 2, 1) B(10, 5, 7). If the y-co-ordinate of P is 4; find its other co-ordinates.

Solution:

Let P(x, y, z) divide the join of A(7, 2, 1) and B(10, 5, 7) in the ratio m : n.

Then $\quad y = \frac{m(5) + n(2)}{m+n}$

But it is given that the y-co-ordinate of P is 4

$\therefore \quad 4 = \frac{5m + 2n}{m+n}$

or $\quad 4m = 4n = 5m + 2n$

or $\quad m = 2n \text{ or } \frac{m}{n} = \frac{2}{1}$

$\therefore \quad x = \frac{m(10) + n(7)}{m+n} = \frac{2(10) + 1(7)}{2+1} = 9;$

$z = \frac{m(7) + n(1)}{m+n} = \frac{2(7) + 1(1)}{2+1} = 5$ **Ans.**

Example 5:

Find the co-rodinates of the point which divides the line joining (1, = 1, 2) and (2, 3, 7) in the ratio 2 : 3.

Solution:

Let (x, y, z,) be the required point, then

$$x = \frac{2(2) + (3)(1)}{2+3} = \frac{7}{5};\ y = \frac{2(3) + 3(-1)}{2+3} = \frac{3}{5}$$

and $$z = \frac{2(7) + 3(2)}{2+3} = \frac{20}{5} = 4.$$

∴ The reuqired point is (7/5, 3/5, 4). **Ans.**

Example 6:

Find the distance of the point (1, 2, 0) from the point where the line joining (2, – 3, 1) and (3, – 4, – 5) cuts the plane 2x + y + z = 7.

Solution:

Let the line joining the point (2, – 3, 1) and (3, – 4, – 5) meet the given plane in (x_1, y_1, z_1). Then as the point (x_1, y_1, z_1) lies on the plane

$$2x + y + z = 7,$$

so we have $$2x_1 + y_1 + z_1 = 7. \quad \text{...(i)}$$

Also let the point (x_1, y_1, z_1) divide the line joining (2, – 3, 1) and (3, = 4, – 5) in the ratio m : n.

Then $$x_1 = \frac{3m + 2n}{m + n},\ y_1 = \frac{-4m - 3n}{m + n},\ z_1 = \frac{-5m + n}{m + n} \quad \text{...(ii)}$$

Substituting these values in (i) we get

$$2(3m + 2n) + (-4m - 3n) + (-5m + n) = 7(m + n)$$

$$\Rightarrow 10m - 5n = 0 \quad \text{or} \quad n = 2m$$

Substituting this value in (ii) we get

$$x_1 = \frac{3m + 4m}{m + 2m} = \frac{7}{3};\ y_1 = \frac{-4m - 6m}{m + 2m} = \frac{-10}{3};\ z_1 = \frac{-5m + 2m}{m + 2m} = -1$$

∴ The point where the line joining (2, – 3, 1) and (3, – 4, – 5) meets the plane 2x + y + z = 7 is (7/3, – 10/3, – 1).

∴ the required distance

= distance between (1, 2, 0) and (7/3, – 10/3, – 1)

$= \sqrt{[\{7/3 - 1\}^2 + \{(-10/3) - 2\}^2 + (-1 - 0)^2]}$

$= \sqrt{[(16/9) + (256/9) + 1]}$

$= \sqrt{(281/9)} = \frac{1}{3}\sqrt{(281)}.$ **Ans.**

Example 7:

Find the ratio in which the co-ordinate planes divide the line joining the points (–2, 4, 7), (3, – 5, 8).

Solution:

Let xy-plane divide the line joining the given points in the ratio m: n.

Then z-co-ordinate of the point where this line meets the xy-plane is zero, as the z-co-ordinate of any point on the xy-plane is zero

$$\therefore \quad 0 = \frac{m(8) + n(7)}{m + n}$$

or $\quad 8m + 7n = 0$

or $\quad m : n = -7 : 8.$ **Ans.**

Similarly, we can find that yz-plane (for every point on which x = 0) and xz-plane (for every point on this plane y = 0) divides the line joining the given points in the ratio 2 : 3 and 4 : 5. **Ans.**

Example 8:

Show that the three points A, B, C whose coordinates are respectively (–2, 3, 5), (1, 2, 3) and (7, 0, – 1) are collinear. Also find the ratio in which point B divides the line AC.

Solution:

Here $\quad AB = \sqrt{[(1 + 2)^2 + (2 - 3)^2 + (3 - 5)^2]}$

$= \sqrt{(9 + 1 + 4)} = \sqrt{(14)};$

$BC = \sqrt{[(7 - 1)^2 + (0 - 2)^2 + (-1 - 3)^2]}$

$= \sqrt{(36 + 4 + 16)} = \sqrt{(56)}$

$= \sqrt{(56)} = 2\sqrt{(14)}$

and $\quad AC = \sqrt{[(7 + 2)^2 + (0 - 3)^2 + (-1 - 5)^2]}$

$= \sqrt{(81 + 9 + 36)}$

$= \sqrt{(126)} = 3\sqrt{(14)}$

$\therefore AB + BC = \sqrt{(14)} + 2\sqrt{(14)}$

$= 3\sqrt{(14)} = AC$

$\therefore$ Point A, B, C are collinear.

Again let B divide AC in the ratio m : n.

The x-coordinate of $B = \dfrac{m.7 + n.(-2)}{m + n}$ or $1 = \dfrac{7m - 2n}{m + n}$

$\Rightarrow \quad m + n = 7m - 2n$

$\Rightarrow \quad 6m = 3n$

$\Rightarrow \quad m : n = 1 : 2$ **Ans.**

Example 9:

From any point (1, – 2, 3) lines are drawn to meet the shpere $x^2 + y^2 + z^2 = 4$ and they are divided in the ratio 2 : 3. Prove that the points of section lie on a sphere.

Solution:

Let any line through (1, –2, 3) meet the given sphere in (x_1, y_1, z_1). Then

$$x_1^2 + y_1^2 + z_1^2 = 4. \quad ...(i)$$

Also let (x_2, y_2, z_2) be the point which divedes the join of (1, – 2, 3) and (x_1, y_1, z_1) in the ratio 2 : 3.

Then $x_2 = \dfrac{2 . x_1 + 3 . 1}{2 + 3}$ or $x_1 = \dfrac{5x_2 - 3}{2}$

$y_2 = \dfrac{2y_1 + 3(-2)}{2 + 3}$ or $y_1 = \dfrac{5y_2 + 6}{2}$

$z_2 = \dfrac{2 . z_1 + 3(3)}{2 + 3}$ or $z_1 = \dfrac{5z_2 - 9}{2}$

Substituting these values of x_1, y_1 and z_1 in (i) we get

$$\frac{1}{4}(5x_2 - 3)^2 + \frac{1}{4}(5y_2 + 6)^2 + \frac{1}{4}(5z_2 - 9)^2 = 4$$

$$\Rightarrow \quad 25x_2^2 + 25y_2^2 + 25z_2^2 - 30x_2 + 60y_2 - 90z_2 + 110 = 0$$

$$\Rightarrow \quad 5x_2^2 + 5y_2^2 + 5z_2^2 - 6x_2 + 12y_2 - 18z_2 + 22 = 0.$$

$\therefore$ The locus of the point of section (x_2, y_2, z_2) is

$5(x^2 + y^2 + z^2) = 6x + 12y - 18z + 22 = 0$, which represents a sphere.

Example 10:

Find (i) the cartesian, (ii) the cylindrical and (iii) the polar equation of the right circular cylinder whose axis is OZ and radius a.

Solution:

Let P(x, y, z) be any point on the cylinder.

Then the radius a = PC = ON

$$= \sqrt{(OA^2 + NA^2)} = \sqrt{(x^2 + y^2)}.$$

$\Rightarrow x^2 + y^2 = a^2$ is the required cartesian equation of the cylinder.

(ii) Let P (u, ϕ, z) be any point on the cylinder.

Then radius a = PC = ON

$\Rightarrow$ a = u *i.e.,* u = a is the required cylindrical equation.

(iii) Let P(r, θ, ϕ) be any point of the cylinder.

Then radius a = PC = OP sin θ

$\Rightarrow a = r \sin\theta$

i.e., $r \sin\theta = \alpha$ is the required polar equation.

Example 11:

Show that the points A(1, 2, 3), B (4, 0, 4) and C(–2, 4, 2) are collinear.

Solution:

$$AB = \sqrt{[(4-1)^2 + (0-2)^2 + (4-3)^2]}$$
$$= \sqrt{(9+4+1)} = \sqrt{(14)}$$
$$BC = \sqrt{[(-2-4)^2 + (4-0)^2 + (2-4)^2]}$$
$$= \sqrt{(36+16+4)}$$
$$= \sqrt{(56)} = 2\sqrt{(14)}$$

and
$$AC = \sqrt{[(-2-1)^2 + (4-2)^2 + (2-3)^2]}$$
$$= \sqrt{(9+4+1)} = \sqrt{(14)}$$

$\therefore AB + BC = \sqrt{(14)} + \sqrt{(14)} = 2\sqrt{(14)} = AC$

∴ The points A, B C are collinear. **Hence proved.**

Example 12:

If A and B are the points (3, 4, 5) and (–1, 3, – 7) then find the locus of P which moves so that $PA^2 - PB^2 = 3$.

Solution:

Let P be (x, y, z). Then

$$PA^2 = (x-3)^2 + (y-4)^2 + (z-5)^2$$

and
$$PB^2 = (x+1)^2 + (y-3)^2 + (z+7)^2.$$

Given $PA^2 - PB^2 = 3$.

$\Rightarrow [(x-3)^2 + (y-4)^2 + (z-5)^2] - [(x+1)^2 + (y-3)^2 + (z+7)^2] = 3$

$\Rightarrow 4x + y + 12z + 6 = 0$ is the required loucs. **Ans.**

Example 13:

Find the locus of the point the difference of whose distances from (2, 0, 0) and (–2, 0, 0) is 1.

Solution:

Do your self. **Ans.** $60x^2 - 4(y^2 + z^2) = 15$.

Example 14:

Find the distance of the point whose spherical polar co-ordinates are $(2\sqrt{2}, 1/4\pi, \pi/6)$ from the point whose cartesian co-ordinates are $(2\sqrt{3}, -1, -4)$.

Soultion:

We have read if (r, θ, ϕ) be the spherical polar co-ordinates of a point (x, y, z), then

$$x = r \sin\theta \cos\phi,$$

$$y = r \sin\theta \sin\phi,$$

$$z = r \cos\theta.$$

$\therefore$ If (x_1, y_1, z_1) be the cartesian co-ordinates of the point whose spherical polar co-ordinates are $\left\{2\sqrt{2}, \frac{1}{4}\pi, \frac{1}{6}\pi\right\}$, then we have

$$x_1 = 2\sqrt{2} \sin \frac{1}{4}\pi \cos \frac{1}{6}\pi$$

$$= 2\sqrt{2}.(1/\sqrt{2}). \frac{1}{2} \sqrt{3} = \sqrt{3}$$

$$y_1 = 2\sqrt{2} \sin \frac{1}{4}\pi \sin \pi/6$$

$$= 2\sqrt{2}.(1/\sqrt{2}) \frac{1}{2} = 1;$$

$$z_1 = 2\sqrt{2} \cos \frac{1}{4}\pi$$

$$= 2\sqrt{2}(1/\sqrt{2}) = 2.$$

$\therefore$ This point in cartesian co-ordinates is $(\sqrt{3}, 1, 2)$.

The other point is $(2\sqrt{3}, -1, -4)$. So the required distance

$$= \sqrt{[(2\sqrt{3} - \sqrt{3})^2 + (-1 - 1)^2 + (-4 - 2)^2]}$$

$$= \sqrt{(3 + 4 + 36)} = \sqrt{(43)}.$$ **Ans.**

Example 15:

Find the locus of a point P which is at a distance r from the point (a, b, c).

Solution:

Let P be (x, y, z). Then distance of P(x, y, z) from the point (a, b, c) = r (given)

i.e., $\sqrt{[(x-a)^2+(y-b)^2+(z-c)^2]} = r$

$\Rightarrow$ $(x-a)^2+(y-b)^2+(z-c)^2 = r^2$ is the required locus. **Ans.**

Example 16:

Find the co-ordinates of the point which is equidistant from the four points O, A, B and C, where O is the origin and A, B, C are the points on the axes of x, y, z respectively at distances a, b, c from the origin.

Solution:

According to the problem, the co-ordinates of the points O, A, B and C are respectively (0, 0, 0), (a, 0. 0), (0, b, 0), and (0, 0 c).

Let P(x, y, z) be the centre of the sphere. Then as O, A, B and C lie on the sphere, so P is equi-distant from O, A, B and C,

i.e., PO = PA = PB = PC

i.e., $PO^2 = PA^2 = PB^2 = PC^2$.

From $PO^2 = PA^2$,

we get $x^2+y^2+z^2$

$$= (x-a)^2+(y-0)^2+(z-0)^2$$

$\Rightarrow$ $2ax = a^2$ or $x = \frac{1}{2}a$.

Similarly from $PO^2 = PB^2$ we get $y = \frac{1}{2}b$

and from $PO^2 = PC^2$, we get $z = \frac{1}{2}c$.

Example 17:

If OA = 2, OB = 3 and OC = 4,

(i) What are the equations to the planes PNAM, PMCK and PNBK?

(ii) What equations are satisfied by the co-ordinates of any ponit on the line PN?

Solution:

(i) The y and z-co-ordinates of points on the plane PNAM differ but all have the same x-co-ordinate *i.e.,* the distance of all points on the plane

PNAM from the yz-plane remains constant and equal to OA *i.e.*, 2. Hence, the equation of the plane PNAM is $x = 2$.

Similarly, the equations of the planes PMCK and PNBK are $z = 4$ and $y = 3$ respectively.

(ii) All points on the line PN are equidistant from yz-plane and each of them is at a distance BN = OA = 2 from yz-plane *i.e.*, for each point on the line PN we have $x = 2$. Similarly each point on the line PN is equidistant from the xz-plane and as such we have $y = OB = 3$. But the z-co-ordinate of all points on the line PN are different. Hence all the points on the line PN satisfy the equations $x = 2$, $y = 3$. **Ans.**

Example 18:

What is the locus of the point (i) whose x-co-ordinate is 5 and (ii) whose x-co-ordinate is 3 and y-co-ordinate is 4?

Solution:

(i) The required locus is $x = 5$ which represents a plane parallel to YOZ plane (or yz-plane) at a distance 5 from it.

(ii) The required locus is $x = 3$ and $y = 4$ which represents line parallel to z-axis.

Example 19:

Find (i) the cartesian, (ii) the cylindrical and (iii) the polar equations of the sphere whose centre is the origin and radius a.

Solution:

(i) Let P be any point (x, y, z) on the sphere.

Then $a = OP = \sqrt{(PN^2 + ON^2)}$,

$$= \sqrt{\{PN^2 + (OA^2 + NA^2)\}} = \sqrt{\{z^2 + (x^2 + y^2)\}}$$

$\Rightarrow \quad a^2 = x^2 + y^2 + z^2$ *i.e.*, $x^2 + y^2 + z^2 = a^2$

is the required cartesian equation.

(ii) Let P be any point on the sphere.

Then $a = OP = \sqrt{(PN^2 + ON^2)}$,

$\Rightarrow \quad a^2 = (z^2 + u^2)$ *i.e.*, $u^2 + z^2 = a^2$,

is the required cylindrical equation

(iii) Let $P(r, \theta, \phi)$ be any point on the sphere.

Then $a = OP = r$

i.e., $r = a$ is the required polar equation.

Example 20:

Transform the cartesian coordinates (1, 2, 3) of a point into spherical polar coordinates.

Solution:

Let (r, θ, ϕ) be the polar co-ordinates of the point whose cartesian co-ordinates are given by $x = 1, y = 2, z = 3$.

we know that

$$r^2 = x^2 + y^2 + z^2 = 1^2 + 2^2 + 3^2 = 14$$

$$\Rightarrow \quad r = \sqrt{(14)}$$

Also $\tan\theta = \dfrac{\sqrt{(x^2 + y^2)}}{z} = \dfrac{\sqrt{(1^2 + 2^2)}}{z} = \dfrac{\sqrt{5}}{2}$

and $\quad \tan\phi = y/x = 2/1 = 2.$

∴ The required polar co-ordinates are given as

$$[\sqrt{(14)}, \tan^{-1}(1/2\ \sqrt{5}), \tan^{-1} 2]$$ **Ans.**

EXERCISES

1. Find the equation of the sphere whose centre is (0, 1,–1) and radius 2.
2. Given three collinear points A(3, 2, –4), B(5, 4, –6), C(9, 8, –10), show that B divides AC in the ratio 1: 2.
3. Find the ratio in which the xz-plane divides the join of (–3, 4, –8) and (5, –6, 4).
4. Find the ratio in which the join of (2, 1, 5) and (3, 4, 3) is divided by the plane $x + y - z = 1/2$.
5. Find the co-ordinates of the point which divides the join of (1, 2, 3) and (3, –5, 6) in the ratio 3 : (–5).
6. Find the ratio in which the line joining the points (2, 4, 5) and (3, 5, – 4) is divided by the yz-plane. Find also the co-ordinates of the point at which the line meets the xy-plane.
7. Find the ratio in which yz-plane divides the line joining the points (–2, 4, 7) and (3, – 5, 8).
8. Find the ratio in which the join of the points (x_r, y_r, z_r) r = 1, 2 is divided by the plane $ax + by + cz + d = 0$
9. The line joining the points (1, 8, – 1) and (4, – 4, 2) meets the zx and xy planes at P and Q respectively. Find the co-ordinates of P and Q.

10. Show that the distance fo the point (1, 2, 3) from the co-ordinate axes are $\sqrt{(13)}$, $\sqrt{(10)}$, $\sqrt{(5)}$.
11. What is the polar equation of the plane which is parallel to the xy-plane and is at a distance c from it? What is the cylindrical equation of the plane parallel to the zx-plane at a distance b from it?
12. Find the polar co-ordinates of the point (3, 4, 5) so that r may be positive.
13. The axes are rectangular and A, B are the points (3, 4, 5) (–1, 3, –7). A variable point P has co-ordinates (x, y, z). Find the locus of P if $PA^2 - PB^2 = 2k^2$.
14. Find the locus of the point P(x, y, z), the difference of whose distances from (0, 0, – 4) and (0, 0, 4) is 4.
15. Prove that the four points A, B, C, D whose coordinates are (1, 1, 1), (– 2, 4, 1), (–1, 5, 5) and (2, 2, 5) are the vertices of a square.
16. Find the locus of the point the sum of whose distances from (4, 0, 0) and (–4, 0, 0) is equal to 10.
17. Show that the point D(– 1/2, 2, 0) is the circumcentre of the triangle formed by the points A(1, 1, 0), B(1, 2, 1) and C (– 2, 2, – 1).
18. A is the point (–2, 2, 3) and B is the point (13, – 3, 13). A point P(x, y, z) moves so that 3 PA = 2 PB. Find the locus of P.
19. Find the locus of the point P(x, y, z) if $PA^2 + PB^2 = 2k^2$, where A and B are the points (3, 4, 5) and (–1, 3, – 7).

2

Direction Cosines and Projection

DIRECTION COSINES OF A LINE

If α, β γ are the angles that a given line AB makes with the positive directions of x, y and z axes then cos α, cos β, cos γ are called the direction cosines (or d. e.'s) of the line AB. Generally the direction conies are represented by l, m, n.

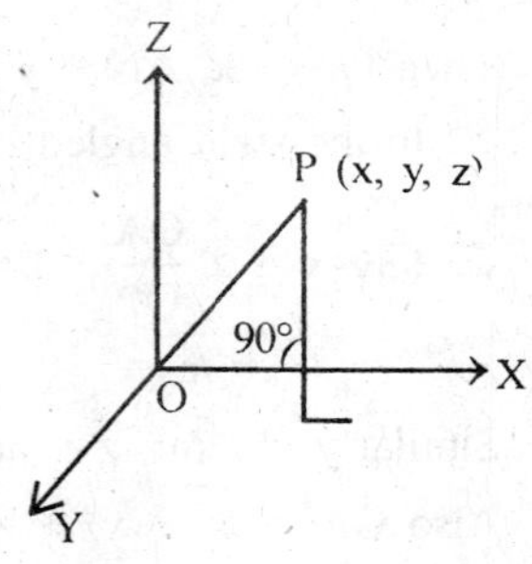

If O be the origin of co-ordinates and (x, y, z). The co-ordinate of a point P then $x = lr$, $y = mr$, $z = nr$.

l, m, n being the direction cosines of the lined OP and r be the length of the segments OP through the point P draw the line P_2 perpendicular to x-axis to that OL = x. From right angel trigangle OPL, we have

$$OL/OP = \cos \angle LOP \Rightarrow x/r = d \Rightarrow x - dr.$$

Similarly, $y = mr$, $z = nr$.

These angle α, β, γ are called the *direction angles* of the line AB.

The direction cosines of the line BA are cos $(\pi - \alpha)$, cos $(\pi - \beta)$ and cos $(\pi - \gamma)$ i.e., $-\cos\alpha$, $-\cos\beta$, $-\cos\gamma$ since the direction angles of the line OP' through O parallel to BA and $\pi - \alpha$, $\pi - \beta$ and $\pi - \gamma$. **(Note)**

Here students should remember that angles α, β and γ are not coplanar.

Cor. d.c.'s of the Co-ordinate Axes

The x-axis makes angles $0, \pi/2$ and $\pi/2$ with x, y and z-axis respectively. So the direction cosines of the x-axis are

$$\cos 0, \cos \pi/2, \cos \pi/2, \textit{ i.e.}, \ 1, 0, 0.$$

Similarly the d.c.'s of y and z-axes are 0, 1, 0 and 0, 0, 1 respectively.

To prove that $l^2 + m^2 + n^2 = 1$, *where* l, m, n *are the direction-cosines of a line AB*

Or

To show that $\cos^2 \alpha + \cos^2 \beta + \cos^2 \gamma = 1$, where $\cos \alpha$, $\cos \beta$, $\cos \gamma$ are the direction cosines of a line AB.

Draw a line OP parallel to AB and of length r (say) so that direction cosines of the line OP are l, m, n or $\cos^2 \alpha$, $\cos^2 \beta$, $\cos^2 \gamma$ (say).

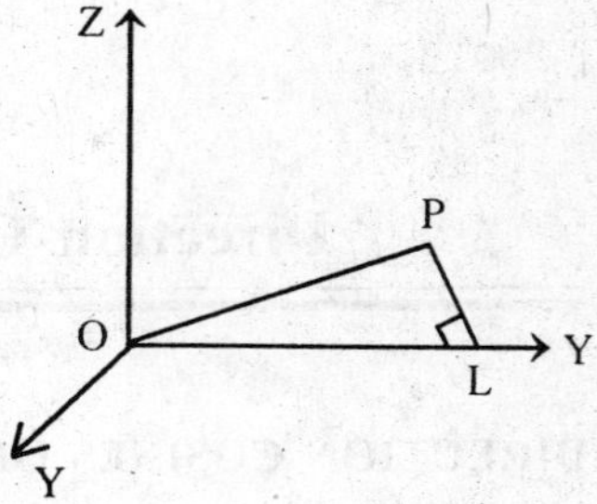

Let the co-ordinates of P be (x, y, z). Draw PN perpendicular to the xy-plane and NA perpendicular to x-axis.

Then OA = x, AN = y and NP = z.

$\therefore$ In the right angled triangle ΔOAP

we have $\cos \alpha \dfrac{OA}{OP}$ or $l = \dfrac{x}{r}$, $\because \cos \alpha = l$

$\Rightarrow$ $x = lr$.

Similarly $y = mr$, $z = nr$.

Also $OA^2 = x^2 + y^2 + z^2$

$\therefore$ $r^2 = (lr)^2 + (mr)^2 + (nr)^2$ or $l^2 + m^2 + n^2 = 1$ ...(i)

$\Rightarrow$ $\cos^2 a + \cos^2 b + \cos^2 g = 1.$ ...(ii)

i.e., The sum of the squares of the direction cosines of a line is equal to unity. *(Remember)*

DIRECTION RATIOS

Definition: If a, b, c, be three numbers proportional to the actual direction cosines l, m, n of a given line then they are defined as direction ratios or direction numbers

Now from the above definition, we have

$$\frac{l}{a} = \frac{m}{b} = \frac{n}{c} \pm \frac{\sqrt{(l^2 + m^2 + n^2)}}{\sqrt{(a^2 + b^2 + c^2)}} = \pm \frac{1}{\sqrt{(a^2 + b^2 + c^2)}}$$

$\because l^2 + m^2 + n^2 = 1$

$$\Rightarrow \quad l = \pm \frac{a}{\sqrt{(a^2 + b^2 + c^2)}}, \quad m = \pm \frac{b}{\sqrt{(a^2 + b^2 + c^2)}},$$

$$n = \pm \frac{c}{\sqrt{(a^2 + b^2 + c^2)}}$$

If AB be the line whose direction ratios are a, b, c then the direction cosines of AB are given by the + sign and those of the line BA by – sin.

Notes:

1. From above we conclude that if direction ratios a, b and c of a line are given we should divide each of them by $\sqrt{(a^2 + b^2 + c^2)}$ to get the corresponding direction cosines.
2. The sum of the squares of the direction ratios of a line is *not equal* to unity.

Example:

Find the direction cosines of a line that makes equal angles with the axes.

Solution:

Let l, m n be the required direction cosines. Thus as the line makes equal angles with the axes, so we have $l = m = n$

(*i.e.*, $\cos\alpha = \cos\beta = \cos\gamma$ or $\alpha = \beta = \gamma$).

Also we know $l^2 + m^2 + n^2 = 1$.

So here we have $l^2 + l^2 + l^2 = 1$

or $3l^2 = 1$ or $l = \pm 1/\sqrt{3}$.

The required direction cosines are $\pm 1/\sqrt{3}$, $\pm 1/\sqrt{3}$, $\pm 1/\sqrt{3}$. **Ans.**

(**Note:** Here *four* such lines are possible, and their d.c.'s are $1/\sqrt{3}$, $1/\sqrt{3}$, $1/\sqrt{3}$; $-1/\sqrt{3}$, $1/\sqrt{3}$, $1/\sqrt{3}$; $1/\sqrt{3}$, $-1/\sqrt{3}$, $1/\sqrt{3}$ and $1/\sqrt{3}$, $1/\sqrt{3}$, $-1/\sqrt{3}$; as l, m, n and $-l$, $-m$, $-n$ are the d.c.'s of the same line).

PROJECTION OF A POINT ON A LINE

The projecting of a point P on a line AB is the foot N of the perpendicular PN from P on the line AB.

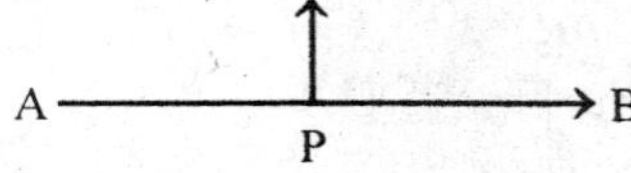

N is also the same point where the line AB meets the plane through P and perpendicular to AB.

PROJECTION OF A SEGMENT OF A LINE ON ANOTHER LINE

The projection of the segment AB of a given line on another line CD is the segment A' B' of CD, where A' and B' are the projections of the points A and B on the line CD.

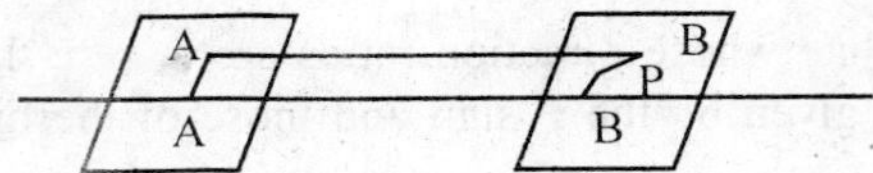

Through A draw the line AN parallel to A'B' meeting the plane through B perpendicular to CD in the point N.

Then we have AN = A'B' ...(i) **(Note)**

If θ be the angle between the lines AB and CD, then AN being parallel to CD we have ∠ BAN = θ.

Also BN is a line through N and is lying in the plane which is perpendicular to the line CD and hence perpendicular to AN. ∴ ∠ ANB = 90°.

∴ From (i) we get A'B' = AN = AB cos θ

Which is the length of the projection A'B'

PROJECTION OF BROKEN LINE ON A GIVEN LINE

Let A, B, C, D, ..., S be any number of points in space and let A', B', C', ... S' be their projections on any line MN. Then as A', B', C', D, ... S' lie on the same straight line MN, so we have

A'B' + B'C' + C'D' + ... + R'S' = A'S'. ...(i)

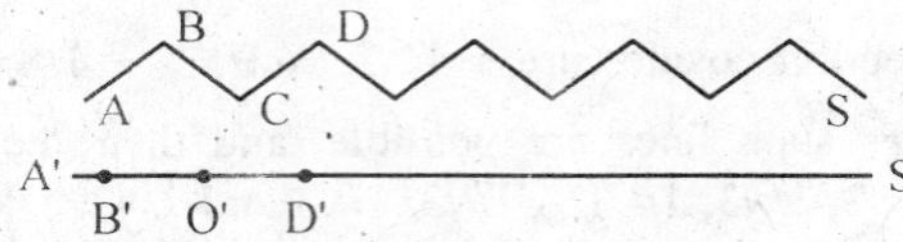

But A'B' is the projection of AB on the line MN.

Hence from (i) we conclude that the sum of projections of AB, BC, CD,...., RS on the line MN = projection A'S' of AS on the line MN.

DIRECTION COSINES OF THE LINE JOINING THE POINTS *P* (x_1, y_1, z_1) *and Q* (x_2, y_2, z_2)

Let N and M be the projections of the points P and Q on x-axis. Then we have ON = x_1 and OM = x_2.

∴ MN = OM – ON = $x_2 - x_1$

Also if OP makes angles α, β and γ with the axes of x, y and z respectively, then we know that

NM = PQ cos α

or $(x_2 - x_1)$ = PQ cos α.

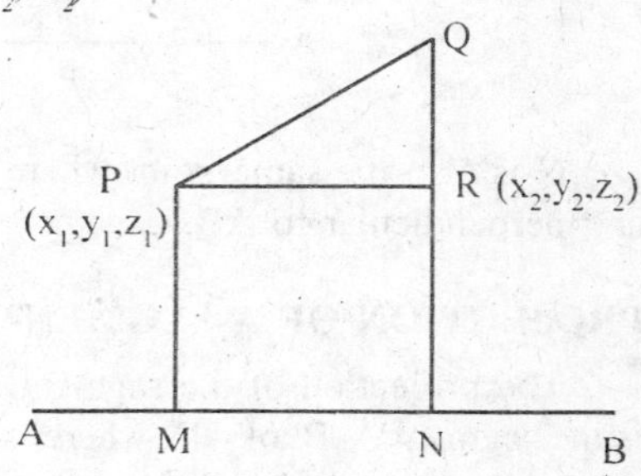

$$\Rightarrow \quad \frac{x_2 - x_1}{\cos \alpha} = PQ$$

Similarly projecting PQ on the y and z-axes we can find that

$$\frac{y_2 - y_1}{\cos \beta} = PQ = \frac{z_2 - z_1}{\cos \gamma}$$

So we have $\dfrac{x_2 - x_1}{\cos\alpha} = \dfrac{y_2 - y_1}{\cos\beta} = \dfrac{z_2 - z_1}{\cos\gamma}$

Hence the direction cosines of the line joining two points P (x_1, y_1, z_1) and Q (x_2, y_2, z_2) are proportional to $x_2 - x_1$, $- y_1$ and $z_2 - z_1$ *i.e.,* direction ratios of PQ are $x_2 - x_1$, $y_2 - y_1$, $z_2 - z_1$ *i.e.,* direction ratios of the line joining two points are difference of their x-co-ordinates, difference of their y-co-ordinates, difference of their z-co-ordinates. *(Remember)*

And actual direction cosines are given as

$$\frac{x_2 - x_1}{PQ}, \frac{y_2 - y_1}{PQ}, \frac{z_2 - z_1}{PQ}$$

where $PQ = \sqrt{\{(x_2 - x_1)^2 + (y_2 - y_1)^2 + (z_2 - z_1)^2\}}$

PROJECTION OF A LINE JOINING THE POINTS P (x_1, y_1, z_1) AND Q (x_2, y_2, z_2) ON ANOTHER LINE WHOSE DIRECTION COSINES ARE l, M AND n.

Draw planes through P and Q parallel to the co-ordinate planes and thus form a rectangular parallelepiped as shown in the figure.

Then we have given

$PN \;\; = x_2 - x_1$

$NK' \; = y_2 - y_1$

and $\quad K'Q. = z_2 - z_1$

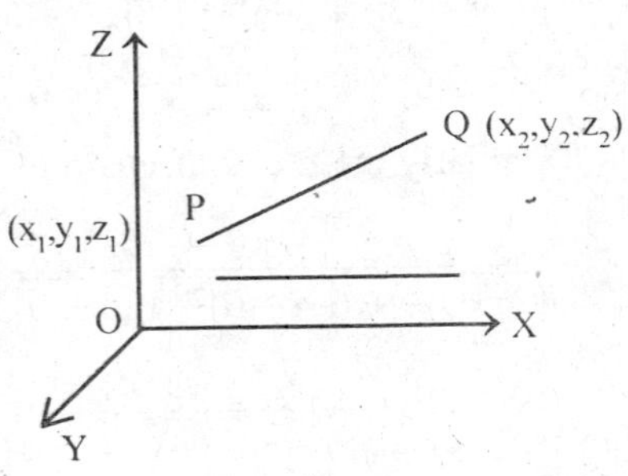

If the line whose direction cosine are given as l, m, n, makes angles α, β and γ with co-ordinate axes, then $l = \cos \alpha$, $m = \cos \beta$, $n = \cos \gamma$.

The projections of PN (which is parallel to x-axis) on the given line

$$= PN. \cos \alpha = (x_2 - x_1) \cos \alpha.$$

Similarly, the projections of NK' and K'Q on the given line are NK' cos β and K' Q cos γ respectively *i.e.,* $(y_2 - y_1) \cos \beta$ and $(z_2 - z_1) \cos \gamma$ respectively.

Also we know that the sum of projections of PN, NK' and K'Q on a line equal tot he projections of PQ on that line and therefore the projections of PQ on the given line.

= sum of projections of PN, NK' and K' Q on that line.

$= (x_2 - x_1) \cos \alpha + (y_2 - y_1) \cos \beta + (z_2 - z_1) \cos \gamma$

$= (x_2 - x_1)\, l + (y_2 - y_1)\, m + (z_2 - z_1)\, n$, from (i).

Cor.: If P is point (x_1, y_1, z_1) then the projection of OP on a line whose direction cosines are l_1, m_1, n_1 is $l_1x_1 + m_1y_1 + n_1z_1$, where O is the origin.

The projection of OP on the given line

= projection of OA + projection of AN + projection of NP on the line

$= x_1l_1 + y_1m_1 + z_1n_1$.

Example:

A and B are (2, 3, –6), (3, –4, 5). Find the direction cosines of OA, OB, BO, and AB where O is (0, 0, 0).

Solution:

The direction ratios of OA are (2, 0), (3 – 0), (–6 –0)

i.e., 2, 3, –6.

∴ The d.c.'s of OA are

$$\frac{2}{\sqrt{[2^2 + 3^2 + (-6)^2]}}, \frac{3}{\sqrt{[2^2 + 3^2 + (-6)^2]}}, \frac{-6}{\sqrt{[2^2 + 3^2 + (-6)^2]}}$$

$\Rightarrow$ 2/7, 3/7, –6/7. **Ans.**

Similarly the d.c.'s of OB are

$$\frac{3-0}{\sqrt{[3-0)^2 + (-4-0)^2 + (5-0)^2]}}, \frac{-4-0}{\sqrt{[(3-0)^2 + (-4-0)^2 + (5-0)^2]}},$$

$$\frac{5-0}{\sqrt{[3-0)^2 + (-4-0)^2 + (5-0)^2]}},$$

$$\Rightarrow \frac{3}{\sqrt{(50)}}, \frac{-4}{\sqrt{(50)}}, \frac{5}{\sqrt{(50)}}, \text{ or } \frac{3}{5\sqrt{2}}, \frac{-4}{5\sqrt{2}}, \frac{1}{5\sqrt{2}}$$ **Ans.**

And the d.c.'s of BO are

$$\frac{3-0}{\sqrt{[(0-3)^2 + (0+4)^2 + (0-5)^2]}}, \frac{0-(-4)}{\sqrt{[(0-3)^2 + (0+4)^2 + (0-5)^2]}},$$

$$\frac{0-5}{\sqrt{[(0-3)^2+(0+4)^2+(0-5)^2]}}$$

$\Rightarrow$ $$\frac{-3}{\sqrt{(50)}}, \frac{4}{\sqrt{(50)}}, \frac{-5}{\sqrt{(50)}}, \text{ or } \frac{-3}{5\sqrt{2}}, \frac{4}{5\sqrt{2}}, \frac{-1}{5\sqrt{2}}$$

Also the d.c.'s of AB are

$$\frac{3-2}{\sqrt{[(3-2)^2+(-4-3)^2+(5+6)^2]}}, \frac{-4-3}{\sqrt{[(3-2)^2+(-4-3)^2+(5+6)^2]}},$$

$$\frac{5-(-6)}{\sqrt{[(3-2)^2+(-4-3)^2+(5+6)^2]}}$$

$\Rightarrow$ $$\frac{1}{\sqrt{[(1+49+121)}}, \frac{-7}{\sqrt{[(1+49+121)}}, \frac{11}{\sqrt{[(1+49+121)}}$$

$\Rightarrow$ $1/\sqrt{(171)}, -7/\sqrt{(171)}, 11/\sqrt{(171)}$. **Ans.**

THE ANGLE BETWEEN TWO LINES

If (l_1, m_1, n_1) and (l_2, m_2, n_2) be the direction cosines of any two lines and θ be the angle between them, then

$$\cos\theta = l_1 l_2 + m_1 m_2 + n_1 n_2.$$

Draw two lines OA and OB parallel respectively to the given lines with d.c.'s (l_1, m_1, n_1) and (l_2, m_2, n_2)

And Let OB = r and the co-ordinates of B be (x_1, y_1, z_1). Then we have

$x_1 = rl_2$, $y_1 = rm_2$ and $z_1 = rn_2$ we know that the projection of OB on OA is given as

$$= l_1 x_1 + m_1 y_1 + n_1 z_1$$

$$= l_1 . rl_2 + m_1 . rm_2 + n_1 . rn_2, \text{ from} \quad ...(i)$$

$$= r(l_1 l_2 + m_1 m_2 + n_1 n_2) \quad ...(ii)$$

Also from the figure 20 above it is evident that the projection of OB on OA

$$= OB \cos\theta = r\cos\theta, \quad \because OB = r. \quad ...(iii)$$

$\therefore$ From (ii) and (iii), we have $r\cos\theta = r(l_1 l_2 + m_1 m_2 + n_1 n_2)$

$\Rightarrow$ $$\mathbf{\cos\theta = l_1 l_2 + m_1 m_2 + n_1 n_2} \quad ...(A)$$

If instead of $(l_1 + m_1 + n_1)$ and $(l_2 + m_2 + n_2)$ the d.c.'s of the given lines are expressed as $\cos\alpha$, $\cos\beta$, $\cos\gamma$ and $\cos\alpha'$, $\cos\beta'$, $\cos\gamma'$ respectively, then the relation (A) takes the form

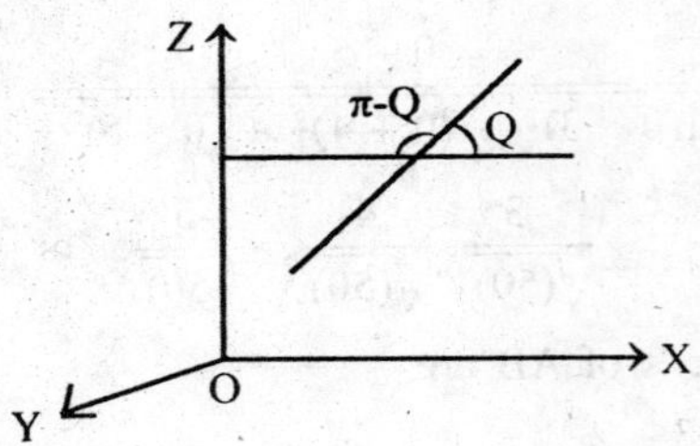

$$\cos\theta = \cos\alpha\cos\alpha' + \cos\beta\cos\beta' + \cos\gamma\cos\gamma' \quad ...(B)$$

Cor. 1: *If (a_1, b_1, c_1) and (a_2, b_2, c_2) be the direction ratios of the given lines, then their actual direction cosines are given by*

$$\frac{a_1}{\sqrt{(a_1^2+b_1^2+c_1^2)}};\ \frac{b_1}{\sqrt{(a_1^2+b_1^2+c_1^2)}};\ \frac{c_1}{\sqrt{(a_1^2+b_1^2+c_1^2)}}$$

and

$$\frac{a_2}{\sqrt{(a_2^2+b_2^2+c_2^2)}};\ \frac{b_2}{\sqrt{(a_2^2+b_2^2+c_2^2)}};\ \frac{c_2}{\sqrt{(a_2^2+b_2^2+c_2^2)}}$$

∴ The angle θ between the line is given by

$$\cos\theta = \frac{a_1a_2+b_1b_2+c_1c_2}{\sqrt{(a_1^2+b_1^2+c_1^2)}\cdot\sqrt{(a_2^2+b_2^2+c_2^2)}} \quad ...(C)$$

Cor. 2 : *Lagrange's Identity we know that*

$$\left(l_1^2+m_2^2+n_1^2\right)\left(l_2^2+m_2^2+n_2^2\right)-\left(l_1l_2+m_1m_2+n_1n_2\right)^2$$

$$\equiv (m_1n_2-m_2n_1)^2+(n_1l_2-n_2l_1)^2+(l_2m_2-l_2m_1)^2$$

Now $\sin^2\theta = 1-\cos\theta$

$= 1-(l_1l_2+m_1m_2+n_1n_2)^2$

$= \left(l_1^2+m_1^2+n_1^2\right)\left(l_2^2+m_2^2+n_2^2\right)-\left(l_1l_2+m_1m_2+n_1n_2\right)^2,$

since $l_1^2+m_1^2+n_1^2 = 1 = l_2^2+m_2^2+n_2^2$

$= (m_1n_2-m_2n_1)^2+(n_1l_2-n_2l_1)^2+(l_1m_2-l_2m_1)^2$, by Lagrange's identity

$$\therefore\ \sin\theta = \pm\sqrt{[(m_1n_2-m_2n_1)^2+(n_1l_2-n_2l_1)^2+(l_1m_2-l_2m_1)^2]} \quad ...(D)$$

$$= \pm\sqrt{\left[\begin{vmatrix} m_1 & m_2 \\ n_1 & n_2 \end{vmatrix}^2+\begin{vmatrix} n_1 & n_2 \\ l_1 & l_2 \end{vmatrix}^2+\begin{vmatrix} l_1 & l_2 \\ m_1 & m_2 \end{vmatrix}^2\right]}$$

and

$$\tan\theta = \frac{\sin\theta}{\cos\theta} = \frac{\pm\sqrt{[\sum(m_1n_2-m_2n_1)^2]}}{l_1l_2+m_1m_2+n_1n_2}$$

substituting values of sin θ and cos θ.

Cor. 3 : *If instead of direction cosines we have the direction ratios of the lines as in cor. 1 above*, then we have

$$\sin\theta = \frac{\pm\sqrt{[(b_1c_2 - b_2c_1)^2 + (c_1a_2 - c_2a_1)^2 + (a_1b_2 - a_2b_1)^2]}}{\sqrt{(a_1^2 + b_1^2 + c_1^2)}\,\sqrt{(a_2^2 + b_2^2 + c_2^2)}}$$

and $$\tan\theta = \frac{\sin\theta}{\cos\theta} = \pm\frac{\sqrt{[\sum(b_1c_2 - b_2c_1)^2]}}{a_1a_2 + b_1b_2 + c_1c_2}$$

Cor. 4: *Condition for Perpendicularity of Two Lines*

If the two given lines whose d.c.'s are (l_1, m_1, n_1) and (l_2, m_2, n_2) are perpendicular to each other, then θ = 90° *i.e.*, cos θ = 0 and from result (A) of we have the required condition as

$$\mathbf{l_1l_2 + m_1m_2 + n_1n_2 = 0} \quad \text{...(E)}$$

If instead of d.c.'s we have the direction ratios of the lines as a_1, b_1, c_1 and a_2, b_2, c_2 then from result (C) above, we have

$$\mathbf{a_1a_2 + b_1b_2 + c_1c_2 = 0} \quad \text{...(F)}$$

Cor. 5 : *Condition for Parallelism of Two Lines*

If two given lines are parallel, then θ = 0 *i.e.*, sin θ = 0 and from result (D) above we have the required condition as

$$(m_1n_2 - m_2n_1)^2 + (n_1l_2 - n_2l_1)^2 + (l_1m_2 - l_2m_1)^2 = 0.$$

But this being the sum of squares of three quantities can be zero only if each of them is separately zero. **(Note)**

i.e., if $m_1m_2 - m_2n_1 = 0,\ n_1l_2 - n_2l_1 = 0$

and $l_1m_2 - l_2m_1 = 0$

i.e., if $$= \frac{m_1}{m_2} = \frac{n_1}{n_2} = \frac{l_1}{l_2} = \frac{\sqrt{(l_1^2 + m_1^2 + n_1^2)}}{\sqrt{(l_2^2 + m_2^2 + n_2^2)}} = 1$$ **(Note)**

i.e., if $l_1 = l_2$, $m_1 = m_2$ and $n_1 = n_2$

i.e., *the d.c.'s of parallel lines are the same* (Remember)

If instead of d.c.'s we have the direction ratios of the given lines as a_1, b_1, c_1 and a_2, b_2, c_2 then proceeding as above with the help of cor. 3 above we get the required condition as

$$\frac{a_1}{a_2} = \frac{b_1}{b_2} = \frac{c_1}{c_2}$$

i.e., *the direction ratios of parallel lines are proportional.*

Example 1:

Find the angles of triangle ABC whose vertices A, B and C are the points (1, 3, 5), (–1, 2, 3) and (3, 4, –2).

Solution:

$\cos^{-1}[3/2\sqrt{(10)}]$, 90°, $\cos^{-1}[3/2\sqrt{2}]$.

Example 2:

If A, B, C are (–1, 3, 2), (2, 3, 5) and (3, 5, –2); find the angles of the triangle ABC).

Solution:

The direction ratios of the line AB are {2 – (–1)}, (3 – 3), (5 – 2) *i.e.,* 3, 0, 3.

∴ The d.c.'s of the line AB are

$$\frac{3}{\sqrt{(3^2+0^2+3^2)}}, \frac{0}{\sqrt{(3^2+0^2+3^2)}}, \frac{3}{\sqrt{(3^2+0^2+3^2)}}$$

i.e., $\frac{3}{3\sqrt{2}}, \frac{0}{3\sqrt{2}}, \frac{3}{3\sqrt{2}}$, i.e. $\frac{1}{\sqrt{2}}, 0\ \frac{1}{\sqrt{2}}$,

$\frac{1}{3\sqrt{6}}, \frac{2}{3\sqrt{6}}, \frac{-7}{3\sqrt{6}}$, and $\frac{2}{3}, \frac{1}{3}, \frac{-2}{3}$ respectively

Similarly d.c.'s of the lines BC are

∴ cos A = cosine of the angle between AB and AC

$$= \text{“}l_1l_2 + m_1m_2 + n_1n_2\text{”}$$

$$= \frac{1}{\sqrt{2}}.\frac{2}{3} + 0.\frac{1}{3} - \frac{1}{\sqrt{2}}.\left(-\frac{2}{3}\right) = 0$$

Therefore $\angle A = 90°$ **Ans.**

cos B = cosine of the angle between BC and BA

$$= \frac{1}{3\sqrt{6}}\left(-\frac{1}{\sqrt{2}}\right) + \frac{2}{3\sqrt{6}}(0) - \frac{7}{3\sqrt{6}}\left(-\frac{1}{\sqrt{2}}\right),$$ Note the d.c.'s of BA

and $\cos B = \frac{6}{3\sqrt{6}.\sqrt{2}} = \frac{1}{\sqrt{3}}$

⇒ $\angle B = \cos^{-1}(1/\sqrt{3})$.

Similarly, we can find that $\angle C = \cos^{-1}(1/3)$ **Ans.**

SOME SOLVED EXAMPLES

Example 1:

If α, β, γ be the angles which a given line makes with the positive directions of the axes then prove that

$$\sin^2 \alpha + \sin^2 \beta + \sin^2 \gamma = 2.$$

Solution:

We know that $\cos^2 \alpha + \cos^2 \beta + \cos^2 \gamma = 1$

$\Rightarrow \quad (1 - \sin^2 \alpha) + (1 - \sin^2 \beta) + (1 - \sin^2 \gamma) = 1$

$\Rightarrow \quad \sin^2 \alpha + \sin^2 \beta + \sin^2 \gamma = 3 - 1 = 2.$

Example 2:

Find the d.c.'s l, m, n of two lines which are connected by the relation $l - 5m + 3n = 0$ and $7l^2 + 5m^2 - 3n^2 = 0$.

Solution:

Given $\quad l - 5m + 3n = 0 \quad$ ...(i)

$7l^2 + 5m^2 - 3n^2 = 0. \quad$ ...(ii)

Substituting the value of l from (i) in (ii) we get

$7(5m - 3n)^2 + 5m^2 - 3n^2 = 0$

$\Rightarrow \quad 180m^2 - 210mn + 60n^2 = 0$

$\Rightarrow \quad 6m^2 - 7mn + 2n^2 = 0$

$\Rightarrow \quad (3m - 2n)(2m - n) = 0$

$\Rightarrow \quad \dfrac{m}{n} = \dfrac{2}{3}, \dfrac{1}{2}$

If m/n = 2/3, we have $\dfrac{m}{2} = \dfrac{n}{3} = \dfrac{5m - 3n}{5.2 - 3.3} = \dfrac{l}{1}$, from (i)

$\Rightarrow \quad \dfrac{l}{1} = \dfrac{m}{2} = \dfrac{n}{3} = \dfrac{\sqrt{(l^2 + m^2 + n^2)}}{\sqrt{(1^2 + 2^2 + 3^2)}} = \dfrac{1}{\sqrt{(14)}}$

∴ The direction cosines of one line are

$1/\sqrt{14},\ 2/\sqrt{14},\ 3/\sqrt{14}.$

If m/n = 1/2, we have $\dfrac{m}{1} = \dfrac{n}{2} = \dfrac{5m - 3n}{5.2 - 3.3} = \dfrac{l}{-1}$, from (i)

$\Rightarrow \quad \dfrac{l}{-1} = \dfrac{m}{1} = \dfrac{n}{2} = \dfrac{\sqrt{(l^2 + m^2 + n^2)}}{\sqrt{[(-1)^2 + (1)^2 + (2)^2)}} = \dfrac{1}{\sqrt{6}}$

∴ The direction cosines of the other line are $-1/\sqrt{6}$, $1/\sqrt{6}$, $2/\sqrt{6}$ **Ans.**

Example 3:

If A, B, C, D are the points (3, 4, 5), (4, 6, 3), (–1, 2, 4) and (1, 0, 5), find the projection of CD on AB.

Solution:

Let l, m, n be the direction cosines of AB.

$$\text{Then } l = \frac{4-3}{\sqrt{[(4-3)^2+(6-4)^2+(3-5)^2]}} = \frac{1}{\sqrt{[(1^2+2^2+2^2)]}} = \frac{1}{3}.$$

$$\text{Similarly } m = \frac{6-4}{\sqrt{[(4-3)^2+(6-4)^2+(3-5)^2]}} = \frac{2}{3}$$

$$\text{and } n = \frac{3-5}{\sqrt{[(4-3)^2+(6-4)^2+(3-5)^2]}} = \frac{-2}{3}$$

∴ The projection of CD on AB

$= [1-(-1)\,l + (0-2)\,m + (5-4)n,$

$= 2\left(\frac{1}{3}\right) - 2\left(+\frac{2}{3}\right) + 1\left(-\frac{1}{3}\right)$

$= -\frac{4}{3} = \frac{4}{3}$, taking numerical value.

Example 4:

Coordinates of four points A, B, C, D are (1, 2, 3), (3, 5, 7), (2, 3, –1) and (3, 4, –3). Find the projection of AB and CD.

Solution:

Let l, m, n be the direction cosines of CD.

$$\text{Then } l = \frac{(3-2)}{\sqrt{[(3-2)^2+(4-3)^2+(-3+1)^2]}} = \frac{1}{\sqrt{6}},$$

$$m = \frac{(4-3)}{\sqrt{[(3-2)^2+(4-3)^2+(-3+1)^2]}} = \frac{1}{\sqrt{6}}$$

$$n = \frac{(-3+1)}{\sqrt{[(3-2)^2+(4-3)^2+(-3+1)^2]}} = \frac{-2}{\sqrt{6}}$$

∴ The required projection of AB on CD.

$= "(x_2 - x_1)\,l + (y_2 - y_1)\,m + (z_2 - z_1)\,n"$

$= (3 - 1)(1/\sqrt{6}) + (5 - 2)(1/\sqrt{6}) + (7 - 3)(-2/\sqrt{6})$

$= (2/\sqrt{6}) + (3/\sqrt{6}) - (8/\sqrt{6}) = -3/\sqrt{6} = -\sqrt{(3/2)}$ **Ans.**

Example 5:

The projection of a line on the co-ordinate axes are 2, 3, 6. Find the length and the direction cosines of the line.

Solution:

Let PQ be the line and its direction cosines be $\cos\alpha$, $\cos\beta$ and $\cos\gamma$. Then the projection of the line PQ on the co-ordinate axes are PQ $\cos\alpha$, PQ $\cos\beta$ and PQ $\cos\gamma$.

Then according to the problem, we have

$$PQ\cos\alpha = 2,\ PQ\cos\beta = 3,\ PQ\cos\gamma = 6. \qquad \text{...(i)}$$

Squaring and adding these, we get

$(PQ)^2[\cos^2\alpha + \cos^2\beta + \cos^2\gamma] = 2^2 + 3^2 + 6^2 = 4 + 9 + 36$

$\Rightarrow \quad (PQ)^2[1] = 49, \quad \because \cos^2\alpha, \cos^2\beta$ and $\cos^2\gamma = 1$

$\Rightarrow \quad PQ = \sqrt{(49)} = 7.$ **Ans.**

$\therefore$ From (i), we have $7\cos\alpha = 2,$

$7\cos\beta = 3,$

$7\cos\gamma = 6$

$\Rightarrow \quad \cos\alpha = 2/7,\ \cos\beta = 3/7$ and $\cos\gamma = 6/7.$ **Ans.**

Example 6:

The co-ordinates of a point A are (2, 3, 6). Find the direction cosines of OA, where O is origin.

Solution:

The direction cosines of OA are

$$\frac{2-0}{\sqrt{(2^2+3^2+6^2)}}, \frac{3-0}{\sqrt{(2^2+3^2+6^2)}}, \frac{6-0}{\sqrt{(2^2+3^2+6^2)}},$$

$\because$ O is (0, 0, 0)

$\Rightarrow \quad 2/7, 3/7, 6/7.$

Example 7:

If a pair of opposite edges of a tetrahedron be perpendicular, then prove that the distances between the mid-point of the other two pairs of opposite edges are equal.

Solution:

Let OABC be the tetrahedron, then with the same notations, we find that if OA is perpendicular to BC, then

$$x_1 (x_3 - x_2) + y_1 (y_3 - y_2) + z_1 (z_3 - z_2) = 0 \quad \ldots(i)$$

Also the mid-points of OB and CA are

$$\left(\frac{1}{2}x_2, \frac{1}{2}y_2, \frac{1}{2}z_2\right) \text{ and } \left(\frac{x_1 + x_3}{2}, \frac{y_1 + y_3}{2}, \frac{z_1 + z_3}{2}\right) \text{ respectively.}$$

$\therefore$ square of the distance between the mid-points of OB and CA

$$= \{\tfrac{1}{2}(x_1 + x_3) - \tfrac{1}{2}x_2\}^2 + \{\tfrac{1}{2}(y_1 + y_3) - \tfrac{1}{2}y_2\}^2 + \{\tfrac{1}{2}(z_1 + z_3) - \tfrac{1}{2}z_2\}^2$$

$$= \tfrac{1}{4}[(x_1 + x_3 - x_2)^2 + (y_1 + y_3 - y_2)^2 + (z_1 + z_3 - z_2)^2] \quad \ldots(ii)$$

Similarly the square of the distance between the mid-points of OC and AB $= \tfrac{1}{4}[(x_1 + x_2 - x_3)^2 + (y_1 + y_2 - y_3)^2 + (z_1 + z_2 - z_3)^2] \quad \ldots(iii)$

If these distances are equal, then from (ii) and (iii) we have

$$(x_1 + x_3 - x_2)^2 + (y_1 + y_3 - y_2)^2 + (z_1 + z_3 - z_2)^2$$
$$= (x_1 + x_2 - x_3)^2 + (y_1 + y_2 - y_3)^2 + (z_1 + z_2 - z_3)^2$$

$$\Rightarrow \quad \{(x_1 + x_3 - x_2)^2 - (x_1 + x_2 - x_3)^2\} + \{(y_1 + y_3 - y_2)^2 - (y_1 + y_2 - y_3)^2\} + \ldots = 0$$

$$\Rightarrow \{2x_1 . 2(x_3 - x_2)\} + \{2y_1 . 2(y_3 - y_2)\} + \{2z_1 . 2(z_3 - z_2)\} = 0$$

$\Rightarrow x_1 (x_3 - x_2) + y_1 (y_3 - y_2) + z_1 (z_3 - z_2) = 0$, which is true by virtue of (i).

Hence proved.

Example 8:

Prove that there concurrent lines with direction cosines (l_1, m_1, n_1), (l_2, m_2, n_2) and (l_3, m_3, n_3) are coplanar, if

$$\begin{vmatrix} l_1 & m_1 & n_1 \\ l_2 & m_2 & n_2 \\ l_3 & m_2 & n_3 \end{vmatrix} = 0$$

Solution:

Let l, m, n be the d.c.'s of the normal to the plane which contains the two concurrent lines with d.c.'s (l_1, m_1, n_1) and (l_2, m_2, n_2). Then we have

$$ll_1 + mm_1 + nn_1 = 0 \quad \ldots(i)$$

and $$ll_2 + mm_2 + nn_2 = 0 \quad \ldots(ii)$$

If the third line with d.c.'s (l_2, m_3, n_3) also lies on this plane then the line wide d.c.'s $(l\ m, n)$ is at right angles to this third line and so we have

$$ll_3,\ mm_3,\ nn_3 = 0 \quad \ldots\text{(iii)}$$

Eliminating l, m, n, from (i), (ii) and (iii) we have

$$\begin{vmatrix} l_1 & m_1 & n_1 \\ l_2 & m_2 & n_2 \\ l_3 & m_2 & n_3 \end{vmatrix} = 0,$$ as the required condition.

Example 9:

If in a tetrahedron OABC, $OA^2 + BC^2 = OB^2 + CA^2 = OC^2 + AB^2$, then its pairs of opposite edges are at right angles.

Solution:

With the same coordinates, we have

$$OA^2 + BC^2 = (x_1^2 + y_1^2 + z_1^2) + [(x_3 - x_2)^2 + (y_3 - y_2)^2 + (z_3 - z_2)^2]$$

$$= (x_1^2 + y_1^2 + z_1^2) + (x_2^2 + y_2^2 + z_2^2) + (x_3^2 + y_3^2 + z_3^2) - 2\,(x_2x_3 + y_2y_3 + z_2z_3)$$

Similarly, we can prove that

$$OB^2 + CA^2 = (x_1^2 + y_1^2 + z_1^2) + (x_2^2 + y_2^2 + z_2^2) + (x_3^2 + y_3^2 + z_3^2) - 2\,(x_1x_3 + y_1y_3 + z_1z_3)$$

$$OC^2 + AB^2 = (x_1^2 + y_1^2 + z_1^2) + (x_2^2 + y_2^2 + z_2^2) + (x_3^2 + y_3^2 + z_3^2) - 2\,(x_1x_2 + y_1y_2 + z_1z_2)$$

If $OA^2 + BC^2 = OB^2 + CA^2$, then we have

$$x_2x_3 + y_2y_3 + z_2z_3 = x_1x_3 + y_1y_3 + z_1z_3$$

$$\Rightarrow x_3\,(x_2 - x_1) + y_3\,(y_2 - y_1) + z_3\,(z_2 - z_1) = 0,$$

which shows that OC is perpendicular to AB *i.e.,* a pair of opposite edges of a tetrahedron OABC are perpendicular. In a similar manner we can prove for other pairs.

Example 10:

Show that the points (0, 4, 1), (2, 3, –1) (4, 5, 0) and (2, 6, 2) are the vertices of a square.

Solution:

Let A, B, C, D be respectively the points (0, 4, 1), (2, 3, –1), (4, 5, 0) and (2, 6, 2).

Find the lengths of sides AB, BC, CD, DA and diagonals AC, BD of the quadrilateral ABCD.

The prove that AB = BC = CD = DA = 3 and AC = BD = $3\sqrt{2}$

$\because$ sides of quad. ABCD are equal in length and its diagonals are also equal in length, so it represents a square.

Example 11:

The vertices of a triangle ABC are the points A (–1, 2, –3), B (5, 0, –6) and C (0, 4, –1) in order. Find the direction ratios of the bisectors of the angle BAC.

Solution:

$$AB = \sqrt{[(5 + 1)^2 + (0 - 2)^2 + (-6 + 3)^2]} = 7;$$

$$AC = \sqrt{[0 + 1)^2 + (4 - 2)^2 + (-1 + 3)^2]} = 3.$$

Let the internal bisector of $\angle$ BAC meet BC in D. Then

BD : DC = BA : CA = 7 : 3 **(Note)**

i.e., D divides BC internally in the ratio 7 : 3

$\therefore$ The co-ordinates of D are

$$\left[\frac{7.0 + 3.5}{7 + 3}, \frac{7.4 + 3.0}{7 + 3}, \frac{7(-1) + 3(-6)}{7 + 3}\right] \text{ i.e. } \left[\frac{3}{2}, \frac{14}{5}, -\frac{5}{2}\right]$$

$\therefore$ The d.c.'s of the line AD are proportional to [(3/2) + 1], [(14/5) – 2], [(–5/2) + 3] *i.e.,* 5/2, 4/5, 1/2 *i.e.,* proportional to (25, 8, 5).

Again if the external bisector of the angle BAC meets BC in E, then E divides BC *externally* in the ratio 7 : –3.

$\therefore$ The coordinates of E are

$$\left[\frac{7.0 - 3.5}{7 - 3}, \frac{7.4 - 3.0}{7 - 3}, \frac{7(-1) - 3(-6)}{7 - 3}\right] \text{ i.e. } \left[\frac{-15}{4}, 7, -\frac{11}{4}\right]$$

$\therefore$ The direction ratios of the line AE are

$$\left(\frac{-15}{4} + 1, 7 - 2, -\frac{11}{4} + 3\right) \text{ i.e. } \left(\frac{-11}{4}, 5, \frac{23}{4}\right)$$

i.e., (–11, 20, 23).m **Ans.**

Example 12:

A plane makes intercepts OA, OB, OC whose measures are a, b, c on the axes OX, OY, OZ. Find the area of the triangle ABC.

Solution:

The co-ordinates of A, B, C are (a, 0, 0), (0, b, 0) and (0, 0, c) respectively.

Now the area of the triangle ABC = 1/2 CA.CB sin ∠ ACB. ...(i)

Now $CA = \sqrt{[(0-a)^2 + (0-0)^2 + (c-0)^2]} = \sqrt{(a^2 + c^2)}$

$CB = \sqrt{[(0-0)^2 + (0-b)^2 + (c-0)^2]} = \sqrt{(b^2 + c^2)}$

Also direction cosines of the lines CA and CB are

$$\frac{0-a,\ 0-0,\ c-0}{\sqrt{\{(0-a)^2 + (0-0)^2 + (c-0)^2\}}} \text{ i.e. } \frac{-a,\ 0,\ c}{\sqrt{(a^2+c^2)}}$$

and
$$\frac{0-0,\ 0-b,\ c-0}{\sqrt{\{(0-0)^2 + (0-b)^2 + (c-0)^2\}}} \text{ i.e. } \frac{0,\ -b,\ c}{\sqrt{(b^2+c^2)}}$$

$\sin ACB =$ "$\sqrt{[\Sigma (m_1n_2 - m_2n_1)^2]}$"

$$= \frac{\sqrt{[(0.c + b.c)^2 + (c.0 + c.a)^2 + (a.b - 0.0)^2}}{\sqrt{(a^2+c^2)}\,.\,\sqrt{(b^2+c^2)}}$$

$$= \frac{\sqrt{(b^2c^2 + c^2a^2 + c^2b^2)}}{\sqrt{(a^2+c^2)}\,.\,\sqrt{(b^2+c^2)}}$$

∴ From (i), the required area of Δ ABC = 1/2 CA.CB sin ∠ ACB

$$= \frac{1}{2}\sqrt{(a^2+c^2)}\,.\,\sqrt{(b^2+c^2)}\ \frac{\sqrt{(b^2c^2 + c^2a^2 + c^2b^2)}}{\sqrt{(a^2+c^2)}\,.\,\sqrt{(b^2+c^2)}}$$

$$= \frac{1}{2}\sqrt{(b^2c^2 + c^2a^2 + c^2b^2)}$$ **Ans.**

Example 13:

If two pairs of opposite edges of a tetrahedron are perpendicular, show that the third pair is also perpendicular.

Solution:

Let OABC be the tetrahedron where, O, A, B and C are (0, 0, 0), (x_1, y_1, z_1), (x_2, y_2, z_2) and (x_3, y_3, z_3) respectively.

Let two pairs of opposite edges *viz.*, OA, BC and OC, AB be perpendicular.

The d.c.'s of OA, OC, BC and AB are proportional to (x_1, y_1, z_1), (x_3, y_3, z_3), $(x_3 - x_2, y_3 - y_2, z_3 - z_2)$ and $(x_2 - x_1, y_2 - y_1, z_2 - z_1)$ respectively.

As OA is perpendicular to BC, so

$$x_1 (x_3 - x_2) + y_1 (y_3 - y_2) + z_1 (z_3 - z_2) = 0 \quad ...(i)$$

As OC is perpendicular to AB, so

$$x_3 (x_2 - x_1) + y_3 (y_2 - y_1) + z_3 (z_2 - z_1) = 0 \quad ...(ii)$$

Adding (i) and (ii) we have $x_2(x_3 - x_1) + y_2(y_3 - y_1) + z_2(z_3 - z_1) = 0$ which shows that the third pair of opposite edges viz. OB, AC are also perpendicular. **Hence proved.**

Example 14:

Find the area of the triangle OAB where, O, A and B are (0, 0, 0), (x_1, y_1, z_1) and (x_2, y_2, z_2) respectively.

Solution:

The area of the triangle OAB $= \frac{1}{2}$ OA.OB sin ∠ OAB. ...(i)

Now $OA = \sqrt{(x_1^2 + y_1^2 + z_1^2)}$, $OB = \sqrt{(x_2^2 + y_2^2 + z_2^2)}$

and $\sin \angle AOB = \dfrac{\sqrt{[(y_1z_2 - y_2z_1)^2 + (z_1x_2 - z_2x_1)^2 + (x_1y_2 - x_2y_1)^2]}}{\sqrt{(x_1^2 + y_1^2 + z_1^2)}\sqrt{(x_2^2 + y_2^2 + z_2^2)}}$

since the d.c.'s of the lines OA and OB are

$$\frac{x_1}{\sqrt{(x_1^2 + y_1^2 + z_1^2)}}, \frac{y_1}{\sqrt{(\Sigma x_1^2)}}, \frac{z_1}{\sqrt{(\Sigma x_1^2)}}$$

and $$\frac{x_2}{\sqrt{(x_2^2 + y_2^2 + z_2^2)}}, \frac{y_2}{\sqrt{(\Sigma x_2^2)}}, \frac{z_2}{\sqrt{(\Sigma x_2^2)}}$$

∴ from (i), the required area of Δ OAB

$$= \frac{1}{2}\sqrt{(x_1^2 + y_1^2 + z_1^2)}.\sqrt{(x_2^2 + y_2^2 + z_2^2)} \times \frac{\sqrt{[\Sigma(y_1z_2 - y_2z_1)^2]}}{\sqrt{(x_1^2 + y_1^2 + z_1^2)}\sqrt{(x_2^2 + y_2^2 + z_2^2)}}$$

$$= \frac{1}{2}\sqrt{[(y_1z_2 - y_2z_1)^2 + (z_1x_2 - z_2x_1)^2 + (x_1y_2 - x_2y_1)^2]}.$$ **Ans.**

Example 15:

Show that the lines where direction cosines are given by the equations $2l + 2m - n$ and $mn + nl + lm = 0$ are at right angles.

Solution:

Eliminating n between given equations we get

$$m(2l + 2m) + (2l + 2m)\, l + lm = 0 \text{ or } 2l^2 + 2m^2 + 5lm = 0$$

$$\Rightarrow \quad 2(l/m)^2 + 5(l/m) + 2 = 0, \text{ dividing each term by } m^2$$

$$\Rightarrow \quad \frac{l}{m} = \frac{-5 \pm \sqrt{[(5)^2 - 4(2)(2)]}}{2(2)} = \frac{-5 \pm 3}{4} = -\frac{1}{2}, -2$$

Let l_1, m_1, n_1 and l_2, m_2, n_2 be the d.c.'s of the two lines, then we have

$$\frac{l_1}{m_1} = -\frac{1}{2} (= -1/2, \quad \frac{l_2}{m_2} = -2 \qquad \text{...(i)}$$

If $\frac{l_1}{m_1} = -\frac{1}{2}$, then $\frac{l_1}{-1} = \frac{m_1}{2} = k_1$ (say)

$$\Rightarrow \quad l_1 = -k_1, \; m_1 = 2k_1$$

Also from $n = 2l + 2m$ we have $n_1 = 2l_1 + 2m_1$

$$\Rightarrow \quad n_1 = 2(-k_1) + 2(2k_1) = 2k_1$$

$\therefore$ we have $l_1 = -k_1$, $m_1 = 2k_1$, $n = 2k_1$ ...(iii)

Again from (i), $\frac{l_2}{m_2} = \frac{-2}{1}$ or $\frac{l_2}{-2} = \frac{m_2}{1} = k_2$ (say)

$$\Rightarrow \quad l_2 = -2k_2 \; m_2 = k_2$$

And from $n = 2l + 2m$

we have $n_2 = 2l_2 + 2m_2$

$$\Rightarrow \quad n_2 = 2(-2k_2) + 2k_2 = -2k_2$$

$\therefore$ we have $l_2 = -2k_2$, $m_2 = k_2$, $n_2 = -2k_2$

$\therefore$ If θ be the angle between the given lines, then

$$\begin{aligned} \cos\theta &= \text{"}a_1a_2 + b_1b_2 + c_1c_2\text{"} \\ &= (-k_1)(-2k_2) + (2k_1)(k_2) + (2k_1)(-2k_2) \\ &= 2k_1k_2 + 2k_1k_2 - 4k_1k_2 = 0 \end{aligned}$$

Hence the given lines are at right angles.

Example 16:

Show that the equation to the right circular cone whose vertex is at the origin, whose axis has d.c.'s $\cos\alpha$, $\cos\beta$, $\cos\gamma$ and whose semi-vertical angle is θ is

$$(y\cos\gamma - z\cos\beta)^2 + (z\cos\alpha - x\cos\gamma)^2 + (x\cos\beta - y\cos\alpha)^2 = (x^2 + y^2 + z^2)\sin^2\theta.$$

Solution:

Let P(x, y, z) be any point on the cone (*i.e.,* on the surface of the cone). The vertex of the cone is given as the origin O(0, 0, 0).

Then the d.c.'s of the generator OP are

$$\frac{x-0}{\sqrt{[(x-0)^2+(y-0)^2+(z-0)^2]}}, \frac{y-0}{\sqrt{[(x-0)^2+(y-0)^2+(z-0)^2]}}, \frac{z-0}{\sqrt{[(x-0)^2+(y-0)^2+(z-0)^2]}}$$

i.e., $$\frac{x}{\sqrt{(x^2+y^2+z^2)}}, \frac{y}{\sqrt{(x^2+y^2+z^2)}}, \frac{z}{\sqrt{(x^2+y^2+z^2)}}.$$

Also the d.c.'s of the axis of the cone are given as $\cos\alpha$, $\cos\beta$, $\cos\gamma$ and the line OP is inclined at an angle θ to the axis.

$$\therefore \sin^2\theta = \Sigma (m_1n_2 - m_2n_1)^2$$

$$= \Sigma\left[\frac{y\cos\gamma - z\cos\beta}{\sqrt{(x^2+y^2+z^2)}}\right]^2 = \frac{\Sigma(y\cos\gamma - z\cos\beta)^2}{(x^2+y^2+z^2)}$$

$$\Rightarrow \quad (x^2+y^2+z^2)\sin^2\theta = (y\cos\gamma - z\cos\beta)^2 + (z\cos\alpha - x\cos\gamma)^2 + (x\cos\beta - y\cos\alpha)^2.$$ **Ans.**

Example 17:

O, A, B, C are four points not necessarily lying in the same plane and such that OA ⊥ BC and OB ⊥ CA. Prove that OC ⊥ AB.

Solution:

Let O, A, B, C be the points (0, 0, 0), (x_1, y_1, z_1), (x_2, y_2, z_2), (x_3, y_3, z_3) respectively.

Then direction ratios of the lines OA, OB and OC are $x_1 - 0, y_1 - 0, z_1 - 0$; $x_2 - 0, y_2 - 0, z_2 - 0$ and $x_3 - 0, y_3 - 0, z_3 - 0$ *i.e.*, x_1, y_1, z_1; x_2, y_2, z_2 and x_3, y_3, z_3 respectively. ...(i)

Also the direction ratios of AB, BC and CA are respectively

$x_2 - x_1, y_2 - y_1, z_2 - z_1$, $x_3 - x_2, y_3 - y_2, z_3 - z_2$

and $x_1 - x_3, y_1 - y_3, z_1 - z_3$ ...(ii)

Now if OA ⊥ BC, then from (i) and (ii), we get

$$x_1(x_3 - x_2) + y_1(y_3 - y_2) + z_1(z_3 - z_2) = 0$$

$$\Rightarrow \quad x_1x_3 - x_1x_2 + y_1y_3 - y_1y_2 + z_1z_3 - z_1z_2 = 0 \quad \text{...(iii)}$$

And if OB ⊥ CA, then from (i) and (ii), we get

$$x_2(x_1 - x_3) + y_2(y_1 - y_3) + z_2(z_1 - z_3) = 0$$

$\Rightarrow \quad x_2x_1 - x_2x_3 + y_2y_1 - y_2y_3 + z_2z_1 - z_2z_3 = 0$ ...(iv)

Adding (iii) and (iv), we get

$$x_1x_3 - x_2x_3 + y_1y_3 - y_2y_3 + z_1z_3 - z_2z_3 = 0$$

$$\Rightarrow \quad x_3(x_2 - x_1) + y_3(y_2 - y_1) + z_3(z_2 - z_1) = 0$$

i.e., the lines whose direction ratios are x_3, y_3, z_3 and $x_2 - x_1, y_2 - y_1, z_2 - z_1$ are perpendicular.

i.e., OC ⊥ AB, from (i) and (ii). **Hence proved.**

Example 18:

Find the angle between the diagonals of a cube

Solution:

We can prove that the d.c.'s of the diagonals OP and CN of the cube are $\left(\frac{1}{\sqrt{3}}, \frac{1}{\sqrt{3}}, \frac{1}{\sqrt{3}}\right)$ and $\left(\frac{1}{\sqrt{3}}, \frac{1}{\sqrt{3}}, \frac{-1}{\sqrt{3}}\right)$ respectively.

Therefore if θ be the angle, between OP and CN, then

$$\cos\theta = "l_1l_2 + m_1m_2 + n_1n_2 = \frac{1}{\sqrt{3}}\cdot\frac{1}{\sqrt{3}} + \frac{1}{\sqrt{3}}\cdot\frac{1}{\sqrt{3}} + \frac{1}{\sqrt{3}}\left(-\frac{1}{\sqrt{3}}\right) = \frac{1}{3}$$

$\Rightarrow \quad \theta = \cos^{-1}\left(\frac{1}{3}\right)$ **Ans.**

Example 19:

Show that the line joining the points (1, 2, 3), (–1, –2, –3) is parallel to the line joining the points (2, 3, 4), (5, 9, 13) and perpendicular to the line joining the points (–2, 1, 5), (3, 3, 2).

Solution:

The direction ratios of the line joining (1, 2, 3) and (–1, –2, –3) are –1 –1, –2 –2, –3, –3 *i.e.*, –2, –4, –6 *i.e.*, 1, 2, 3 ...(i)

The direction ratios of the line joining the points (2, 3, 4) and (5, 9, 13) are 5 –2, 9 –3, 13 –4 *i.e.*, 3, 6, 9 *i.e.*, 1, 2, 3 ...(ii)

From (i) and (ii) we conclude that these two lines are parallel.

Also the direction ratios of the line joining the points (–2, 1, 5) and (3, 3, 2) are 3 + 2, 3 –1, 2 –5 *i.e.*, 5, 2, –3 ...(iii)

Now as (1) (5) + (2) (2) + (3) (–3) = 5 + 4 – 9 = 0

So the lines whose d.c.'s are given by (i) and (iii) are perpendicular.

Example 20:

A line makes angles α, β, γ, δ with the four diagonals of a cube; prove that $\cos^2 \alpha + \cos^2 \beta + \cos^2 \gamma + \cos^2 \delta = 4/3$.

Solution:

We can prove that the direction ratios of the diagonals OP, CN, AE and BD of the cube (whose edges are all equal *i.e.,* a = b = c) are (a, a, a); (a, a, –a) and (–a, a, a) respectively.

$\therefore$ d.c.'s of diagonals are

$$\frac{a}{\sqrt{(a^2 + a^2 + a^2)}}, \frac{a}{\sqrt{(a^2 + a^2 + a^2)}}, \frac{a}{\sqrt{(a^2 + a^2 + a^2)}} \text{ i.e. } \frac{1}{\sqrt{3}}, \frac{1}{\sqrt{3}}, \frac{1}{\sqrt{3}}$$

In a similar way the d.c.'s of the diagonals CN, AE and BD are

$$\left(\frac{1}{\sqrt{3}}, \frac{1}{\sqrt{3}}, \frac{-1}{\sqrt{3}}\right); \left(\frac{-1}{\sqrt{3}}, \frac{1}{\sqrt{3}}, \frac{1}{\sqrt{3}}\right) \text{ and } \left(\frac{1}{\sqrt{3}}, \frac{-1}{\sqrt{3}}, \frac{1}{\sqrt{3}}\right)$$

Let l, m, n be d.c.'s of the line which makes angles α, β, γ, δ with the diagonals of the cube.

Then $\cos \alpha = l\,(1/\sqrt{3}) + m(1/\sqrt{3}) + n(1/\sqrt{3}) = (l + m + n)/\sqrt{3}$

$\Rightarrow \quad \cos^2 \alpha = \frac{1}{3}\,(l + m + n)^2$ ...(i)

Similarly $\cos \beta = l\,(1/\sqrt{3}) + m(1/\sqrt{3}) + n(-1/\sqrt{3}) = (l + m - n)/\sqrt{3}$

$\Rightarrow \quad \cos^2 \beta = \frac{1}{3}\,(l + m - n)^2$ ...(ii)

And in a similar manner we can prove that

$\cos^2 \gamma = \frac{1}{3}\,(-l + m + n)^2$

and $\quad \cos^2 \delta = \frac{1}{3}\,(l - m + n)^2$ ...(iii)

From (i), (ii) and (iii),

we get $\cos^2 \alpha + \cos^2 \beta + \cos^2 \gamma + \cos^2 \delta$

$= \frac{1}{3}\,(l + m + n)^2 + (l + m - n)^2 + (-l + m + n)^2 + (l - m + n)^2]$

$= \frac{1}{3}\,[4\,(l^2 + m^2 + n^2)]$, on simplifying

$= 4/3$, since $l^2 + m^2 + n^2 = 1$. **Hence proved.**

Example 21:

Find the area of the triangle included between the plane $x + y + z = 5$ and the co-ordinate planes.

Solution:

If the given plane x + y + z = 5 or $\frac{x}{5}+\frac{y}{5}+\frac{z}{5}=1$ meet the coordinate area in A, B, and C, then we have A (5, 0, 0), B (0, 5, 0) and C (0, 0, 5).

Here a = b = c = 5. **Ans.** $(25\sqrt{3})/2$ sq. units.

Example 22:

If the edges of a rectangular parallepiped be a, b, c show that the angles between the four diagonals are given by

$$\cos^{-1}\left[\frac{\pm a^2 \pm b^2 \pm c^2}{a^2+b^2+c^2}\right].$$

Solution:

Let one corner O of the rectangular parallelepiped be taken as origin and the three coterminus edges OA, OB and OC be taken as coordinate axes. Let OA = a, OB = b and OC = c.

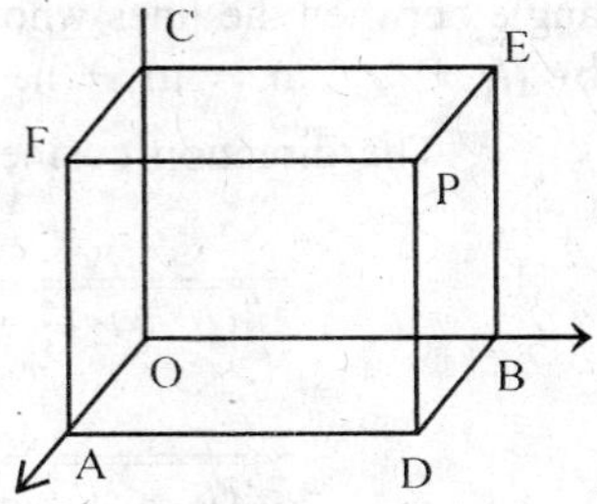

Then the co-ordinates of O, A, B, C, D, P, N and E are respectively (0, 0, 0), (a, 0, 0) (0, b, 0), (0, 0, c), (a, 0, c), (a, b, c), (a, b, 0) and (0, b, c) as is evident from the adjoining figure.

Here OP, CN, AE and BD are the diagonals.

The direction ratios of OP are a – 0, b – 0, c – 0

i.e., a, b, c

Again direction ratios of CN are (a – 0), (b – 0), (0 – c),

i.e., a, b, –c.

Similarly we can find that the direction ratios of AE and BD are –a, b, c and a, –b, c respectively.

If θ be the angle between OP and CN, then

$$\cos\theta = \frac{a_1a_2+b_1b_2+c_1c_2}{\sqrt{(a_1^2+b_1^2+c_1^2)}\cdot\sqrt{(a_2^2+b_2^2+c_2^2)}}$$

$$\Rightarrow \cos\theta = \frac{a.a+b.b+c(-c)}{\sqrt{(a^2+b^2+c^2)}\sqrt{[a^2+b^2+(-c)^2]}} = \frac{a^2+b^2-c^2}{a^2+b^2+c^2}$$

Similarly, we can find the angle between other pairs of diagonals and we have six such pairs out of these four diagonals and all these angles are given by

$$\cos^{-1}\left[\frac{\pm a^2 \pm b^2 \pm c^2}{a^2 + b^2 + c^2}\right]$$ **Ans.**

Example 23:

Find the direction ratios of the lines bisecting the angles between the lines whose direction ratios are l_1, m_1, n_1 and l_2, m_2, n_2 and the angle between these lines is θ.

Solution:

We have proved that the direction ratios of the internal bisector of the angle between the lines whose d.c.'s are (l_1, m_1, n_1) and (l_2, m_2, n_2) are given by $[l_1 + l_2 + m_1 + m_2 + n_1 + n_2]$.

∴ The direction cosines of the internal bisector are

$$\frac{l_1 + l_2}{\sqrt{[(l_1 + l_2)^2 + (m_1 + m_2)^2 + (n_1 + n_2)^2]}}$$

$$\frac{m_1 + m_2}{\sqrt{[(l_1 + l_2)^2 + (m_1 + m_2)^2 + (n_1 + n_2)^2]}}, \frac{n_1 + n_2}{\sqrt{[\ldots]}}$$

Now $\sqrt{(l_1 + l_2)^2 + (m_1 + m_2)^2 + (n_1 + n_2)^2]} = \sqrt{[\Sigma l_1^2 + \Sigma l_2^2 + 2\,\Sigma l_1 l_2]}$

$$= \sqrt{(1 + 1 + 2\cos\theta)}, \because \Sigma\, l_1 l_2 = \cos\theta$$

$$= \sqrt{[2(1 + \cos\theta)]} = 2\cos\frac{1}{2}\theta$$

The d.c.'s of the internal bisector are

$$\frac{l_1 + l_2}{2\cos\frac{1}{2}\theta}, \frac{m_1 + m_2}{2\cos\frac{1}{2}\theta}, \frac{n_1 + n_2}{2\cos\frac{1}{2}\theta}$$ **Ans.**

In a similar manner we can prove that the d.c.'s of the external bisector are

$$\frac{l_1 - l_2}{2\sin\frac{1}{2}\theta}, \frac{m_1 - m_2}{2\sin\frac{1}{2}\theta}, \frac{n_1 - n_2}{2\sin\frac{1}{2}\theta}$$ **Ans.**

where $\sqrt{[(l_1 - l_2)^2 + (m_1 - m_2)^2 + (n_1 - n_2)^2]}$

$$= \sqrt{[\Sigma l_1^2 + \Sigma l_2^2 + 2\,\Sigma l_1 l_2]}$$

$$= \sqrt{[1 + 1 - 2 \cos \theta]}$$

$$= \sqrt{[2(1 - \cos \theta)]} = 2 \sin \frac{1}{2} \theta.$$

Example 24:

Prove that the three lines drawn from a point with direction cosines proportional to (1, –1, 1), (2, –3, 0) and (1, 0, 3) are coplanar.

Solution:

Let PA, PB and PC be three given lines with d.c.'s proportional to (1, –1, 1), (2, –3, 0) and (1, 0, 3).

Let l, m, n be the direction ratios of the normal to the plane APB. Then as PA and PB are perpendicular to this normal so we have

$$l \, . \, 1 + m(-1) + n \, . \, 1 = 0 \qquad \ldots(i)$$

and

$$l \, . \, 2 + m(-3) + n \, . \, 0 = 0 \qquad \ldots(ii)$$

Solving (i) and (ii), we have

$$\frac{l}{(-1)\,0 - 1.(-3)} = \frac{m}{1.2 - 1.0} = \frac{b}{1.(-3) - (-1)2} \text{ i.e. } \frac{l}{3} = \frac{m}{2} = \frac{n}{-1} \qquad \ldots(iii)$$

If PC also lies in this plane PAB, then PC must be at right angles to this normal whose direction ratios are given by (iii), the condition of the same as

$$\text{"}a_1a_2 + b_1b_2 + c_1c_2 = 0\text{"}$$

Here $a_1a_2 + b_1b_2 + c_1c_2 = l \, . \, 1 + m \, . \, 0 + n \, . \, 3$

$$= 3.1 + 2.0 + (-1) \, . \, 3, \text{ from (iii)}$$

$$= 0.$$

Hence proved.

Example 25:

(l_1, m_1, n_1) and (l_2, m_2, n_2) are the d.c.'s of two concurrent lines, show that the d.c.'s of two lines bisecting the angles between them are proportional to $(l_1 \pm l_2, m_1 \pm m_2, n_1 \pm n_2)$.

Solution:

Draw two lines OP and OQ through the origin O parallel to the given concurrent lines.

Mark off OP = r and OQ = r. Then as (l_1, m_1, n_1) and (l_2, m_2, n_2) are the d.c.'s of the lines OP and OQ, therefore the co-ordinates of P and Q are (l_1r, m_1r, n_1r) and (l_2r, m_2r, n_2r) respectively.

Produce QO and mark OR = OQ = r, then the co-ordinates of R are $(-l_2r, -m_2r, -n_2r)$

Let A and B be the mid-points of PQ and PR then OA and OB are the internal and external bisectors of the ∠QOP.

The co-ordinates of A and B are

$$[\frac{1}{2}(l_1 r + l_2 r),\ \frac{1}{2}(m_1 r + m_2 r),\ \frac{1}{2}(n_1 r + n_2 r)]$$

and $$[\frac{1}{2}(l_1 r - l_2 r),\ \frac{1}{2}(m_1 r - m_2 r),\ \frac{1}{2}(n_1 r - n_2 r)]$$ respectively

∴ The direction ratios of OA and OB, where O is the origin (0, 0, 0) are

$$[\frac{1}{2}(l_1 + l_2)r,\ \frac{1}{2}(m_1 + m_2)r,\ \frac{1}{2}(n_1 + n_2)r]$$

and $$[\frac{1}{2}(l_1 - l_2)r,\ \frac{1}{2}(m_1 - m_2)r,\ \frac{1}{2}(n_1 - n_2)r]$$

i.e., d.c.'s of OA and OB are proportional to

$[(l_1 + l_2), (m_1 + m_2), (n_1 + n_2)]$ and $[(l_1 - l_2), (m_1 - m_2), (n_1 - n_2)]$ respectively.

Example 26:

Find the direction cosines of the line which is perpendicular to the lines whose direction cosines are proportional to 3, –1, 1 and –3, 2, 4.

Solution:

Do your self. **Ans.** $\frac{-2}{\sqrt{(30)}}, \frac{-5}{\sqrt{(30)}}, \frac{1}{\sqrt{(30)}}$.

Example 27:

(l_1, m_1, n_1) and (l_2, m_2, n_2) are the d.c.'s of two intersecting lines. Show that all lines through the intersection of these two whose direction cosines are proportional to $l_1 + \lambda l_2$, $m_1 + \lambda m_2$ + $n_1 + \lambda n_2$ are coplanar with them.

Solution:

Let l, m, n be the d.c.'s os the normal to the plane containing the two given intersecting lines. Then we have $l\, l_1 + m\, m_1 + n\, n_1 = 0$ ...(i)

and $l\, l_2 + m\, m_2 + n\, n_2 = 0$ ...(ii)

Multiplying (ii) by λ and adding to (i), we get

$l(\lambda l_2 + l_1) + m(\lambda m_2 + m_1) + n(\lambda n_2 + n_1) = 0$, which shows that the lines with d.c.'s (l, m, n) and $(l_1 + \lambda l_2, m_1 + \lambda m_2, n_1 + \lambda n_2)$ are perpendicular and as such all lines having d.c.'s $(l_1 + \lambda l_2, m_1 + \lambda m_2, n_1 + \lambda n_2)$ are coplanar with the given two intersecting lines. **Hence proved.**

Example 28:

The direction ratios of two lines are 1, –2, –2 and 0, 2, 1. Find the direction cosines of the line perpendicular to the above lines.

Solution:

Let a, b, c be the direction ratios of the line whose direction cosines are required. Then as this line is perpendicular to the given lines so we have

$$a(1) + b(-2) + c(-2) = 0$$

and

$$a\,(0) + b(2) + c(1) = 0$$

Solving these simultaneously, we get

$$\frac{a}{(-2)(1)-(-2)(2)} = \frac{b}{(-2)(0)-(1)(1)} = \frac{c}{(1)(2)-(0)(-2)}$$

$\Rightarrow$ $a/2 = b/2{-}1 = c/2$ *i.e.,* $a : b : c = 2 : -1 : 2$

$\therefore$ The required direction cosines are

$$\frac{2}{\sqrt{(2^2+1^2+2^2)}}, \frac{-1}{\sqrt{(2^2+1^2+2^2)}}, \frac{2}{\sqrt{(2^2+1^2+2^2)}} \text{ i.e. } \frac{2}{3}, \frac{-1}{3}, \frac{2}{3}$$

Ans.

Example 29:

Find the direction cosines of the line perpendicular to a pair of mutually perpendicular lines with their direction cosines as (l_1, m_1, n_1) *and* (l_2, m_2, n_2) *respectively.*

Solution:

Let l, m, n be the required direction cosines of the line which is perpendicular to given lines, then $l\,l_1 + m\,m_1 + n\,n_1 = 0$ ...(i)

and $l\,l_2 + m\,m_2 + n\,n_2 = 0$ ...(ii)

Also as the given lines are mutually perpendicular so we have

$$l_1\,l_2 + m_1\,m_2 + n_1\,n_2 = 0 \quad \text{...(iii)}$$

Solving (i) and (ii) simultaneously we get

$$\frac{l}{m_1n_2 - m_2n_1} = \frac{m}{n_1l_2 - n_2l_1} = \frac{n}{l_1m_2 - l_2m_1}$$

$$= \frac{\sqrt{(l^2 + m^2 + n^2)}}{\sqrt{[(m_1n_2 - m_2n_1)^2 + (n_1l_2 - n_2l_1)^2 + (l_1m_2 - l_2m_1)^2]}}$$

$$= \frac{1}{\sqrt{[(l_1^2 + m_1^2 + n_1^2)(l_2^2 + m_2^2 + n_2^2) - (l_1 l_2 + m_1 m_2 + n_1 n_2)^2]}}$$

$\because$ $l^2 + m^2 + n^2 = 1$ and by Lagrange's Identity

$$= \frac{1}{\sqrt{[(1)(1) - (0)]}}, \text{ from (iii) and } \Sigma l_1^2 = 1 = \Sigma l_2^2$$

$= 1.$

$\therefore$ $l = m_1 n_2 - m_2 n_1$,

$m = n_1 l_2 - n_2 l_1$, $n = l_1 m_2 - l_2 m_1$. **Ans.**

Example 30:

Lines OA, OB are drawn from O with d.c.'s proportional to (1, –2, –1), (3, –2, 3). Find the d.c.'s of the normal to the plane AOB.

Solution:

Let a, b, c be the direction ratios of the required normal to the plane AOB.

Then as OA lies in this plane so it is perpendicular to the normal to this plane and consequently we have $a(1) + b(-2) + c(-1) = 0$ ...(i)

Similarly OB is also perpendicular to this normal and so we have

$$a(3) + b(-2) + c(3) = 0 \quad \text{...(ii)}$$

Solving (i) and (ii) simultaneously, we have

$$\frac{a}{(-2)(3) - (-1)(-2)} = \frac{b}{(-1)(3) - (1)(3)} = \frac{c}{(1)(-2) - (-2)(3)}$$

$$\Rightarrow \frac{a}{-8} = \frac{b}{-6} = \frac{c}{4} \text{ or } a : b : c = 4 : 3 : -2$$

Also $4^2 + 3^2 + (-2)^2 = 29$.

$\therefore$ Required d.c.'s are $4/\sqrt{(29)}$, $3/\sqrt{(29)}$, $-2/\sqrt{(29)}$. **Ans.**

Example 31:

The direction cosines of a variable line in two adjacent positions are l, m, n; l + δl, m + δm, n + δn, show that the small angle dθ between the two positions is given by $(\delta\theta)^2 = (\delta l)^2 + (\delta m)^2 + (\delta n)^2$.

Solution:

As (l, m, n) and $(l + \delta l, m + \delta m, n + \delta n)$ are the d.c.'s of lines so we have

$$l^2 + m^2 + n^2 = 1 \quad \text{...(i)}$$

and

$$(l + \delta l)^2 + (m + \delta m)^2 + (n \ \delta n)^2 = 1 \quad \text{...(ii)}$$

Subtracting (i) from (ii) we get

$$2(l\,\delta l) + m\,\delta m + n\,\delta n) + [(\delta l)^2 + (\delta m)^2 + (\delta n)^2] = 0$$

$$\Rightarrow \quad 2\Sigma\,(l\,\delta l) = -\,\Sigma\,(\delta l)^2 \qquad \text{...(iii)}$$

Also as $\delta\,\theta$ is the angle between these lines so we have

$$\cos\delta\theta = l\,(l + \delta l) + m(m + \delta m) + n\,(n + \delta n)$$

$$= (l^2 + m^2 + n^2) + (l\,\delta l + m\,\delta m + n\,\delta n)$$

$$= 1 - \frac{1}{2}\,[\,\Sigma\,(\delta l)^2], \text{ from (i) and (iii)}$$

$$\Rightarrow \quad \frac{1}{2}[\,\Sigma\,(\delta l)^2] = 1 - \cos\delta\theta$$

$$= 2\sin^2(\frac{1}{2}\delta\theta)] = 2(\frac{1}{2}\delta\theta)^2, \qquad \because \sin(\frac{1}{2}\delta\theta) = \frac{1}{2}\delta\theta$$

$$\Rightarrow \quad (\delta l) + (\delta m)^2 + (\delta n)^2 = (\delta\theta)^2$$ **Hence proved.**

Example 32:

If (l_1, m_1, n_1), (l_2, m_2, n_2), (l_3, m_3, n_3) are d.c.'s of three mutually perpendicular lines then prove that the line whose direction cosines are proportional to $l_1 + l_2 + l_3$, $m_1 + m_2 + m_3$, $n_1 + n_2 + n_3$ makes equal angles with them.

Solution:

As (l_1, m_1, n_1), (l_2, m_2, n_2) and (l_3, m_3, n_3) are d.c.'s of three mutually perpendicular lines, therefore we have

$$l_1l_2 + m_1m_2 + n_1n_2 = 0 \qquad \text{...(i)}$$

$$l_2l_3 + m_2m_3 + n_2n_3 = 0 \qquad \text{...(ii)}$$

and $$l_1l_3 + m_1m_3 + n_1n_3 = 0. \qquad \text{...(iii)}$$

Also l_1, m_1, n_1 being d.c.'s of a line we have $l_1^2 + m_1^2 + n_1^2 = 1$.

Similarly $l_2^2 + m_2^2 + n_2^2 = 1$ and $l_3^2 + m_3^2 + n_3^2 = 1$ ...(iv)

Again $\sqrt{[(l_1 + l_2 + l_3)^2 + (m_1 + m_2 + m_3)^2 + (n_1 + n_2 + n_3)^2]}$

$= \sqrt{[(l_1^2 + m_1^2 + n_1^2) + (l_2^2 + m_2^2 + n_2^2) + (l_3^2 + m_3^2 + n_3^2)}$

$+ 2(l_1l_2 + m_1m_2 + n_1n_2) + 2(l_2l_3 + m_2m_3 + n_2n_3) + 2(l_3l_1 + m_3m_1 + n_3n_1)$

expanding and rearranging terms

$= \sqrt{[1 + 1 + 1]}$, from (i), (ii), (iii) and (iv)

$= \sqrt{3}$.

$\therefore$ The d.c.'s of the given line are

$$\frac{l_1 + l_2 + l_3}{\sqrt{3}}, \frac{m_1 + m_2 + m_3}{\sqrt{3}}, \frac{n_1 + n_2 + n_3}{\sqrt{3}}$$

If θ be the angle between this line and the line with d.c.'s l_1, m_1, n_1, then cos

$$\theta = l_1\left(\frac{l_1 + l_2 + l_3}{\sqrt{3}}\right) + m_1\left(\frac{m_1 + m_2 + m_3}{\sqrt{3}}\right) + n_1\left(\frac{n_1 + n_2 + n_3}{\sqrt{3}}\right)$$

$= (1/\sqrt{3})\,[(l_1^2 + m_1^2 + n_1^2) + (l_1l_2 + m_1m_2 + n_1n_2) + (l_1l_3 + m_1m_3 + n_1n_3)]$

$\Rightarrow \cos\theta\ (1/\sqrt{3})\ [1 + 0 + 0]$ ...from (i), (iii) and (iv)

$\Rightarrow \theta = \cos^{-1}(1/\sqrt{3})$

In a similar way we can prove that the line with d.c.'s $(l_1 + l_2 + l_3)/\sqrt{3}$, $(m_1 + m_2 + m_3)/\sqrt{3}$, $(n_1 + n_2 + n_3)/\sqrt{3}$ is inclined to line with d.c.'s l_2, m_2, n_2 and l_3, m_3, n_3 also at angle $\cos^{-1}(1/\sqrt{3})$. **Hence proved.**

Example 33:

Show that the lines whose direction cosines are given by $l + m + n = 0$, $2mn + 3ln - 5lm = 0$ are perpendicular to one another.

Solution:

Given $l + m + n = 0$ ...(i)

$2mn + 3ln - 5lm = 0$ ...(ii)

Eliminating n between (i) and (ii), we have

$2m(-l - m) + 3l(-l - m) - 5lm = 0$

$\Rightarrow 3l^2 + 10lm + 2m^2 = 0$

$\Rightarrow 3l\,(l/m)^2 + 10\,(l/m) + 2 = 0$

$\Rightarrow \frac{l}{m} = \frac{-10 \pm \sqrt{(100 - 24)}}{6} = \frac{-5 \pm \sqrt{(19)}}{3}$ ...(iii)

Let l_1, m_1, n_1 and l_2, m_2, n_2 be the d.c's of the two lines, then from (iii)

we have $\frac{l_1}{m_1} = \frac{-5 + \sqrt{(19)} - 5}{3}$ and $\frac{l_2}{m_2} = \frac{-5 - \sqrt{(19)}}{3}$

If $\frac{l_1}{m_1} = \frac{\sqrt{(19)} - 5}{3}$; then $\frac{l_1}{\sqrt{(19)} - 5} = \frac{m_1}{3} = k_1$ (say)

$\therefore l_1 = k_1\,[\sqrt{(19)} - 5]$; $m_1 = 3k_1$

Also from (i) $n_1 = -l_1 - m_1 = k_1(5 - \sqrt{19}) - 3k_1 = k_1(2 - \sqrt{19})$.

$$\therefore \quad \frac{l_1}{\sqrt{(19)} - 5} = \frac{m_1}{3} = \frac{n_1}{2 - \sqrt{(19)}} \quad \text{...(iv)}$$

Similarly taking $\dfrac{l_2}{m_2} = \dfrac{-5 - \sqrt{(19)}}{3}$ we can find that

$$\frac{l_1}{-\sqrt{(19)} - 5} = \frac{m_2}{3} = \frac{n_2}{2 + \sqrt{(19)}} \quad \text{...(v)}$$

From (iv) and (v) we have the direction ratios of two lines and if they are at right angles then we must have

"$a_1a_2 + b_1b_2 + c_1c_2 = 0$"

Here $a_1a_2 + b_1b_2 + c_1c_2 = [\sqrt{(19)} - 5][-(19) - 5) + 3 \times 3$
$+ [2 - \sqrt{(19)}][2 + \sqrt{(19)}] = (-5)^2 - 19 + 9 + (2)^2 - 19 = 38 - 38 = 0$

Hence the two lines are at right angles. **Hence proved.**

Example 34:

If points P, Q are (2, 3, –6) and (3, –4, 5), find the angle that OP makes with OQ.

Solution:

The direction ratios of OP are 2 – 0, 3 – 0, –6 – 0, *i.e.*, 2, 3, –6 and those of OQ are 3 – 0, –4 – 0, 5 – 0 *i.e.*, 3, –4. 5

∴ If θ be the required angle, then

$$\cos\theta = \frac{a_1a_2 + b_1b_2 + c_1c_2}{\sqrt{(a_1^2 + b_1^2 + c_1^2)} \cdot \sqrt{a_2^2 + b_2^2 + c_2^2)}}$$

$$= \frac{2.3 + 3(-4) + (-6)5}{\sqrt{[2^2 + 3^2 (-6)^2]} \cdot \sqrt{[3^2 + (-4)^2 + 5^2}} = \frac{6 - 12 - 30}{\sqrt{(49)}\sqrt{(50)}} = \frac{-36}{7 \times 5\sqrt{2}}$$

$\Rightarrow \quad \theta = \cos^{-1}[-18\sqrt{2}/35]$ **Ans.**

Example 35:

If l_1, m_1, n_1 and l_2, m_2, n_2 are the d.c.'s of two lines, then the direction ration of another which is perpendicular to both the given lines are $(m_1n_2 - m_2n_1)$, $(n_1l_2 - n_2l_1)$, $(l_1m_2 - l_2m_1)$.

Prove further if the given lines are at right angles to each other then these direction ratios are the actual direction cosines.

Solution:

Let l, m, n be the required d.c.'s of the line.

Then as this line is perpendicular to both the given lines, so we have

$$ll_1 + mm_1 + nn_1 = 0 \quad \text{and} \quad ll_2 + mm_2 + nn_2 = 0$$

Solving these, we get

$$\frac{l}{m_1n_2 - m_2n_1} = \frac{m}{n_1l_2 - n_2l_1} = \frac{n}{l_1m_2 - l_2m_1},$$ which give the required direction ratios.

∴ The d.c.'s of this line are

$$\frac{m_1n_2 - m_2n_1}{\sqrt{\left[\sum(m_1n_2 - m_2n_1)^2\right]}}, \frac{n_1l_2 - n_2l_1}{\sqrt{\left[\sum(m_1n_2 - m_2n_1)^2\right]}}, \frac{l_1m_2 - l_2m_1}{\sqrt{\left[\sum(m_1n_2 - m_2n_1)^2\right]}}, \quad \text{...(i)}$$

If θ be the angle between the two given lines whose d.c.'s are (l_1, m_1, n_1) and (l_2, m_2, n_2) then we know

$$\sin\theta = \sqrt{[\Sigma\,(m_1n_2 - m_2n_1)^2]}$$

It these lines are at right angles, then θ = 90° or sin θ = 1 and therefore

$$\sqrt{[\Sigma\,(m_1n_2 - m_2n_1)^2]} = 1.$$

∴ From (i) the actual d.c.'s are

$(m_1n_2 - m_2n_1)$, $(n_1l_2 - n_2l_1)$ and $(l_1m_2 - l_2m_1)$. **Hence proved.**

Example 36:

Find the angle between the tow lines whose direction cosines are cos α, cos β, cos γ and cos α′, cos β′, cos γ′. Deduce the condition of perpendicularity and parallelism of the above lines.

Solution:

If θ be the required angle, we can prove that

$$\cos\theta = \cos\alpha\cos\alpha' + \cos\beta\cos\beta' + \cos\gamma\cos\gamma' \quad \text{...(i)}$$

If these two lines are perpendicular, then θ = 90° and from (1) we get the *condition of perpendicularity* as

$$\cos\alpha\cos\alpha' + \cos\beta\cos\beta' + \cos\gamma\cos\gamma' = 0.$$

Again if two lines are parallel, then $\theta = 0$ and from we get the *condition of perpendicularity* as

$$\cos\alpha \cos\alpha' + \cos\beta \cos\beta' + \cos\gamma \cos\gamma' = 1. \qquad \textbf{Ans.}$$

Otherwise, if given lines are parallel, the their d.cosines are equal and so we have

$$\cos\alpha = \cos\alpha', \cos\beta = \cos\beta', \cos\gamma = \cos\gamma'$$

$$\Rightarrow \quad \alpha = \alpha', \beta = \beta', \gamma = \gamma'$$

Example 37:

If (l_1, m_1, n_1), (l_2, m_2, n_2), (l_3, m_3, n_3) are the d.c.'s three mutually perpendicular lines then prove that the line whose direction cosines are proportional to $l_1 + l_2 + l_3$, $m_1 + m_2 + m_3$, $n_1 + n_2 + n_3$ makes equal angles with them.

Solution:

As (l_1, m_1, n_1), (l_2, m_2, n_2), and (l_3, m_3, n_3) are d.c.'s of three mutually perpendicular lines, therefore we have

$$l_1 l_2 + m_1 m_2 + n_1 n_2 = 0 \qquad ...(i)$$

$$l_2 l_3 + m_2 m_3 + n_2 n_3 = 0 \qquad ...(ii)$$

and $$l_1 l_3 + m_1 m_3 + n_1 n_3 = 0. \qquad ...(iii)$$

Also l_1, m_1, n_1 being d.c.'s of a line we have $l_1^2 + m_1^2 + n_1^2 = 1$.

Similarly $l_2^2 + m_2^2 + n_2^2 = 1$ and $l_3^2 + m_3^2 + n_3^2 = 1$ ...(iv)

Again $\sqrt{[l_1 + l_2 + l_3) + (m_1 + m_2 + m_3)^2 + (n_1 + n_2 + n_3)^2]}$

$$= \sqrt{\left(l_1^2 + m_1^2 + n_1^2\right) + \left(l_2^2 + m_2^2 + n_2^2\right) + \left(l_3^2 + m_3^2 + n_3^2\right)}$$

$$+ 2(l_1 l_2 + m_1 m_2 + n_1 n_2) + 2(l_2 l_3 + m_2 m_3 + n_2 n_3)$$

$$+ 2(l_3 l_1 + m_3 m_1 + n_3 n_1)$$

expanding and rearranging terms

$$= \sqrt{[1 + 1 + 1]}, \text{ from (i), (ii), (iii) and (iv)}$$

$$= \sqrt{3}.$$

$\therefore$ The d.c.'s of the given line are

$$\frac{l_1 + l_2 + l_3}{\sqrt{3}}, \frac{m_1 + m_2 + m_3}{\sqrt{3}}, \frac{n_1 + n_2 + n_3}{\sqrt{3}}$$

If θ be the angle between this line and the line with d.c.'s l_1, m_1, n_1, then

$$\cos\theta = l_1\left(\frac{l_1 + l_2 + l_3}{\sqrt{3}}\right) + m_1\left(\frac{m_1 + m_2 + m_3}{\sqrt{3}}\right) + n_1\left(\frac{n_1 + n_2 + n_3}{\sqrt{3}}\right)$$

$$= (1/\sqrt{3})\,[(l_1^2 + m_1^2 + n_1^2) + (l_1 l_2 + m_1 m_2 + n_1 n_2) + (l_1 l_3 + m_1 m_3 + n_1 n_3)]$$

$$\Rightarrow \quad \cos\theta = (1/\sqrt{3})\,[1 + 0 + 0] \text{...from (i), (iii) and (iv)}$$

$$\Rightarrow \quad \theta = \cos^{-1}(1/\sqrt{3})$$

In a similar way we can prove that the line with d.c.'s $(l_1 + l_2 + l_3)/\sqrt{3}$, $(m_1 + m_2 + m_3)/\sqrt{3}$, $(n_1 + n_2 + n_3)/\sqrt{3}$ in inclined to line with d.c.'s l_2, m_2, n_2 and l_3, m_3, n_3 also at angle $\cos^{-1}(1/\sqrt{3})$. **Hence proved.**

Example 38:

Find the angle between the lines whose direction ratios are (2, 3, 4) and (1, – 2, 1).

Solution:

If θ be the angle between the given lines, then

$$\cos\theta = \frac{a_1 a_2 + b_1 b_2 + c_1 c_2}{\sqrt{(a_1^2 + b_1^2 + c_1^2)}\,.\,\sqrt{(a_2^2 + b_2^2 + c_2^2)}}$$

$$= \frac{2.1 + 3.(-2) + 4.1}{\sqrt{(2^2 + 3^2 + 4^2)}\,.\,\sqrt{[1^2 + (-2)^2 + 1^2]}} = \frac{0}{\sqrt{(20)}\,\sqrt{(6)}} = 0$$

$\Rightarrow \quad \theta = 90^\circ$ *i.e.,* lines are at right angles.

Example 39:

Prove that the straight lines whose direction cosines are given by relations al + bm + cn = 0 abd fmn + gnl + hlm = 0 are perpendicular

if $\frac{f}{a} + \frac{g}{b} + \frac{h}{c} = 0$ *and parallel if* $\sqrt{(af)} \pm \sqrt{(bg)} \pm \sqrt{(ch)} = 0$

Solution:

Let the d.c.'s of the two lines be (l_1, m_1, n_1) and (l_2, m_2, n_2).

Eliminating n between the given relations, we get

$$fm[-(al + bm)/c] + gl[-(al + bm)/c] + hlm = 0$$

$$\Rightarrow \quad -afm - bfm^2 - agl^2 - bglm + chlm = 0$$

$$\Rightarrow \quad ag\,(l/m)^2 + (af + bg - ch)\,(l/m) + bf = 0.$$

dividing each term by m^2.

Its roots are l_1/m_1 and l_2/m_2

$$\therefore \quad \frac{l_1}{m_1} \cdot \frac{l_2}{m_2} = \text{Product of the roots} = \frac{bf}{ag}$$

$$\Rightarrow \quad \frac{l_1 l_2}{bf} = \frac{m_1 m_2}{ag} \text{ or } \frac{l_1 l_2}{(f/a)} = \frac{m_1 m_2}{(g/b)} = \frac{n_1 n_2}{(h/c)}, \text{ by symmetry}$$

If the lines are perpendicular, then $l_1 l_2 + m_1 m_2 + n_1 n_2 = 0$

$\Rightarrow \quad (f/a) + (g/b) + (h/c) = 0.$ **Hence proved.**

If the lines are parallel, then their d.c.'s must be the same *i.e.*, the roots of (i) must be equal, the condition for the same is "$b^2 = 4ac$"

i.e., $\quad (af + bg - ch)^2 = 4ag.\ bf$ ...(ii)

$\Rightarrow \quad af + bg - ch = \pm 2\sqrt{(af)}\sqrt{(bg)}$

$\Rightarrow \quad af + bg \pm 2\sqrt{(af)}\sqrt{(bg)} = ch$

$\Rightarrow \quad [\sqrt{(af)} \pm \sqrt{(bg)}]^2 = ch = [\sqrt{(ch)}]^2$

$\Rightarrow \quad \sqrt{(af)} \pm \sqrt{(bg)} \pm \sqrt{(ch)} = 0$ is the required condition.

Also From (ii), we get $a^2f^2 + b^2g^2 + c^2h^2 + 2abfg - 2acfh - 2bcgh = 4abfg$.

$\Rightarrow \quad a^2f^2 + b^2g^2 + c^2h^2 - 2bcgh - 2cafh - 2abfg = 0.$ **Ans.**

Example 40:

Find the acute angle between the lines whose direction cosines are proportional to (2, 3, –6), and (3, – 4, 5).

Solution:

If θ be the angle between the given lines, then

$$\cos\theta = \frac{a_1a_2 + b_1b_2 + c_1c_2}{\sqrt{(a_1^2 + b_1^2 + c_1^2)} \cdot \sqrt{(a_2^2 + b_2^2 + c_2^2)}}$$

$$= \frac{2.3 + 3.(-4) + (-6).5}{\sqrt{(2^2 + 3^2 + (-6)^2)} \cdot \sqrt{[3^2 + (-4)^2 + 5^2]}} = \frac{-36}{35\sqrt{(2)}}$$

= negative. Hence θ is obtuse angle.

$\therefore$ Required acute angle $= \pi - \theta$, where $\cos\theta = (-36/35)\sqrt{(1/2)}$ **Ans.**

Example 41:

Find the angle between the two lines whose d.c.'s (l, m, n) satisfy the equatios $l + m + n = 0$ and $l^2 + m^2 + n^2 = 0$.

Solution:

Eliminating n between the given relations, we get

$$l^2 + m^2 + (-l - m)^2 = 0$$

$$\Rightarrow \quad l^2 + m^2 + lm = 0$$

$$\Rightarrow \quad (l/m)^2 + (l/m) + 1 = 0, \text{ dividing each term by } m^2$$

Let its roots be l_1/m_1 and l_2/m_2, then

$$\frac{l_1}{m_1} \cdot \frac{l_2}{m_2} = \text{product of the roots} = -\frac{1}{1}$$

$$\Rightarrow \quad \frac{l_1 l_2}{1} = \frac{m_1 m_2}{-1} \qquad \text{...(i)}$$

Again eliminating m between the given relations we get as above

$$l^2 + ln + n^2 = 0$$

$$\Rightarrow \quad (l/n)^2 + (l/n) + 1 = 0, \text{ dividing each term by } n^2$$

If its roots be l_1/n_1 and l_2/n_2, then

$$\frac{l_1}{n_1} \cdot \frac{l_2}{n_2} = \text{product of the roots} = -\frac{1}{1}$$

$$\Rightarrow \quad \frac{l_1 l_2}{1} = \frac{n_1 n_2}{-1} \qquad \text{...(ii)}$$

$\therefore$ From (i) and (ii)

we get $\dfrac{l_1 l_2}{1} = \dfrac{m_1 m_2}{-1} = \dfrac{n_1 n_2}{-1}$

$\therefore$ If θ be the required angle, then $\cos\theta = l_1 l_2 + m_1 m_2 + n_1 n_2$

$$\Rightarrow \quad \cos\theta = 1 - 1 - 1 = -1 \quad \text{or} \quad \theta = \pi.$$ **Ans.**

Example 42:

If A, B, C, D are the points (3, 4, 5), (4, 6, 3), (–1, 2, 4) and (1, 0, 5), find the angle between CD and AB.

Solution:

The direction rations of AB are 4 – 3, 6 – 4, 3 – 5 or 1, 2, – 2

And the direction ratios of CD are 1 + 1, 0 – 2, 5 – 4 or 2, – 2, 1

$\therefore$ If θ be the required angle between AB and CD, then

$$\cos\theta = \frac{a_1 a_2 + b_1 b_2 + c_1 c_2}{\sqrt{(a_1^2 + b_1^2 + c_1^2)} \cdot \sqrt{(a_2^2 + b_2^2 + c_2^2)}}$$

$$= \frac{1.2 + 2.(-2) + (-2).1}{\sqrt{\left(1^2 + 2^2 + (-2)^2\right)} \cdot \sqrt{\left[2^2 + (-2)^2 + 1^2\right]}} = \frac{-4}{9}$$

= negative. Hence $\theta = \cos^{-1}(-4/9)$ and is an obtuse angle.

Example 43:

Prove that the acute angle between the lines whose direction cosines are given by the relations $l + m + n = 0$ and $l^2 + m^2 - n^2 = 0$ is $\pi/3$.

Solution:

Eliminating n between the given relations, we get

$$l^2 + m^2 - (-l - m)^2 = 0 \text{ or } lm = 0$$

∴ either $l = 0$ or $m = 0$

If $l = 0$, then from $l + m + n = 0$ we get $m = -n$

∴ we have $l = 0$, $m = -n$ or $\frac{l}{0} = \frac{m}{-1} = \frac{n}{1}$

∴ Direction cosines of one line are

$$\frac{0}{\sqrt{[0^2 + (-1)^2 + 1^2]}} = -\frac{-1}{\sqrt{[0^2 + (-1)^2 + 1^2]}} = \frac{1}{\sqrt{[0^2 + (-1)^2 + 1^2]}}$$

⇒ $0, -1/\sqrt{2}, 1/\sqrt{2}$

If $m = 0$, then from $l + m + n = 0$ we get $l = -n$

∴ we have $\frac{l}{1} = \frac{m}{0} = \frac{n}{-1}$

∴ Direction cosines of the second line as above are $1/\sqrt{2}, 0, -1/\sqrt{2}$

∴ If θ be the angle between these two lines, then

$$\cos\theta = "l_1 l_2 + m_1 m_2 + n_1 n_2"$$

$$= 0 \,.\, (1/\sqrt{2}) + (-1/\sqrt{2}) \,.\, 0 + (1/\sqrt{2})(-1/\sqrt{2})$$

⇒ $\cos\theta = -1/2$ or $\theta = 120°$

∴ Required acute angle between the lines

$= 180° - 120° = 60°$ *i.e.*, $\pi/3$. **Hence proved.**

Example 44:

Find the angle between the lines whose d.c.'s (l, m, n) satisfy the equations $l + m + n = 0$ and $2lm + 2nl - mn = 0$.

Solution:

Eliminating n between the given relations, we get .

$$2lm + 2(-l - m)l - m(-l - m) = 0$$

$$\Rightarrow \quad 2l^2 - lm - m^2 = 0$$

$$\Rightarrow \quad (2l + m)(l - m) = 0 \qquad \text{...(i)}$$

If $2l + m = 0$, then from $l + m + n = 0$, we get $n = l$

$\therefore$ we have $2l = -m = 2n$ or $\frac{l}{1} = \frac{m}{-2} = \frac{n}{1}$.

$\therefore$ The direction cosines of one line are

$$\frac{1}{\sqrt{[1^2 + (-2)^2 + 1^2]}}, \frac{-2}{\sqrt{[1^2 + (-2)^2 + 1^2]}}, \frac{1}{\sqrt{[1^2 + (-2)^2 + 1^2]}}$$

$$\Rightarrow \quad 1/\sqrt{6}, -2/\sqrt{6}, 1/\sqrt{6}. \qquad \text{...(ii)}$$

If $l - m = 0$, then from $l + m + n = 0$, we get $n = -2m$.

$\therefore$ we have $l = m = -\frac{n}{2}$ or $\frac{l}{1} = \frac{m}{1} = \frac{n}{-2}$.

$\therefore$ The direction cosines of the second line are

$$\frac{1}{\sqrt{[1^2 + 1^2 + (-2)^2]}}, \frac{-2}{\sqrt{[1^2 + 1^2 + (-2)^2]}}, \frac{1}{\sqrt{[1^2 + 1^2 + (-2)^2]}}$$

$$\Rightarrow \quad 1/\sqrt{6}, 1/\sqrt{6}, -2/\sqrt{6}.$$

$\therefore$ If θ be the required angle, then

$$\cos\theta = "l_1 l_2 + m_1 m_2 + n_1 n_2"$$

$$= \frac{1}{\sqrt{6}} \cdot \frac{1}{\sqrt{6}} + \left(-\frac{2}{\sqrt{6}}\right) \cdot \frac{1}{\sqrt{6}} + \frac{1}{\sqrt{6}} \cdot \left(\frac{-2}{\sqrt{6}}\right) = \frac{-3}{6}$$

$$\Rightarrow \quad \cos\theta = -1/2 \quad \text{or} \quad \theta = 120°.$$ **Ans.**

Example 45:

Show that the straight line whose direction cosines are given by the equations : $ul + vm + wn = 0$ $al^2 + bm^2 + cn^2 = 0$ are (α) perpendicular if $u^2(b + c) + v^2(c + a) + w^2(a + b) = 0$ and (β) parallel, if $(u^2/a) + (v^2/b) + (w^2/c) = 0$.

Solution:

The d.c.'s of the lines are given by

$$al + vm + wn = 0 \text{ and } al^2 + bm^2 + cn^2 = 0$$

Eliminating n between these, we get

$$al^2 + bm^2 + c\,[-(ul + vm)/w]^2 = 0$$

$$\Rightarrow \quad (aw^2 + cu^2)l^2 + (bw^2 + cv^2)\,m^2 + 2cuvlm = 0$$

$$\Rightarrow \quad (aw^2 + cu^2)\,((l/m)^2 + 2cuv\,(l/m) + (bw^2 + cv^2) = 0, \quad \ldots(i)$$

dividing each term by m^2.

(α) Its two roots are l_1/m_1 and l_2/m_2, if the d.c.'s of the two lines be taken as (l_1, m_1, n_1) and (l_2, m_2, n_2).

$\therefore$ From (i), we have

$$\frac{l_1}{m_1} \cdot \frac{l_2}{m_2} = \text{product of the roots} = \frac{bw^2 + cv}{cu^2 + aw^2}$$

$$\Rightarrow \quad \frac{l_1 l_2}{bw^2 + cv^2} = \frac{m_1 m_2}{cu^2 + aw^2} = \frac{n_1 n_2}{av^2 + bu^2}, \text{ by symmetry.}$$

$\therefore$ If the two lines are perpendicular, then we have

$l_1 l_2 + m_1 m_2 + n_1 n_2 = 0$ *i.e.,* $(bw^2 + cv^2) + (cu^2 + aw^2) + (av^2 + bu^2) = 0.$

$$\Rightarrow \quad u^2(b + c) + v^2(c + a) + w^2(a + b) = 0.$$ **Hence proved.**

(β) If the two lines are parallel, then their d.c.'s are equal are consequently the roots of (i) are equal, the condition for the same being

"$b^2 = 4ac$" *i.e.,* $(2cuw)^2$

$$= 4(aw^2 + cu^2)(bw^2 + cv^2)$$

$$\Rightarrow \quad c^2u^2v^2 = abw^4 + acw^2v^2 + bcu^2w^2 + c^2u^2v^2$$

$$\Rightarrow \quad abw^4 + acw^2v^2 + bcu^2w^2 = 0$$

$$\Rightarrow \quad abw^2 + acv^2 + bcu^2 = 0$$

$$\Rightarrow \quad \frac{w^2}{c} + \frac{v^2}{b} + \frac{u^2}{a} = 0, \text{ dividing each term by abc.}$$

EXERCISES

1. Obtain the direction cosines of the diagonal of a cube through one corner, taking the coordinate axes along the edges of the cube through that corner.
2. Find the angle between the lines whose direction ratios are 1, 1, 2 and $\sqrt{3} - 1, -\sqrt{3} - 1, 4$.
3. If θ is the angle between the lines whose direction cosines are proportional to 5, –12, 13 and –3, 4, 5; then find the value of $\tan \theta$.
4. If points A and B are (2, 3, 4) and (1, –2, 1) respectively, then prove that OA is perpendicular to OB, where O is (0, 0, 0).

5. Show that the pair of lines whose direction cosines are given by the equations $l + 2m + 3n = 0$ and $mn - 4nl + 3lm = 0$ are at right angles.
6. The condition of orthogonality for the two lines whose d.c.'s are l_1, m_1, n_1 and l_2, m_2, n_2 is
   (i) $l_1^2 + m_1^2 + n_1^2$ and $l_2^2 + m_2^2 + n_2^2 = 1$;
   (ii) $l_1l_2 + m_1m_2 + n_1n_2 = 0$;
   (iii) $l_1/l_2 = m_1/m_2 = n_1/n_2$;
   (iv) $l_1l_2 + m_1m_2 + n_1n_n = 1$.
7. Which are the direction cosines of a line?
   (i) (2/3, 2/3, 2/3);
   (ii) (1/3, 1/3, 1/3);
   (iii) $(1/\sqrt{3}, 1/\sqrt{3}, 1/\sqrt{3})$;
   (iv) $(1/\sqrt{3}, 1/\sqrt{3}, 2/\sqrt{3})$.
8. l_r, m_r, n_r (r = 1, 2, 3) are the direction cosines of three mutually perpendicular lines and also.
   $$\frac{a}{l_1} + \frac{b}{m_1} + \frac{c}{n_1} = 0, \frac{a}{l_2} + \frac{b}{m_2} + \frac{c}{n_2} = 0,$$
   then prove that $\frac{a}{l_3} + \frac{b}{m_3} + \frac{c}{n_3} = 0$.
9. The direction ratios of a line are 2, 3, 4. What are its direction cosines?
10. Choose the correct answer:
    The direction cosines of the line joining P (x_1, y_1, z_1) and Q (x_2, y_2, z_2) is
    (i) $\frac{x_1 + x_2}{PQ}, \frac{y_1 + y_2}{PQ}, \frac{z_1 + z_2}{PQ}$,
    (ii) $\frac{x_2 - x_1}{PQ}, \frac{y_2 - y_1}{PQ}, \frac{z_2 - z_1}{PQ}$
    (iii) $x_2 - x_1, y_2 - y_1, z_2 - z_1$,
    (iv) none of these.
11. If A, B, are (2, 3, 5) (–1, 3, 2), find the direction cosines of AB.
12. The projections of a line on the axes are 12, 4, 3. Find the length of the line and its directions cosines.
13. A (3, 1, 2), B (5, 1, 2), C (0, 1, 5) and D(0, –1, –5) are four points. Show that AB is perpendicular to CD.

14. Find the direction cosines of the line joining the origin to the point (3, 12, 4).

15. Find the angles of triangle ABC whose vertices A, B, and C are the points (1, 3, 5), (–1, 2, 3) and (3, 4, –2).

16. Prove that the equation to the right circular cone whose vertex is the point (2, –3, 5) and whose axis is the line which makes equal angles with the co-ordinate axes and semivertical angle 30° is

$$4[(y - z + 8)^2 + (z - x - 3)^2 + (z - y - 5)^2] = 3[(x - 2)^2 + (y + 3)^2 + (x - 5)^2]$$

17. Find the angle between the lines whose direction cosines are given by the equations $3l + m + 6n = 0$, $6mn - 2ln + lm = 0$.

18. Find the angle between the lines whose direction cosines are give by the relations $3l + m + 5n = 0$ and $6mn - 2nl + 5lm = 0$.

3

The Straight Lines

REPRESENTATION OF LINE

In this chapter it is proposed to discuss the manner in which a straight line can be represented. We introduce the method and lytically a follows.

Consider any two of the co-ordination planes say YOZ and ZOX whose equations are $X = 0$ and $Y = 0$ respectively. These two planes intersect in Z-axis.

A point (x, y, z) lies on the Z plane.

$\Rightarrow$ The point (x, y, z) lies on the YOZ plane and the point (x, y, z) lies on the ZOX plane $X = 0$ and $Y = 0$, thus, we see that a point (x, y, z) lies on the Z-axis of and only if, we simultaneously, have $X = 0$, $Y = 0$, we are thus led to say that $X = 0$, $Y = 0$ are the two equation on of Z-axis.

Consider now any line what so ever any two planes through the line. Let $ax + by + cz + d = 0$ and $a_1x + b_1x + c_1z + d_1 = 0$ be the equations of these two planes. Clearly, we have the following statement.

A point (x, y, z) lies on the given line if and only if we simultaneously have.

$$ax + by + cz + d = 0$$

and

$$ax + b_1y + c_1z + d_1 = 0.$$

Thus we say that

$$ax + by + cz + d = 0$$

and

$$ax + b_1y + c_1z + d_1 = 0.$$

are the two equation of the line.

It follows that a straight lone in represented by two equations of the first degree in x, y, z.

SYMMETRICAL FORM OF THE EQUATIONS OF A LINE

Let the line pass through a fixed point A (α, β, γ) and have direction cosines l, m, n. If P (x, y, z) is any point on the line at a distance r from A, then the projection of AP on the x-axis is $x - \alpha$. Also it is lr. Hence, we have $x - \alpha = lr$.

Similarly, we have $y - \beta = mr$ and $z - \gamma = nr$.

$\therefore$ The co-ordinates of any point (x, y, z) on the line satisfy the equations

$$\frac{x-\alpha}{l} = \frac{y-\beta}{m} = \frac{z-\gamma}{n} (= r), \quad \text{...(i)}$$

which are the equations of a straight line passing through the point A (α, β, γ) and l, m, n are the direction-cosines of the line.

Note: From the equations (i) we have

$$x = \alpha + lr,\ y = \beta + mr$$

and $\quad z = \gamma + nr,$

which are the general coordinates of any point on the line in terms of r upon the varishing or ceter base of any of l, m, n.

Example 1:

Find the point in which the line $-(x + 1) = \frac{1}{5}(y - 12) = \frac{1}{2}(z - 7)$ *cuts the surface* $11x^2 - 5y^2 + z^2 = 0$.

Solution:

Any point on the line $\frac{x+1}{-1} = \frac{y-12}{5} = \frac{z-7}{2} = r$ is

$$(-1 - r, 12 + 5r, 7 + 2r). \quad \text{...(i)}$$

If it lies on the surface $11x^2 - 5y^2 + z^2 = 0$, then we have

$$11(-1 - r)^2 - 5(12 + 5r)^2 + (7 + 2r)^2 = 0$$

$$\Rightarrow \quad 11(1 + 2r + r^2) - 5(144 + 120r + 25r^2) + (49 + 28r + 4r^2) = 0$$

$$\Rightarrow \quad 110r^2 + 550r^2 + 660 = 0 \quad \text{or} \quad r^2 + 5r + 6 = 0$$

$$\Rightarrow \quad r = -2, -3.$$

Substituting these values of r in (i), we have required points as (1, 2, 3) and (2, – 3, 1). **Ans.**

Example 2:

Find where the line $\frac{x-1}{2} = \frac{2-y}{3} = \frac{z+3}{4}$ *meets the plane*

$$2x + 4y - z - 1 = 0.$$

Solution:

The equations of the line are $\frac{x-1}{2} = \frac{y-2}{-3} = \frac{z+3}{4} = r$, say ...(i) and the plane is $2x + 4y - z - 1 = 0$ ...(ii)

Any point on the line (i) is $(1 + 2r, 2 - 3r, -3 + 4r)$ ...(iii)

This point lies on (ii), if

$$2(1 + 2r) + 4(2 - 3r) - (-3 + 4r) - 1 = 0 \quad \text{...(iv)}$$

$$\Rightarrow \quad -12r + 12 = 0 \quad \Rightarrow \quad r = 1.$$

Substituting this value of r in the coordinates of the point given by (iii) we have the required point as $(1 + 2, 2 - 3, -3 + 4)$ or $(3, -1, 1)$. **Ans.**

Line Through Two Points

Let two given points be P (x_1, y_1, z_1) and Q (x_2, y_2, z_2).

Then the d.c.'s of the line PQ are $x_2 - x_1, y_2 - y_1, z_2 - z_1$.

$\therefore$ The equations of the line PQ are $\dfrac{x - x_1}{x_2 - x_1} = \dfrac{y - y_1}{y_2 - y_1} = \dfrac{z - z_1}{z_2 - z_1}$

Also we know that if there be a point R which divides PQ internally in the ratios λ: 1, then, the coordinates of R are

$$\left(\frac{\lambda x_2 + x_1}{\lambda + 1}, \frac{\lambda y_2 + y_1}{\lambda + 1}, \frac{\lambda z_2 + z_1}{\lambda + 1}\right). \quad \text{...(i)}$$

Example 1:

Find the co-ordinates of the point of intersection of the line $\dfrac{x+1}{1} = \dfrac{y+3}{3} = \dfrac{z-2}{-2}$ *wirh the plane* $3x \mid 4y + 5z = 5$.

Solution:

Let $\dfrac{x+1}{1} = \dfrac{y+3}{3} = \dfrac{z-2}{-2} = r$ so that two point $(r - 1, 3r - 3, -2r + 2)$ is a point on the given lie for all valges of r. It it also lies on the given plane, we have

$$3r - 3 + 12r = 12 - 10r + 10 = 5 \Rightarrow r = 2$$

Hence the required point of intersection is $(1, 3, -2)$. Its distance from the point $(-1, -3, 2)$ is $\sqrt{56}$ which is different from the value 2 or r.

Example 2:

Find the equations of the lines through the points (a, b, c) and (a', b', c') and prove that it passes through the origin, if aa' + bb' + cc' = rr', where r and r' are the distances of these points from the origin.

Solution:

The equations of the line through (a, b, c) and (a', b', c') are

$$\frac{x-a}{a'-a}=\frac{y-b}{b'-b}=\frac{z-c}{c'-c} \quad \text{...(i)}$$

If (i) passes through the origin *i.e.*, (0, 0, 0), then

$$\frac{0-a}{a'-a}=\frac{0-b}{b'-b}=\frac{0-c}{c'-c} \text{ or } \frac{a'-a}{a}=\frac{b'-b}{b}=\frac{c'-c}{c} \quad \textbf{(Note)}$$

$$\Rightarrow \quad \frac{a'}{a}-1=\frac{b'}{b}-1=\frac{c'}{c}-1 \text{ or } \frac{a'}{a}=\frac{b'}{b}=\frac{c'}{c}$$

From the, we can get $ab' - a'b = 0$, $bc' - b'c = 0$, $ca' - c'a = 0$ ...(ii)

Also we have $r = \sqrt{(a^2+b^2+c^2)}$, $r' = \sqrt{(a'^2+b'^2+c'^2)}$ ...(iii)

Now from Lagrange's Identity, we have

$$(a^2 + b^2 + c^2)(a'^2 + b'^2 + c'^2) = (aa' + bb' - cc')^2$$

$$= (ab' - a'b)^2 + (bc' - b'c)^2 + (ca' - c'a)^2$$

$$\Rightarrow \quad r^2 (r')^2 - (aa' + bb' + cc')^2 = 0, \text{ from (ii) and (iii)}$$

$$rr' = aa' + bb' + cc'.$$ **Hence proved.**

Transformation from General Form to Symmetric Form

To transfor the equation

$$a_1x + b_1y + c_1z + d_1 = 0$$

and $\quad a_2x + b_2y + c_2z + d_2 = 0$

of a line to the symmatrical form.

Let the equations of a straight line be given as

$$\left.\begin{aligned} a_1x+b_1y+c_1z+d_1 &= 0 \\ \text{and} \quad a_2x+b_2y+c_2z+d_2 &= 0 \end{aligned}\right\} \quad \text{...(i)}$$

In order to transform the general form of the straight line given above by (i) we should find out the d.c.'s of the line and the co-ordinates of some point on it.

Let *l*, m, n be the d.c.'s of this line. Then as this line lies on both the planes given by (i), so that it is perpendicular to the normals to these planes. The direction ratios of the normals to the planes given by (i) are a_1, b_1, c_1 and a_2, b_2, c_2 respectively.

∴ We have $a_1 l + b_1 m + c_1 n = 0;$

$a_2 l + b_2 m + c_2 n = 0.$

Solving these, we get $\dfrac{l}{b_1c_2 - b_2c_1} = \dfrac{m}{c_1a_2 - c_2a_1} = \dfrac{n}{a_1b_2 - a_2b_1}$ ...(ii)

which give the direction ratios of the line.

Now the co-ordinates of any point on the line can be calculated in many ways. One of them is that we choose the point as the one where the line meets z = 0.

(*i.e.*, xy-plane).

Putting z = 0 in (i), we get

$$a_1x + b_1y + d_1 = 0$$

and $a_2x + b_2y + d_2 = 0.$

Solving these, $\dfrac{x}{b_1d_2 - b_2d_1} = \dfrac{y}{d_1a_2 - d_2a_1} = \dfrac{1}{a_1b_2 - a_2b_1}$

$$\Rightarrow \quad x = 0\frac{b_1d_2 - b_2d_1}{a_1b_2 - a_2b_1};\ y = \frac{d_1a_2 - d_2a_1}{a_1b_2 - a_2b_1}$$

∴ The co-ordinates of the point where the line meets the plane z = 0 is given by

$$\left[\frac{b_1d_2 - b_2d_1}{a_1b_2 - a_2b_1}, \frac{d_1a_2 - d_2a_1}{a_1b_2 - a_2b_1}, 0\right] \qquad ...(iii)$$

∴ from (ii) and (iii) we can write the equations of the line given by (i) in the symmetric form as

$$\frac{x - \left(\dfrac{b_1d_2 - b_2d_1}{a_1b_2 - a_2b_1}\right)}{b_1c_2 - b_2c_1} = \frac{y - \left(\dfrac{d_1a_2 - d_2a_1}{a_1b_2 - a_2b_1}\right)}{c_1a_2 - a_2b_1} = \frac{z - 0}{a_1b_2 - a_2b_1}$$

Note: If $a_1b_2 - a_2b_1 = 0$, then we should choose the point where the line meets the plane x = 0 or y = 0.

Example 1:

Find the symmetrical form of the equ. of line 3x – y + z + 1 = 0, 5x + y + 3z = 0. Also find the equation to the plane through (2, 1, 4) and perpendicular to the given line.

Solution:

Let l, m, n be the direction rations of the given line then we know

$$3l = m + n = 0$$

$$5l + m + 3n = 0$$

$$l/1 = m/1 = n/-2$$

Again supose the given line interset the plane z = 0 at $(x_1, y_1, 0)$. Then

$$3x_1 - y_1 + 1 = 0 \text{ and } 5x_1 + y_1 = 0$$

$$x_1 = -1/8,\ y_1 = 5/8$$

Example 2:

Find the symmetric form of the equations of the line x + y + z + 1 = 0 and 4x + y – 2z + 2 = 0 and hence find the equation to the plane through (1, 1, 1) and perpendicular to the given line.

Solution:

Let l, m, n be the d.c.'s of the line. Then as this line lies on both the given planes, so it is perpendicular to the normal to these planes.

∴ We have $l + m + n = 0,\ 4l + m - 2n = 0.$ **(Note)**

From these we hence $\frac{l}{1+2} = \frac{m}{-2-4} = \frac{n}{4-1}$ or $\frac{l}{1} = \frac{m}{-2} = \frac{n}{1}$ ...(i)

Putting z = 0 in the given equations we obtain

$$x + y = -1,\ 4x + y = -2$$

Solving these we get $x = -\frac{1}{3},\ y = -\frac{2}{3}$

∴ The required line meets the plane z = 0 at $\left(-\frac{1}{3}, -\frac{2}{3}, 0\right)$ ...(ii)

∴ From (i) and (ii) the required equations in symmetric form are

$$\frac{x+\frac{1}{3}}{1} = \frac{y+\frac{2}{3}}{-2} = \frac{z-0}{1}$$

or $$\frac{3x+1}{3} = \frac{3y+2}{-6} = \frac{z}{1}$$

And its direction ratios are 3, – 6, 1.

Let the plane through (1, 1, 1) be A(x – 1) + B(y – 1) + C(z – 1)= 0

If this plane is perpendicular to the above line, then are how

$A/1 = B/(-2) = C/1$ and so its equation is

given as $(x - 1) - 2(y - 1) + (z - 1) = 0$ or $x - 2y + z = 0$. **Ans.**

Example 3:

Prove the join of (2, 3, 4), (3, 4, 5) is normal to the plane through (–2, –3, 6), (4, 0, –3), (0, –1, 2) the axes being rectangular.

Solution:

The equation of any plane through (–2, – 3, 6) is given as

$$A(x + 2) + B(y + 3) + C(z - 6) = 0 \qquad ...(i)$$

If this plane passes through (4, 0, –3) and (0, –1, 2) then we have

$$A(4 + 2) + B(0 + 3) + C(-3 - 6) = 0$$

and $\quad A(0 + 2) + B(-1 + 3) + C(2 - 6) = 0$

$\Rightarrow \quad 2A + B - 3C = 0$ and $A + B - 2C = 0$

Solving these we have $A - C = 0$

$\Rightarrow \quad A = C. \quad \therefore \quad B = C$

$\therefore$ From (i) we have the plane through three given points (–2, – 3, 6), (4, 0, – 3) and (0, – 1, 2) as

$$C(x + 2) + C(y + 3) + C(z - 6) = 0 \quad \text{or } x + y + z - 1 = 0 \quad ...(ii)$$

Also the d.r.'s of the line joining (2, 3, 4), (3, 4, 5) are

$$3 - 2, 4 - 3, 5 - 4 \text{ i.e., } 1, 1, 1$$

And the d.r.'s of the line normal to the plane (ii) are 1, 1, 1 which are the same as those of the line joining the points (2, 3, 4) and (3, 4, 5) *i.e.*, the join of the points (2, 3, 4), (3, 4, 5) is perpendicular to the plane (ii).

Hence proved.

Example 4:

Find the equation of the plane through the point (–1, 3, 1) and perpendicular to the line 2x + 3y + 4z = 5, 3x + 4y + 5z = 6.

Solution:

Let l, m, n be the d.c.'s of the line of intersection of the given planes, then we have

$$2l + 3m + 4n = 0$$

and $\quad 3l + 4m + 5n = 0$

Solving these we get $\dfrac{l}{1} = \dfrac{m}{-2} = \dfrac{n}{1} \qquad ...(i)$

The equation of any plane through (–1, 3, 1) is

$$A(x+1)+B(y-3)+C(z-1)=0 \quad \text{...(ii)}$$

If this plane is perpendicular to the line whose direction ratios are given by (i), then its normal is parallel to the line and as such we have

$$A/l = B/m = C/n = 0 \quad \text{...(iii)}$$

∴ From (i) and (iii) we have

$$\frac{A}{1}=\frac{B}{-2}=\frac{C}{1}=k \text{ (say)}$$

i.e., $A = k,\ B = -2k,\ C = k.$

Substituting these value in (iii), the required equation is given as

$$k(x+1)-2k(y-3)+k(z-1)=0$$

or $x - 2y + z + 6 = 0.$ **Ans.**

Example 5:

Find the equation to the plane through the points (2, –1, 0), (3, – 4, 5) parallel to the line 2x = 3y = 4z.

Solution:

The equation of any plane through (2, – 1, 0) is

$$A(x-2)+B(y+1)+C(z-0)=0 \quad \text{...(i)}$$

If it passes through (3, – 4, 5), then we have

$$A(3-2)+B(-4+1)+C(5-0)=0$$

or $A - 3B + 5C = 0$ ...(ii)

Now the given line is x/6 = y/4 = z/3. If it is parallel to the plane (i), then this line is perpendicular to the normal to the plane given by (i) and so we have

$$A.6 + B.4 + C.3 = 0$$

or $6A + 4B + 3C = 0$ ...(iii)

Eliminating A, B, C from (i), (ii) and (iii), we get

$$\begin{vmatrix} x-2 & y+1 & z \\ 1 & -3 & 5 \\ 6 & 4 & 3 \end{vmatrix} = 0$$

$$\Rightarrow \quad (x-2)(-29)(y+1)(-27)+z(22)=0$$

$$\Rightarrow \quad 29x - 27y - 22z - 85 = 0.$$ **Ans.**

TO FIND THE PERPENDICULAR DISTANCE OF A POINT P (x_1, y_1, z_1) FROM A GIVEN LINE WHEN ITS EQUATIONS ARE GIVEN IN THE SYMMETRIC FORM

Let the equations of the line be $\frac{x-\alpha}{l} = \frac{y-\beta}{m} = \frac{z-\gamma}{n}$ ...(i)

Any point on this line can be taken as N $(\alpha + lr, \beta + mr, \gamma + nr)$. ...(ii)

If N be the foot of the perpendicular from P (x_1, y_1, z_1) to (i) then the line PN is perpendicular to (i). Then

The direction ratios of the line PN are given as

$$\alpha + lr - x_1, \beta + mr - y_1, \gamma + nr - z_1.$$

And is the line PN is perpendicular to line (i), so we have given as

$$l(\alpha + lr - x_1) + m(\beta + mr - y_1) + n(\gamma + nr - z_1) = 0$$

or $(l^2 + m^2 + n^2)\, r = (lx_1 + my_1 + nz_1) - (l\alpha + m\beta + n\gamma)$

or $r = [l(x_1 - \alpha) + m(y_1 - \beta) + n(z_1 - \gamma)]/(l^2 + m^2 + n^2)$.

Substituting this value of r in (ii), we can determine the coordinates of N, the foot of the perpendicular and then PN can be easily calculated.

TO FIND THE EQUATIONS OF THE PERPENDICULAR LINE FROM THE POINT P (x_1, y_1, z_1) TO A GIVEN LINE WHOSE EQUATIONS ARE GIVEN (A) IN GENERAL FORM AND (B) IN SYMMETRIC FORM

(a) General Form

Let the equations of the line be given as

$$ax + by + cz + d = 0, \; a'x + b'y + c'z + d' = 0 \quad \text{...(i)}$$

If l, m, n be the d.c.'s of this line, then we have

$$al + bm + cn = 0 \text{ and } a'l + b'm + c'n = 0.$$

Solving these we get $\frac{l}{bc'-b'c} = \frac{m}{ca'-c'a} = \frac{n}{ab'-a'b}$ ...(ii)

Now the equations of any plane through the line (i) is given as

$$(ax + by + cz + d) + \lambda(a'x + b'y + c'z + d') = 0 \quad \text{...(iii)}$$

If this plane passes through P (x_1, y_1, z_1), then we have

$$(ax_1 + by_1 + cz_1 + d) + \lambda(a'x_1 + b'y_1 + c'z_1 + d') = 0 \quad \text{...(iv)}$$

∴ The equation of the plane through P (x_1, y_1, z_1) and the line is λ-eliminant of (iii) and (iv).

Also the equation of any plane through P (x_1, y_1, z_1) is given as

$$A(x - x_1) + B(y - y_1) + C(z - z_1) = 0 \quad \text{...(v)}$$

If this is perpendicular to the line (i), then we have

$$A/l = B/m = C/n,$$

where l, m, n are given by (ii), since the normal to the plane (v) is parallel to the line (i).

∴ From (v) the equation of the plane perpendicular to the line (i) and passing through P (x_1, y_1, z_1) is $l(x - x_1) + m(y - y_1) + n(z - z_1) = 0$...(vi)

∴ The equations of perpendicular line from P (x_1, y_1, z_1) to (i) are given by (vi) and λ-eliminant of (iii) and (iv).

(b) Symmetric Form

We can find the d.c.'s l_1, m_1, n_1 (say) of line PN.

Also P is (x_1, y_1, z_1). Therefore the equations of the perpendicular from P (x_1, y_1, z_1) to the given line are

$$\frac{x - x_1}{l_1} = \frac{y - y_1}{m_1} = \frac{z - z_1}{n_1}.$$

Example 1:

The equations of a given line AB are x/2 = y/–3 = z/5. Through a point P (1, 2, 3), PN is drawn perpendicular to the line AB and PQ is drawn parallel to the plane 3x + 4y + 5z = 0 to meet AB in Q. Find the coordinates of N and Q and the equations of PN and PQ.

Solution:

Any point on the line AB is N (2r, – 3r, 5r) ...(i)

∴ The d.r.'s of the line PN are proportional to

$$2r - 1, -3r - 2, 5r - 3. \quad \text{...(ii)}$$

∴ If PN is perpendicular to AB (whose d.r.'s are 2, – 3, 5), then we have

$$2(2r - 1) - 3(-3r - 2) + 5(5r - 3) = 0 \text{ or } r = 11/38$$

∴ From (i) the point N is (11/19, – 33/38, 55/38). **Ans.**

And from (ii) the d.r.'s of PN are –8/19, – 109/38, – 59/38

⇒ 16, 109, 59.

The equations of the line PN which passes through P (1, 2, 3) and whose direction ratios are 16, 109, 59 is $\dfrac{x-1}{16} = \dfrac{y-2}{109} = \dfrac{z-3}{59}$ **Ans.**

Again as before any point Q on the line AB is (2r, – 3r, 5r).

∴ The d.r.'s of the line PQ are 2r – 1, – 3r – 2, 5r – 3.

If PQ is parallel to the plane 3x + 4y + 5z = 0 then PQ is perpendicular to the normal to this plane and so we have

$$3(2r-1)+4(-3r-2)+5(5r-3) \text{ or } r = 26/19$$

∴ The point Q is (52/19, – 78/19, 130/19) and d.r.'s of the line PQ are

$$\frac{52}{19}-1, -\frac{78}{19}-2, \frac{130}{19}-3$$

or $\quad 33, -116, 73.$

∴ The equations of PQ the line passing through P (1, 2, 3) are

$$(x-1)/(33) = (y-2)/(-116) = (z-3)/73.$$ **Ans.**

Example 2:

Find the distance of the point A (1, – 2, 3) from the line PQ drawn through P (2, –3, 5) making equal angles with the axes.

Solution:

Since the line PQ makes equal angles with the axes, so its direction ratios are 1, 1, 1.

∴ The equations of the line PQ are $\dfrac{x-2}{1}=\dfrac{y+3}{1}=\dfrac{z-5}{1}$

∴ Any point N on this line is (2 + r, – 3 + r, 5 + r) ...(i)

∴ The direction ratios of the line AN are given as

$$(2+r)-1, (-3, +r)-(-2), (5+r)-3.$$

i.e., $\quad 1+r, -1+r, 2+r$

Now if N be the foot of the perpendicular drawn from A to the line PQ, then AN is perpendicular to PQ and so we have

$$(1+r)1+(-1+r)1+(2+r)1=0$$ **(Note)**

$\Rightarrow \quad 3r+2=0 \quad \text{or} \quad r=-2/3.$

∴ From (i), the co-ordinates of N are given by

$$\left(2-\frac{2}{3}, -3-\frac{2}{3}, 5-\frac{2}{3}\right) \text{ i.e. } \left(\frac{4}{3}, -\frac{11}{3}, \frac{13}{3},\right)$$

∴ The required distance = AN

$$= \sqrt{\left[\left(1-\frac{4}{3}\right)^2 + \left(-2+\frac{11}{3}\right)^2 + \left(3-\frac{13}{2}\right)^2\right]} = \sqrt{\frac{14}{3}}.$$ **Ans.**

PROJECTION OF A LINE ON A GIVEN PLANE

Definition 1: *If A be the point of intersection of the given line and the given plane and B be the foot of the perpendicular from any point on this line to this plane, then AB is called the projection of the given line on the given plane.*

Definition 2 : *The line of intersection of the given plane with another plane through the given line and perpendicular to the given plane is called the projection of the given line on the given plane.*

Example 1:

Find the projection of the line $3x - y + 2z = 1$, $x + 2y - z = 2$ on the plane $3x + 2y + z = 0$ in the symmetric form.

Solution:

Any plane through the given line is given as

$$(3x - y + 2z - 1) + \lambda\,(x + 2y - z - 2) = 0$$

$$\Rightarrow \quad (3 + \lambda)\,x + (2\lambda - 1)\,y + (2 - \lambda)\,z - (1 + 2\lambda) = 0. \qquad ...(i)$$

If this plane is perpendicular to the given plane $3x + 2y + z = 0$. ...(ii)

Then we have $3\,(3 + \lambda) + 2\,(2\lambda - 1) + 1\,(2 - \lambda) = 0$ or $\lambda = -\,3/2$.

∴ From (i) the equation of the plane through the given line and perpendicular to the given plane (ii) is given by

$$\left(3-\frac{3}{2}\right)x + (-\,3 - 1)\,y + \left(2-\frac{3}{2}\right)z - (1 - 3) = 0$$

$$\Rightarrow \quad 3x - 8y + 7z + 4 = 0 \qquad ...(iii)$$

The projection of the given line on the given plane is the line of intersection of the planes (ii) and (iii).

∴ The equations of the projection of the given line on the given plane are given as

$$3x + 2y + z = 0,\ 3x - 8y + 7z + 4 = 0. \qquad ...(iv)$$

Let l, m, n be the d.c.'s of this line, the as this line lies on both the planes given by (iv), so it is perpendicular to the normal to these planes.

∴ We have $3l + 2m + n = 0$ $3l - 8m + 7n = 0$

From above equation we have

$$\frac{l}{14+8}=\frac{m}{3-21}=\frac{n}{-24-6}$$

or $$\frac{l}{11}=\frac{m}{-9}=\frac{n}{-15} \quad \text{...(v)}$$

Again putting z = 0 in (iv),

we get 3x + 2y = 0, 3x – 8y + 4 = 0

Solving these, we get x = – 4/15, y = 2/5

∴ The line (iv) meets the plane z = 0 at (– 4/15, 2/5, 0)

∴ From (v) and (vi), the required equations are given as

$$\frac{x+(4/15)}{11}=\frac{y-(2/5)}{-9}=\frac{z-0}{-15}=0$$

i.e., $$\frac{15x+4}{11}=\frac{5y-2}{-3}=\frac{z}{-1}.$$ **Ans.**

COPLANAR LINES

To find the condition that two lines whose equations are

$$\frac{x-\alpha_1}{l_1}=\frac{x-\alpha_1}{m_1}=\frac{x-\alpha_3}{n_1}$$

and $$\frac{x-\alpha_2}{l_2}=\frac{y-\beta_2}{m_2}=\frac{z-x}{n_2}$$

intersect (i.e., are coplanar) and equation of the plane in which they lie.

The equation of the given lines may be given in three ways:

(a) both in symmetric form, (b) one in symmetric form the other in general form and (c) both in general form.

(a) Both Lines Given in Symmetric Form

Let the lines be given as

$$\frac{x-\alpha_1}{l_1}=\frac{y-\beta_1}{m_1}=\frac{z-\gamma_1}{n_1} \quad \text{...(i),}$$

$$\frac{x-\alpha_2}{l_2}=\frac{y-\beta_2}{m_2}=\frac{z-\gamma_2}{n_2} \quad \text{...(ii)}$$

The equation of any plane through the line (i) is given as

$$A(x-\alpha_1)+B(y-\beta_1)+C(z-\gamma_1)=0, \quad \text{...(iii)}$$

where $\quad Al_1+Bm_1+Cn_1=0. \quad$ ...(iv)

If the two lines given by (i) and (ii) intersect, then the line (ii) must lie on the plane (iii) *i.e.*, the normal to the plane (iii) must be at right angles to (ii), consequently we get

$$Al_2+Bm_2+Cn_2=0. \quad \text{...(v)}$$

Also from (ii) it is evident that $(\alpha_2, \beta_2, \gamma_2)$ is a point on (ii).

If the line (ii) lies on the plane (iii), the this point $(\alpha_2, \beta_2, \gamma_2)$ must lie on (iii) and so we have

$$A(\alpha_2-\alpha_1)+B(\beta_2-\beta_1)+C(\gamma_2-\gamma_1)=0 \quad \text{...(vi)}$$

Hence the given line (i) and (ii) intersect *i.e.*, they are coplanar, then eliminating A, B, C from (iv), (v) and (vi) we have the required condition which is.

Given as $\begin{vmatrix} \alpha_2-\alpha_1 & \beta_2-\beta_1 & \gamma_2-\gamma_1 \\ l_1 & m_1 & n_1 \\ l_2 & m_2 & n_2 \end{vmatrix}=0$

Also the equation of the plane in which the given line (i) and (ii) lie is obtained by eliminating A, B, C from (iii), (iv) and (v) is given

as $\begin{vmatrix} x-\alpha_1 & y-\beta_1 & z-\gamma_1 \\ l_1 & m_1 & n_1 \\ l_2 & m_2 & n_2 \end{vmatrix}=0$

TO FIND THE POINT OF INTERSECTION OF THE LINES (I) AND (II)

Let any point on the line (i) is $(\alpha_1+l_1r_1, \beta_1+m_1r_1, \gamma_1+n_1r_1)$ and any point on the line (ii) is $(\alpha_2+l_2r_2, \beta_2+m_2r_2, \gamma_2+n_rr_2)$.

If the two lies intersect then for some values of r_1, and r_2 these points must coincide *i.e.*, we have

$$\alpha_1+l_1r_1=\alpha_2+l_2r_2,\ \beta_1+m_1r_1$$
$$=\beta_2+m_2r_2,\ \gamma_1+n_1r_1=\gamma_2+n_2r_2$$

$$\Rightarrow \quad l_1r_1-l_2r_2+(\alpha_1-\alpha_2)=0 \quad \text{...(i)}$$

$$m_1r_1-m_2r_2+(\beta_1-\beta_2)=0 \quad \text{...(ii)}$$

and $\quad n_1r_1-n_2r_2+(\gamma_1-\gamma_2)=0. \quad$ ...(iii)

Solving (i) and (ii) we can obtain the values of r_1 and r_2 and if they satisfy (iii) also then two lines intersect and we can find the point of intersection by substituting the value of r_1 in $(\alpha_1 + l_1 r_1, \beta_1 + m_1 r_1, \gamma_1 + n_1 r_1)$ of the value of r_1 in $(\alpha_2 + l_2 r_2, \beta_2 + m_2 r_2, \gamma_2 + n_2 r_2)$.

(b) One Line in Symmetric Form, the Other in General Form

Let the equations of the lines be $\dfrac{x-\alpha}{l} = \dfrac{y-\beta}{m} = \dfrac{z-\gamma}{n}$ ...(i)

and $ax + by + cz + d = 0 = a'x + b'y + c'z + d'$ ...(ii)

The equation of the plane through the line (ii) is given by

$$(ax + by + cz + d) = \lambda (a' x + b'y + c'z + d') = 0 \quad ...(iii)$$

$$\Rightarrow \quad (a + \lambda a') x + (b + \lambda b') y + (c + \lambda c') z + (d + \lambda d') = 0 \quad ...(iv)$$

If this plane is parallel to the line (i), then we have

$$l (a + \lambda a') + m (b + \lambda b') + n (c + \lambda c') = 0$$

$$\Rightarrow \quad \lambda (a' l + b' m + c' n) = -(al + bm + cn)$$

$$\Rightarrow \quad \lambda = -(al + bm + cn)/(a' l + b' m + c' n)$$

Putting this value of λ in (iii), the equation of the plane through the line (ii) and parallel to the line (i) is

$$(a' l + b' m + c' n) (ax + by + cz + d)$$
$$= (al + bm + cn) (a'x + b' y + c' z + d') \quad ...(iv)$$

If the line (i) lies in this plane then the point (α, β, γ) on the line (i) must satisfy (iv) and so condition for the lines (i) and (ii) to be coplanar is

$$(a'l + b'm + c'n) (a\alpha + b\beta + c\gamma + d)$$
$$= (al + bm + cn) (a'\alpha + b'\beta + c'\gamma + d')$$

$$\Rightarrow \frac{a\alpha + b\beta + c\gamma + d}{al + bm + cn} = \frac{a' \alpha + b'\beta + c'\gamma + d'}{a' l + b'm + c'n}$$

Also from (iv) putting $\dfrac{a'l + b'm + c'n}{al + bm + cn} = \dfrac{a'\alpha + b'\beta + c'\gamma + d'}{a\alpha + b\beta + c\gamma + d}$

equation of the plane in which (i) an (ii) lie is

$$\frac{ax + by + cz + d}{a\alpha + b\beta + c\gamma + d} = \frac{a'x + b'y + c'z + d'}{a'\alpha + b'\beta + c'\gamma + d}.$$

BOTH THE LINES IN GENERAL FORM

Let the equations of the lines be given as

$$a_1x + b_1y + c_1z + d_1 = 0 = a_2x + a_2x + b_2y + c_2z + d_2 \quad ...(i)$$

$$a_3x + b_3y + c_3z + d_3 = 0 = a_4x + b_4y + c_4z + d_4$$

If these two lines are coplanar, then they intersect and let (x_1, y_1, z_1) be their point of intersection.

If (x_1, y_1, z_1) is their point of intersection, then it satisfies all the four planes given by (i) and which intersect in the two lines.

$\therefore$ we have

$$a_1x_1 + b_1y_1 + c_1z_1 + d_1 = 0$$
$$a_2x_1 + b_2y_1 + c_2z_1 + d_2 = 0$$
$$a_3x_1 + b_3y_1 + c_3z_1 + d_3 = 0$$
$$a_4x_1 + b_4y_1 + c_4z_1 + d_4 = 0.$$

Eliminating x_1, y_1 and z_1 from these we have the required condition as

$$\begin{vmatrix} a_1 & b_1 & c_1 & d_1 \\ a_2 & b_2 & c_2 & d_2 \\ a_3 & b_3 & c_3 & d_3 \\ a_4 & b_4 & c_4 & d_4 \end{vmatrix} = 0$$

TO OBTAIN THE EQUATIONS OF A STRAIGHT INTERSECTING TWO GIVEN LINES

Case I : Given Lines in Symmetric Form

Let the given lines be $\dfrac{x-\alpha}{l} = \dfrac{y-\beta}{m} = \dfrac{z-\gamma}{n} = r$ (say) ...(i)

and $\dfrac{x-\alpha'}{l'} = \dfrac{y-\beta'}{m'} = \dfrac{z-\gamma'}{n'} = r'$ (say) ...(ii)

Any point on the line (i) is $(\alpha + lr, \beta + mr, \gamma + ar)$ and any point on line (ii) is $(\alpha' + l' r, \beta' + m' r', \gamma' + n' r')$.

The required line is the line which joins these two points for some values of r and r' which will be obtained from other given conditions.

Case II : Given Lines in General Form

Let the given lines be $u = 0$, $v = 0$

and $u' = 0$, $v' = 0$

Then the equations of the required line are

$$u = kv = 0$$

and $u' + k' v' = 0$,

where k and k' will be obtained by another condition.

Example 1:

Show that the equations of the line through (a, b, c) which is parallel to the plane lx + my + nz = 0 and intersects the line

$$A_1x + B_1y + C_1z + D_1 = 0 = A_2x + B_2y + C_2z + D_2$$

are $$l(x-a) + m(y-b) + n(z-c) = 0$$

and $$\frac{A_1x+B_1y+C_1z+D_1}{A_1a+B_1b+C_1c+D_1} = \frac{A_2x+B_2y+C_2z+D_2}{A_2a+B_2b+C_2c+D_2}.$$

Solution:

The required line is the line of intersection of the two planes as stated below:

(i) The plane through the point (a, b, c) and parallel to the given plane $lx + my + nz = 0$ is given by

$$l(x-a) + m(y-b) + n(z-c) = 0 \qquad ...(i)$$

(ii) The plane through the point (a, b, c) and the line

$$(A_1x + B_1y + C_1z + D_1) + \lambda(A_2x + B_2y + C_2z + D_2) = 0, \qquad ...(ii)$$

where λ is calculated from the condition that the plane (ii) passes through (a, b, c).

$\therefore$ From (ii), $(A_1a + B_1b + C_1c + D_1) + \lambda(A_2a + B_2b + C_2c + D_2) = 0$,

$\Rightarrow \lambda = -(A_1a + B_1b + C_1c + D_1)/(A_2a + B_2b + C_2c + D_2)$

$\therefore$ From (ii) putting this value of λ the equation of the plane through (a, b, c) and the given line is

$$\frac{A_1x+B_1y+C_1z+D_1}{A_1a+B_1b+C_1c+D_1} = \frac{A_2x+B_2y+C_2z+D_2}{A_2a+B_2b+C_2c+D_2} \qquad ...(iii)$$

Here the equations of the required line are given by (i) and (iii).

Example 2:

Find the S.D. between the line $x = 0, (y/2) + (z/3) = 1$ *and* $y = 0, (x/4) - (z/3) = 1$.

Solution:

Equations of the plane through the given lines are written as

$$x + \lambda\left[\frac{1}{2}y + \frac{1}{3}z - 1\right] = 0$$

or $$6x + \lambda(3y + 2z - 6) = 0$$

and $$y+\mu\left[\frac{1}{4}x-\frac{1}{3}z-1\right]=0$$

$$12\,y + \mu\,(3x - 4z - 12) = 0$$

i.e., $$6x + 3\lambda + 2\lambda z - 6\lambda = 0 \qquad ...(i)$$

and $$3\mu x + 12y - 4\mu z - 12\mu = 0. \qquad ...(ii)$$

If the plane (i) and (ii) are parallel, then we have

If the plane (i) and (ii) are parallel, then we have

$$\frac{6}{3\mu}=\frac{3\lambda}{12}=\frac{2\lambda}{-4\mu}$$, comparing coefficients of x, y , z

From $$\frac{6}{3\mu}=\frac{2\lambda}{-4\mu}$$ we get $\lambda = -4$

And form $$\frac{6}{3\mu}=\frac{3\lambda}{12}$$ we get $8 = \lambda\,\mu$

$\Rightarrow$ $\mu = -2$, putting the value of λ.

$\therefore$ Form (i) and (ii) substituting values of λ and μ we get the parallel planes as $3x - 6y - 4z + 12 = 0$ and $3x - 6y - 4z - 12 = 0$

Now any point on the plane $3x - 6y - 4z + 12 = 0$ is (0, 0, 3)

$\therefore$ The required length of S.D.

= length of perpendicular from (0, 0, 3) on the plane $3x - 6y - 4z = 12$

$$=\frac{3.0-6.0-4.3-12}{\sqrt{[3^2+(-6)^2+(-4)^2]}}=\frac{-24}{\sqrt{(61)}}=\frac{24}{\sqrt{(61)}}$$, numerically. **Ans.**

Example 3:

Show that the lines

$$\frac{1}{2}(x-1)=\frac{1}{3}(y-2)=\frac{1}{4}(z-3)$$

and $4x - 3y + 1 = 0 = 5x - 3z + z$ *are coplanar. Also find their point of intersection.*

Solution:

Any plane through the second line is given as

$$(4x - 3y + 1) + \lambda\,(5x - 3z + 2) = 0$$

$$\Rightarrow \quad (4 + 5\lambda)\,x - 3y - 3\lambda\,z + (1 + 2\lambda) = 0. \qquad ...(i)$$

If it is parallel to the line $\frac{x-1}{2}=\frac{y-2}{3}=\frac{z-3}{4}$, then we have

$$2.(4 + 5\lambda) + 3.(-3) + 4.(-3\lambda) = 0$$

or $\quad -2\lambda - 1 = 0$

or $\quad \lambda = -\frac{1}{2}$.

Hence from (i) the equation of the plane through the second line and parallel to the first line is

$$[4 - (5/2)]\, x - 3y + (3/2)\, z + [1 - 2.(1/2)] = 0$$

$\Rightarrow \quad (3/2)\, x - 3y + (3/2)\, z = 0$

or $\quad x - 2y + z = 0.$ ...(ii)

Also from the equations of the first line it is evident that (1, 2, 3) is a point on this line. And we find from (ii) that the point (1, 2, 3) lies on this plane (ii). Hence the given lines are coplanar *i.e.,* they intersect.

To find the point of intersection: Any point on the first line is

$$(1 + 2r\ 2 + 3r,\ 3 + 4r). \quad \text{...(iii)}$$

As the two given lines intersect therefore for some value of r the point (iii) lies on the second line and so we have

$$4\,(1 + 2r) - 3\,(2 + 3r) + 1 = 0$$

and $\quad 5\,(1 + 2r) - 3\,(3 + 4r) + 2 = 0.$

Both of these give r = – 1 and therefore from (iii) the required point of intersection is (1 – 2, 2 – 3, 3 – 4) or (– 1, – 1, – 1). **Ans.**

Example 4:

Find the S.D. between the lines

$$\frac{x-1}{2}=\frac{y-2}{3}=\frac{z-3}{4} \quad \text{...(i)}$$

and $$\frac{x-2}{3}=\frac{y-3}{4}=\frac{z-4}{5} \quad \text{...(ii)}$$

Solution:

Let *l*, m, n be the d.c.'s of the S.D. Since the lines of S.D. is perpendicular to both the given lines thus we have

$$2.l + 3.m + 4.n = 0 \quad \text{...(iii)}$$

and $\quad 3l + 4m + 5n = 0.$ ...(iv)

Solving (iii) and (iv) we get

$$\frac{l}{-1}=\frac{m}{2}=\frac{n}{-1}=\frac{\sqrt{(l^2+m^2+n^2)}}{\sqrt{\left[(-1)^2+2^2+(-1)^2\right]}}=\frac{1}{\sqrt{6}}$$

$\Rightarrow \quad l = -\left(1/\sqrt{6}\right),\ m = \left(2/\sqrt{6}\right),\ n = -\left(1/\sqrt{6}\right)$ ...(iv)

Also from the given equations it is evident that A (1, 2, 3) is a point on (i) and B (2, 3, 4) is a point on (ii).

Now S. D. is the projection of AB on the line whose d.c.'s are given by (iii) and so we have

S.D. = $l(1-2) + m(2-3) + n(3-4)$, where l, m, n are given by (iii)

$$= -\frac{1}{\sqrt{6}}(-1)+\frac{2}{\sqrt{6}}(-1)-\frac{1}{\sqrt{6}}(-1) = \left(1/\sqrt{6}\right)(1-2+1) = 0.$$

Example 4:

Show that the equation to a right circular cone whose vertex is the origin, the semi-vertical angle θ and whose axis has direction cosines l, m, n is $\Sigma[yn - zm]^2 = (x^2 + y^2 + z^2)\sin^2\theta$.

Solution:

Let P (x, y, z) be a point on the one whose vertex V is (0, 0, 0) and semi-vertical angle is θ.

From P draw PN perpendicular to its axis VC, whose d.c.'s are l, m, n.

The equation of VC is $\frac{x}{l}=\frac{y}{m}=\frac{z}{n}$.

Then $PN^2 = \begin{vmatrix} x & y \\ l & m \end{vmatrix}^2 + \begin{vmatrix} y & z \\ m & n \end{vmatrix}^2 + \begin{vmatrix} z & x \\ n & l \end{vmatrix}^2$

$= (xm - yl)^2\ (yn - zm)^2 + (zl - nx)^2 = \Sigma\,(yn - zm)^2$ ...(i)

And VP = distance between V (0, 0, 0) and P (x, y, z)

$\Rightarrow \quad VP = \sqrt{(x^2+y^2+z^2)}$ or $VP^2 = x^2 + y^2 + z^2$ ...(ii)

Also from Δ VPN it is evident that

$$PN = VP \sin\theta \text{ or } PN^2 = VP^2 \sin^2\theta$$

$\Rightarrow \quad \Sigma\,(yn - zm)^2 = (x^2 + y^2 + z^2 \sin^2\theta$, from (i) and (ii). **Hence proved.**

Example 5:

Find the equation to the right circular cone whose vertex is (2, 3, 5) the semi-vertical angle is 30°, and the axis is a line equally inclined to the coordinate axes.

Solution:

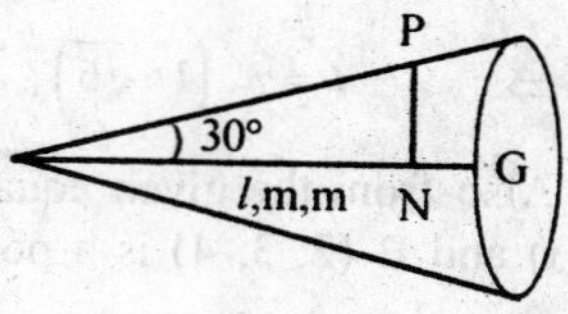

If the axis VC of the cone [Here V is (2, – 3, 5)] is equally inclined to the coordinate axes, then its direction cosines are $l, l, l,$ where

$$\sqrt{(l^2+l^2+l^2)} = 1$$

or $$l = 1/\sqrt{3}$$

∴ The direction cosines of the axis VC of the cone are given as $1/\sqrt{3}$, $1/\sqrt{3}$, $1/\sqrt{3}$ and V is (2, –3, 5).

∴ The equations of the axis VC are $\dfrac{x-2}{1/\sqrt{3}} = \dfrac{y+3}{1/\sqrt{3}} = \dfrac{z-5}{1/\sqrt{3}}$

∴ If P (x, y, z) be any point on the cone, then the length of the perpendicular PN from P to the axis VC is given as

$$PN^2 = \begin{vmatrix} x-2 & x+3 \\ 1/\sqrt{3} & 1/\sqrt{3} \end{vmatrix}^2 + \begin{vmatrix} y+3 & z+5 \\ 1/\sqrt{3} & 1/\sqrt{3} \end{vmatrix}^2 + \begin{vmatrix} z-5 & x-2 \\ 1/\sqrt{3} & 1/\sqrt{3} \end{vmatrix}^2$$

$$= \frac{1}{3}\left[\begin{vmatrix} x-2 & y+3 \\ 1 & 1 \end{vmatrix}^2 + \begin{vmatrix} y-3 & z-5 \\ 1 & 1 \end{vmatrix}^2 + \begin{vmatrix} z-5 & x-2 \\ 1 & 1 \end{vmatrix}^2\right]$$

$$= \frac{1}{3}\ [(x-y-5)^2 + (y-z+8)^2 + (z-x-3)^2]$$

$$= \frac{1}{3}[2x^2+2y^2+2z^2-2xy-2yz-2xz-4x+26y-22z+98] \quad ...(i)$$

Also VP^2 = Square of the distance between V(2, – 3, 5) and P(x, y, z)

$$= (x-2)^2 + (y+3)^2 + (z-5)^2$$

$$= x^2 + y^2 + z^2 - 4x + 6y - 10z + 38 \quad ...(ii)$$

Also from Δ VPN we get

$$PN = VP \sin\theta = VP \sin 30^\circ$$

$$\Rightarrow \quad PN^2 = VP^2 \left(\frac{1}{2}\right)^2$$

$$\Rightarrow \quad (2/3)\ [x^2 + y^2 + z^2 - xy - yz - zx - 2x + 13y - 11z + 49]$$

$\Rightarrow \quad = (1/4)\,[x^2 + y^2 + z^2 - 4x + 6y - 10z + 38]$, from (i) and (ii)

$\Rightarrow \quad 5x^2 + 5y^2 + 5z^2 - 8xy - 8yz - 8zx - 4x + 86y - 58z + 278 = 0.$

TO FIND THE PERPENDICULAR DISTANCE OF A POINT FROM A LINE AND THE CO-ORDINATES OF THE FOOT OF THE PERPENDICULAR

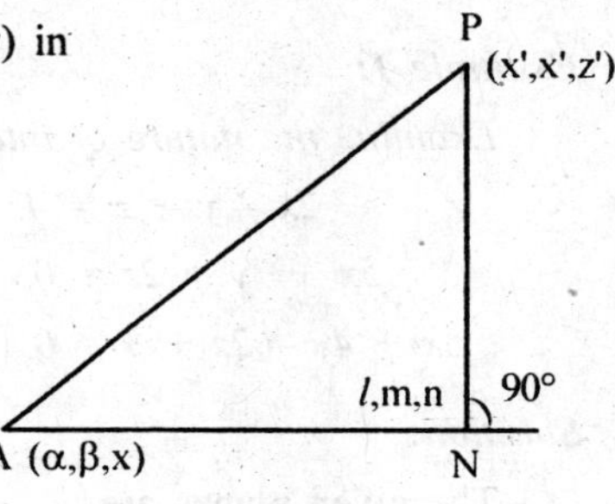

Let the equations of the line AB (say) in the symmetric form be given as

$$\frac{x-\alpha}{l} = \frac{y-\beta}{m} = \frac{z-\gamma}{n}$$

Again let the point be P (x', y', z').

From P draw PN perpendicular to AB. Join AP.

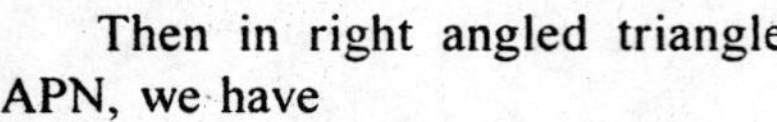

Then in right angled triangle APN, we have

$$PN^2 = AP^2 - AN^2 \qquad \text{...(i)}$$

Now $\quad AP^2 = (x' - \alpha)^2 + (y' - \beta)^2 + (z' - \gamma)^2 \qquad$ ...(ii)

And AN = the projection of AP on the line whose d.c.' are l, m, n

$\Rightarrow \quad AN = l\,(x' - \alpha) + m\,(y' - \beta) + n\,(z' - \gamma)$

$\therefore$ From (i), (ii) and (iii) we have PN^2

$= [(x'-\alpha)^2 + (y' - \beta)^2 + (z' - \gamma)^2] - [l\,(x' - \alpha) + m(y' - \beta) + n(z' - \gamma)^2]$

$= [(x' - \alpha)^2 + (y' - \beta)^2 + (z' - \gamma)^2]\,(l^2 + m^2 + n^2)$

$\qquad - [l\,(x' - \alpha) + m\,(y' - \beta) + n\,(z' - \gamma)]^2, \because l^2 + m^2 + n^2 = 1$

$= \{m\,(z' - \gamma) - n\,(y' - \beta)\}^2 + \{n\,(x' - \alpha) - l\,(z' - \gamma)\}^2$

$\qquad + \{l\,(y' - \beta) - m\,(x' - \alpha)\}^2.$

using Larange's Identity

$$= \begin{vmatrix} m & n \\ y'-\beta & z'-\gamma \end{vmatrix}^2 + \begin{vmatrix} n & l \\ z'-\gamma & x'-\alpha \end{vmatrix}^2 + \begin{vmatrix} l & m \\ x'-\alpha & y'-\beta \end{vmatrix}^2$$

Coordinates of N, the Foot of Perpendicular

Let AN = r, then the coordinates of N, which is a point on AD at a distance r from A are (α, + lr, β + mr, γ + nr). ...(iii)

$\therefore$ The direction ratios of PN are $\alpha + lr - x'$, $\beta + mr - y'$, $\gamma + nr - z'$.

As PN is perpendicular to AB, whose d.c.'s are l, m, n so we have

$$l(\alpha + lr - x') + m(\beta + mr - y') + n(\gamma + nr - z') = 0$$

$$\Rightarrow \quad (l^2 + m^2 + n^2)\, r = l(x' - \alpha) + m(y' - \beta) + n(z' - \gamma)$$

$$\Rightarrow \quad r = l(x' - \alpha) + m(y' - \beta) + n(z' - \gamma), \text{ since } l^2 + m^2 + n^2 = 1.$$

Substituting this value of r in (iv) we can find the coordinates of N, the foot of the perpendicular from P to the line AB.

Example 1:

Examine the nature of intersection of planes.

$$2x - y + z = 4,$$
$$5x + 7y + 2z = 0,$$
$$3x + 4y - 2z + 3 = 0.$$

Solution:

The given planes are

$$2x - y + z - 4 = 0 \qquad \text{...(i)}$$
$$5x + 7y + 2z + 0 = 0 \qquad \text{...(ii)}$$
$$3x + 4y - 2z + 3 = 0 \qquad \text{(iii)}$$

∴ The 'rectangular array' is given as

$$\begin{Vmatrix} 2 & -1 & 1 & -4 \\ 5 & 7 & 2 & 0 \\ 3 & 4 & -2 & 3 \end{Vmatrix}$$

$$\therefore \text{"}\Delta_4\text{"} = \begin{vmatrix} 2 & -1 & 1 \\ 5 & 7 & 2 \\ 3 & 4 & -2 \end{vmatrix} = \begin{vmatrix} 0 & 0 & 1 \\ 1 & 9 & 2 \\ 7 & 2 & -2 \end{vmatrix}$$

subtracting 2 times third column from first and adding third column to second.

$$\Rightarrow \quad \Delta_4 = -61 \neq 0.$$

Hence the given planes intersect in a point.

Solving (i), (ii) and (iii) we obtain

$$\frac{x}{\begin{vmatrix} -1 & 1 & -4 \\ 7 & 2 & 0 \\ 4 & -2 & 3 \end{vmatrix}} = \frac{-y}{\begin{vmatrix} 2 & 1 & -4 \\ 3 & 2 & 0 \\ 3 & -2 & 3 \end{vmatrix}} = \frac{z}{\begin{vmatrix} 2 & -1 & -4 \\ 5 & 7 & 0 \\ 3 & 4 & 3 \end{vmatrix}} = \frac{-1}{\begin{vmatrix} 2 & -1 & 1 \\ 5 & 7 & 2 \\ 3 & 4 & -2 \end{vmatrix}}$$

which give x = 1, y = – 1, z = 1.

Hence the planes meet in the point (1, – 1, 1). **Ans.**

Example 2:

Prove that the planes $2x - 3y - 7z = 0$, $2x - 14y - 13z = 0$, $8x - 31y - 33z = 0$ pass through one line and find its equation.

Solution:

The rectangular array is given as

$$\left\|\begin{matrix} 2 & -3 & -7 & 0 \\ 3 & -14 & -13 & 0 \\ 8 & -31 & -33 & 0 \end{matrix}\right\|$$

$$\therefore \Delta_4 = \begin{vmatrix} 2 & -3 & -7 \\ 3 & -14 & -13 \\ 8 & -31 & -33 \end{vmatrix} = \begin{vmatrix} 2 & 3 & 7 \\ 3 & 14 & 13 \\ 8 & 31 & 33 \end{vmatrix}$$

taking – 1 common from 2nd and 3rd columns.

$$= \begin{vmatrix} 2 & 1 & 1 \\ 3 & 11 & 4 \\ 8 & 23 & 9 \end{vmatrix},$$ sutracting Ist column from 2nd and 3 times Ist column from 3rd respectively.

$$= \begin{vmatrix} 0 & 0 & 1 \\ -5 & 7 & 1 \\ -10 & 14 & 9 \end{vmatrix},$$ subtracting 3rd column once from 2nd and twice from first,

= 0, expanding with respect to first row and evaluating.

$$\text{Also } \Delta_1 = \begin{vmatrix} -3 & -7 & 0 \\ -14 & -13 & 0 \\ -31 & -33 & 0 \end{vmatrix} = 0$$

Since $\Delta_4 = 0$ and $\Delta_1 = 0$, therefore the given planes meet in a line.

The equations of the line of intersection are given by

$$2x - 3y - 7z = 0,$$

$$3x - 14y - 13z = 0$$

(taking any two of the given planes)

These may be written in the symmetric form as

$$\frac{x}{39 - 94} = \frac{y}{-21 + 26} = \frac{z}{-28 + 9}$$

or $$\frac{x}{59} = \frac{y}{-5} = \frac{z}{19}.$$ **Ans.**

Solving these we get $\dfrac{l}{b_1c_2 - b_2c_1} = \dfrac{m}{c_2a_2 - c_2a_1} = \dfrac{n}{a_2b_2 - a_2b_1}$

If the this is parallel to the plane (iii), then this must be at right angles to the normal to the plane (iii) and so we get

$$a_3\,(b_1c_2 - b_2c_1) + b_3\,(c_1a_2 - c_2a_1) + c_3\,(a_1b_2 - a_2b_1) = 0$$

$$\Rightarrow \quad \begin{vmatrix} a_1 & b_1 & c_1 \\ a_2 & b_2 & c_2 \\ a_3 & b_3 & c_3 \end{vmatrix} = 0 \qquad \text{...(vi)}$$

The symmetry of this result shows that this is the condition for the line of intersection of any two planes to be parallel to the third plane.

Hence the planes (i), (ii) and (iii) form a triangular prism if (vi) is satisfied and other determinants of (iv) *do not vanish.*

Case III : *The Planes Intersecting in a Point*

Solving the equations (i), (ii) and (iii) by Crammer's Rule we get

$$\frac{x}{\begin{vmatrix} b_1 & c_1 & d_1 \\ b_2 & c_2 & d_2 \\ b_3 & c_3 & d_3 \end{vmatrix}} = \frac{-y}{\begin{vmatrix} a_1 & c_1 & d_1 \\ a_2 & c_2 & d_2 \\ a_3 & c_3 & d_3 \end{vmatrix}} = \frac{z}{\begin{vmatrix} a_1 & b_1 & d_1 \\ a_2 & b_2 & d_2 \\ a_3 & b_3 & d_3 \end{vmatrix}} = \frac{-1}{\begin{vmatrix} a_1 & b_1 & c_1 \\ a_2 & b_2 & c_2 \\ a_3 & b_3 & c_3 \end{vmatrix}}$$

$$\Rightarrow \quad \frac{x_1}{\Delta_1} = \frac{-y}{\Delta_2} = \frac{z}{\Delta_3} = \frac{-1}{\Delta_4}$$

where $\Delta_1 = \begin{vmatrix} b_1 & c_1 & d_1 \\ b_2 & c_2 & d_2 \\ b_3 & c_3 & d_3 \end{vmatrix}$,

$\Delta_2 = \begin{vmatrix} a_1 & c_1 & d_1 \\ a_2 & c_2 & d_2 \\ a_3 & c_3 & d_3 \end{vmatrix}$,

$\Delta_3 = \begin{vmatrix} a_1 & b_1 & d_1 \\ a_2 & b_2 & d_2 \\ a_3 & b_3 & d_3 \end{vmatrix}$

and $\Delta_4 = \begin{vmatrix} a_1 & b_1 & c_1 \\ a_2 & b_2 & c_2 \\ a_3 & b_3 & c_3 \end{vmatrix}$

$$\Rightarrow \qquad x = -\frac{\Delta_1}{\Delta_4},\ y = \frac{\Delta_2}{\Delta_4},\ z = -\frac{\Delta_3}{\Delta_2}.$$

Hence, if the planes intersect in a point at a finite distance then we must have $\Delta_4 \neq 0$, which is the required condition.

Working Rule : Let the three planes be given by equations (i), (ii) and (iii) and let us Δ_1, Δ_2, Δ_3 and Δ_4 as in case III above.

Now we proceed as follows:

(a) Evaluate Δ_4 and if $\Delta_4 \neq 0$, then the planes intersect in a point and the coordinates of this point can be obtained by solving the given equations.

(b) If $\Delta_4 = 0$, then evaluate any one of Δ_1, Δ_2 and Δ_3 and we have:

 (i) If none of Δ_1, Δ_2, and Δ_3 is zero, then the planes form triangular prism and

 (ii) If any of Δ_1, Δ_2 and Δ_3 is zero, then the planes intersect in a line.

Note: If $\Delta_4 = 0$ and $\Delta_1 \neq 0$,

then $\Delta_2 \neq 0$, $\Delta_3 \neq 0$.

And if $\Delta_4 = 0$ and $\Delta_1 = 0$,

then $\Delta_2 = 0$, $\Delta_3 = 0$.

Hence, evaluation of Δ_4 and only one of the determines out of Δ_1, Δ_2, Δ_3 is sufficient.

INTERSECTION OF THREE PLANES

Let the planes be $a_1x + b_1y + c_1x + d_i = 0$, $i = 1, 2, 3$.

Now if we take two equations at a time, we get three lines of intersection of the above three plane and following three cases arise:

(i) *Three lines* of intersection of these planes may coincide and then the three given *planes intersect in a common line.*

(ii) *Three lines* of intersection of three planes may be *parallel* and then the three given *planes form a triangular prism,* and

(iii) *Three lines* of intersection of three planes may *intersect in a point* and then the three given *planes intersect in a point.*

Now we are to *find the condition* that the planes.

$$u_1 \equiv a_1x + b_1y + c_1z + d_1 = 0 \qquad \text{...(i)}$$

$$u_2 \equiv a_2x + b_2y + c_2z + d_2 = 0 \quad \text{...(ii)}$$

$$u_3 \equiv a_3x + b_3y + c_3z + d_3 = 0 \quad \text{...(iii)}$$

(i) may intersect in a common line.

(ii) may form a triangular prism, and

(iii) may intersect in a point.

Case I : *Planes Intersecting in a Common Line.*

The equation of any plane through the intersection of the planes (i) and (ii) is $u_1 + \lambda u_2 = 0$

i.e., $(a_1x + b_1y + c_1z + d_1) + \lambda (a_2x + b_2y + c_2z + d_2) = 0$

$$\Rightarrow \quad (a_1 + \lambda a_2) x + (b_1 + \lambda b_2) y + (c_1 + \lambda c_2) z + (d_1 + \lambda d_2) = 0 \quad \text{...(iv)}$$

If the planes (i), (ii) and (iii) intersect in a common line, then by properly choosing the value of λ the plane (iv) can be made to represent the plane (iii).

Hence, comparing the coefficients of x, y, z and the constant terms in (iii) and (iv) we get

$$\frac{a_1 + \lambda a_2}{a_3} = \frac{b_1 + \lambda b_2}{b_3} = \frac{c_1 + \lambda c_2}{c_3} = \frac{d_1 + \lambda d_2}{d_3} = k \text{ (say)}$$

i.e.,

$$a_1 + \lambda a_2 - ka_3 = 0$$

$$b_1 + \lambda b_2 - kb_3 = 0$$

$$c_1 + \lambda c_2 - kc_3 = 0$$

$$d_1 + \lambda d_2 - kd_3 = 0.$$

Now, we are to eliminate two variables λ and k and for this we require only three equations whereas there are four equations. Therefore we take any three equations and obtain one condition by eliminating λ and k between them. Now we can choose a set of three equations from the above four equations in four ways and so it appears that we shall get four different conditions. But we can easily find that any two sets of three equations are equivalent to the remaining two sets and consequently we shall get only two independent conditions.

The necessary conditions are expressed by the rectangular array

$$\left\| \begin{matrix} a_1 & b_1 & c_1 & d_1 \\ a_2 & b_2 & c_2 & d_2 \\ a_3 & b_3 & c_3 & d_3 \end{matrix} \right\| = 0, \quad \text{...(v)}$$

The notation signifies that any of the four determinants of third order obtained by omitting one vertical column are zero and thus we get two independent conditions.

Case II : *The Plane Forming a Triangular Prism.*

The planes will form triangular prism provided the three lines of intersection of the planes taken two at a time are parallel.

Let l, m, n be the direction ratios of the line of intersection of the planes (i) and (ii), then we have

$$a_1 l + b_1 m + c_1 n = 0$$

and $$a_2 l + b_2 m + c_2 n = 0,$$

as this line is perpendicular to the normals to these planes.

TO FIND THE SHORTEST DISTANCE (OR S.D.) BETWEEN TWO GIVEN LINES AND TO OBTAIN THE EQUATION OF THIS SHORTEST DISTANCE

Definition: Two lines are said to be *skew lines or non-intersecting lines* if they do not lie in the same plane and the straight line which is perpendicular to each of these two non-intersecting lines is called the line of shortest distance or S. D. The length of this line intercepted between the given lines is called the length of the shortest distance.

Lengths and Equations of S.D.

Method I : *Length of S.D.*

Let the equations of the given lines be

$$\frac{x-\alpha_1}{l_1} = \frac{y-\beta_1}{m_1} = \frac{z-\gamma_1}{n_1} \qquad ...(i)$$

and $$\frac{x-\alpha_2}{l_2} = \frac{y-\beta_2}{m_2} = \frac{z-\gamma_2}{n_2} \qquad ...(ii)$$

Let PQ be the line which is perpendicular to both the given lines AB and CD. Let l, m, n be the direction ratios of the line PQ.

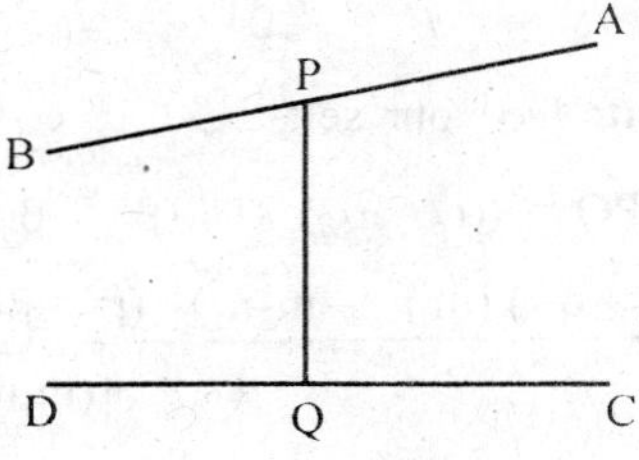

Then as PQ is perpendicular to the lines (i) and (ii), so we have $ll_1 + mm_1 + nn_1 = 0$

and $ll_2 + mm_2 + nn_2 = 0.$

Solving these, we get

$$\frac{l}{m_1n_2 - m_2n_1} = \frac{m}{n_1l_2 - n_2l_1} = \frac{n}{l_2m_2 - l_2m_2} \quad ...(iii)$$

∴ If λ, μ, ν be the actual d.c.'s of the line PQ, then we have

$$\lambda = \frac{(m_1n_2 - m_2n_1)}{\sqrt{\left[(m_1n_2 - m_2n_1)^2 + (n_1l_2 - n_2l_1)^2 + (l_1m_2 - l_2m_1)^2\right]}}$$

$$\Rightarrow \lambda = \frac{m_1n_2 - m_2n_1}{\sqrt{\left[\Sigma(m_1n_2 - m_2n_1)^2\right]}}$$

Similarly $\mu = \dfrac{n_1l_2 - n_2l_1}{\sqrt{\left[\Sigma(m_1n_2 - m_2n_1)^2\right]}}$ and $\nu = \dfrac{l_1m_2 - l_2m_1}{\sqrt{\left[\Sigma(m_1n_2 - m_2n_1)^2\right]}}$ ...(iv)

From figure above it is evident that PQ is the projection of AC [where A and C are points $(\alpha_1, \beta_1, \gamma_1)$ and $(\alpha_2, \beta_2, \gamma_2)$ on the line (i) and (ii) respectively] on the line PQ whose d.c.'s are (λ, μ, ν) given by (iv).

Example 1:

Find the magnitude and the equations of the line of S. D. between the lines

$$\frac{x-8}{3} = \frac{y+9}{-16} = \frac{z-10}{7} \quad an \quad \frac{x-15}{3} = \frac{y-29}{8} = \frac{z-5}{-5}$$

Hint: Do your self. **Ans.** 14; $\dfrac{x-9}{2} = \dfrac{y-13}{3} = \dfrac{z-15}{6}$.

Example 2:

Find the S.D. between the lines

$$\frac{x-3}{1} = \frac{y-8}{-1} = \frac{z-3}{1}$$

and $$\frac{x+3}{1} = \frac{y+7}{2} = \frac{z-6}{4}$$

Hint: Do your self. **Ans.** $5\sqrt{6}$.

∴ PQ = $(\alpha_1 - \alpha_2)\lambda + (\beta_1 - \beta_2)\mu + (\gamma_1 - \gamma_2)\nu$.

$$= \frac{(\alpha_1 - \alpha_2)(m_1n_2 - m_2n_1) + (\beta_1 - \beta_2)(n_1l_2 - n_2l_1) + (\gamma_1 - \gamma_2)(l_1m_2 - l_2m_1)}{\sqrt{\left[(m_1n_2 - m_2n_1)^2\right]}}$$

$$\begin{vmatrix} \alpha_1-\alpha_2 & \beta_1-\beta_2 & \gamma_1-\gamma_2 \\ l_1 & m_1 & n_1 \\ l_2 & m_2 & n_2 \end{vmatrix} \bigg/ \sqrt{[\Sigma(m_1n_2-m_2n_1)^2]} \quad \textbf{Note} \qquad ...(v)$$

Note : If the given lines are coplanar, then

S.D. = 0 *i.e.,* $\Sigma(\alpha_1-\alpha_2)(m_1n_2-m_2n_1)=0$

Equations of S.D.

From the figure 4 above it is evident that the line PQ (which represents S.D.) is the line of intersection of the plane containing the lines AB and PQ and the plane containing the lines CD and PQ.

Now the equation of the plane containing lines AB and PQ is

$$\begin{vmatrix} x-\alpha_1 & y-\beta_1 & z-\gamma_1 \\ l_1 & m_1 & n_1 \\ l & m & n \end{vmatrix}=0 \qquad ...(iv)$$

and the equations of the plane containing the lines CD and PQ is

$$\begin{vmatrix} x-\alpha_2 & y-\beta_2 & z-\gamma_2 \\ l_2 & m_2 & n_2 \\ l & m & n \end{vmatrix}=0 \qquad ...(vii)$$

The line of shortest distance PQ is the line of intersection of these two planes given by (vi) and (vii), hence the equations of the line of the S.D. are given by (vi) and (vii).

Method II : The general coordinates of points on the two lines given by (i) and (ii) above are $(\alpha_1 + l_1r_1, \beta_1 + m_1r_1, \gamma_1 + n_1r_1)$ say point P

and $(\alpha_1 + l_2r_2, \beta_2 + m_2r_2, \gamma_2 + n_2r_2)$ say point Q.

If these are the points where the line of S.D. meets the given lines (i) and (ii) respectively, then line PQ must be at right angles to both the lines (i) and (ii). Find the direction ratios of the lines PQ and apply the conditions that PQ is perpendicular to the lines (i) and (ii).

Solving the two equations so obtained, we can find the values of r_1 and r_2. Hence the coordinates of P and Q and also the d.c.'s of the line PQ are known, which enable us to find the distance between P and Q the equation of the line PQ.

Note: This method is useful when the coordinates of P and Q are required.

Method III : The shortest distance can also be obtained if we use the fact that it is equal to the length of the perpendicular from any point on one of the lines to the plane through the other line parallel to the first.

Note: This method is generally used when the equations of one line is given in general form while those of the other are in the symmetric form.

Method IV : If the equations of both the lines are given in general form as $u_1 = 0 = v_1$

and $u_2 = 0 = v_2$.

Then the equations of any plane through the first line is

$$u_1 + k_1 v_1 = 0 \qquad \text{...(A)}$$

And the equation of any plane through the second line is

$$u_2 + k_2 v_2 = 0. \qquad \text{...(B)}$$

Choose k_1 and k_2 in such a manner that planes (A) and (B) are parallel then the required S.D. is distance between these two parallel planes and the equations of S.D. will be given by planes through each of the given planes and perpendicular to these parallel planes.

Note: For convenience we generally reduce the given equations to the symmetric form (if they are not so) and use method I.

Example 3:

Show that the shortest distance between the lines

$$\frac{x-1}{2}=\frac{y-2}{3}=\frac{z-3}{4};\ \frac{x-2}{3}=\frac{y-4}{4}=\frac{z-5}{5}\ \textit{is}\ \frac{1}{\sqrt{6}}$$

and that its equations are $11x + 2y - 7z + 6 = 0,\ 7x + y - 5z + 7 = 0.$

Solution:

The given lines are

$$\frac{x-1}{2}=\frac{y-2}{3}=\frac{z-3}{4} \qquad \text{...(i)}$$

and

$$\frac{x-2}{3}=\frac{y-4}{4}=\frac{z-5}{5} \qquad \text{...(ii)}$$

Any point P on the line (i) is $(1 + 2r_1, 2 + 3r_1, 3 + 4r_1)$ ...(iii)

and any point Q on the line (ii) is $(2 + 3r_2, 4 + 4r_2, 5 + 5r_2)$ ...(iv)

$\therefore$ The direction ratios of the line PQ are

$$(1 + 2r_1) - (2 + 3r_2),\ (2 + 3r_1) - (4 + 4r_2),\ (3 + 4r_1) - (5 + 5r_2)$$

$\Rightarrow \quad 2r_1 - 3r_2 - 1,\ 3r_1 - 4r_2 - 2,\ 4r_1 - 5r_2 - 2$ ...(v)

If PQ is the required S. D. then PQ is perpendicular to both the given lines and as such we have

∴ The equations of the S.D. are

$$\frac{x-(5/3)}{1} = \frac{y-3}{-2} = \frac{z-(13/3)}{1}$$

$$\Rightarrow \quad \frac{3x-5}{1} = \frac{3y-9}{-2} = \frac{3z-13}{1}$$

$$2(2r_1 - 3r_2 - 1) + 3(3r_1 - 4r_2 - 2) + 4(4r_1 - 5r_2 - 2) = 0$$

and $\quad 3(2r_1 - 3r_2 - 1) + 4(3r_1 - 4r_2 - 2) - 5(4r_1 - 5r_2 - 2) = 0$

$\Rightarrow \quad 29r_1 - 38r_2 - 16 = 0$

and $\quad 32r_1 - 50r_2 - 21 = 0.$

Solving these, we get $r_1 = 1/3$ and $r_2 - 1/6$.

Substituting these values of r_1 and r_2 in (iii), (iv) and (v), we have the coordinates of P and Q as $\left(\frac{5}{3}, 3, \frac{13}{3}\right)$ and $\left(\frac{3}{2}, \frac{10}{3}, \frac{25}{6}\right)$.

And the d.r.'s of the line PQ are $\frac{1}{6}, \frac{-1}{3}, \frac{1}{6}$ or 1, – 2, 1.

∴ The length of S.D.

$$= PQ = \sqrt{\left[\left(\frac{5}{3}-\frac{3}{2}\right) + \left(3-\frac{10}{3}\right)^2 + \left(\frac{13}{3}-\frac{25}{6}\right)^2\right]}$$

$$= \sqrt{\left[\frac{1}{36}+\frac{1}{9}+\frac{1}{36}\right]} = \frac{1}{\sqrt{6}}.$$ **Hence proved.**

Also the line PQ is the line of intersection of the plane containing the line (i) and PQ and the plane containing the line (ii) and PQ.

The equation of the plane containing the line (i) and PQ is

$$\begin{vmatrix} x-1 & y-2 & z-3 \\ 2 & 3 & 4 \\ 1 & -2 & 1 \end{vmatrix} = 0$$

$$\Rightarrow \quad \begin{vmatrix} x-z+2 & y+2z-8 & z-3 \\ -2 & 11 & 4 \\ 0 & 0 & 1 \end{vmatrix} = 0,$$

adding twice third column to second and substracting third column from first.

$\Rightarrow$ $11(x - z + 2) + 2(y + 2z - 8) = 3$

$\Rightarrow$ $11x + 2y - 7z + 6 = 0$ ...(vi)

And the equation of the plane containing the line (ii) and PQ is

$$\begin{vmatrix} x-2 & y-4 & z-5 \\ 3 & 4 & 5 \\ 1 & -2 & 1 \end{vmatrix} = 0$$

$$\Rightarrow \begin{vmatrix} x-z+3 & y+2z-14 & z-5 \\ -2 & 14 & 5 \\ 0 & 0 & 1 \end{vmatrix} = 0,$$

adding twice third column to second and subtracting third column from first.

$\Rightarrow$ $14(x - z + 3) + 2(y + 2z - 14) = 0$

$\Rightarrow$ $7x + y + 5z + 7 = 0$ ...(vii)

$\therefore$ The equations of S.D. are given by (vi) and (vii).

Another Method for Equations of S.D.

The line PQ is the line through P (5/3, 3, 13/3) and having d.r.'s as 1, – 2, 1.

Example 4:

Find the length and the equations of the common perpendicular to the two lines

$$\frac{x+3}{-4} = \frac{y-6}{3} = \frac{z}{2} \quad ...(i) \quad and \quad \frac{x+2}{-4} = \frac{y}{1} = \frac{z-7}{1} \quad ...(ii)$$

(Avadh 95, 91, 90; Gorakhpur 91; Purvanchal 95)

Solution:

Let l, m, n be the d.c.'s of the line of the common perpendicular (or S.D.) to the two given lines. Then we have

$$-4.l + 3m + 2n = 0,$$

$$-4l + m + n = 0$$

Solving these get $\dfrac{l}{3-2} = \dfrac{m}{-8+4} = \dfrac{n}{-4+12}$

$$\Rightarrow \frac{l}{1} = \frac{m}{-4} = \frac{n}{8} = \frac{\sqrt{(l^2 + m^2 + n^2)}}{\sqrt{[1^2 + (-4)^2 + 8^2]}} = \frac{1}{9}$$

∴ The d.c.'s of the S.D. are (1/9), – (4/9), (8/9) ...(iii)

and the d.r.'s of the S.D. are 1, – 4, 8.

Also A (–3, 6, 0) is a point on the line (i) and B (–2, 0, 7) is point on the line (ii).

∴ The length of S.D. = projection of join of A and B on the line whose d.c.'s are given by (iii)

$$= (1/9)\,[(-3) - (-2)] - (4/9)\,[(6) - (0)] + (8/9)\,[(0) - (7)]$$

$$= -\,(1/9) - (24/9) - (56/9) = 9 \text{ (numerically)}.$$

Equations of S.D.:

The equation of the plane through the line (ii) and S.D. is

$$\begin{vmatrix} x+3 & y-6 & z \\ -4 & 3 & 2 \\ 1 & -4 & 8 \end{vmatrix} = 0 \text{ or } 32x + 34y + 13z = 108 \qquad \text{...(iv)}$$

And the equation of the plane through the line (ii) and S.D. is

$$\begin{vmatrix} x+2 & y & z-7 \\ -4 & 1 & 1 \\ 1 & -4 & 8 \end{vmatrix} = 0 \text{ or } 4x + 11y + 5z - 27 = 0 \qquad \text{...(v)}$$

∴ From (iv) and (v) the equations of the S.D. are

$$32x + 34y + 13z - 108 = 0$$

and $\quad 4x + 11y + 5z - 27 = 0.$ **Ans.**

Example 5:

Find the S.D. between the lines

$$\frac{x-3}{1} = \frac{y-5}{-2} = \frac{z-2}{1} \qquad \text{...(i)}$$

and $$\frac{x-1}{7} = \frac{y+1}{-6} = \frac{z+1}{1} \qquad \text{...(ii)}$$

Solution:

Let l, m, n be the d.c.'s of the S.D. of the given lines.

Then we have $l - 2m + n = 0,\ 7l - 6m + n = 0$

Solving these, we get $\dfrac{l}{-2+6} = \dfrac{m}{7-1} = \dfrac{n}{-6+14}$

$$\Rightarrow \frac{l}{2}=\frac{m}{3}=\frac{n}{4}=\frac{\sqrt{(l^2+m^2+n^2)}}{\sqrt{(2^2+3^2+4^2)}}=\frac{1}{\sqrt{(29)}}$$

∴ The direction cosines of S.D. are $\frac{2}{\sqrt{(29)}}, \frac{3}{\sqrt{(29)}}, \frac{4}{\sqrt{(29)}}$ ...(iii)

Also as A (3, 5, 2) is a point on the line (i) and B (1, –1, –1) is a point on the line (ii).

∴ The length of S.D. = projection of join of A and B on the line whose d.c.'s are given by (iii)

$$= \frac{2}{\sqrt{(29)}}[3-1]+\frac{3}{\sqrt{(29)}}[5-(-1)]+\frac{4}{\sqrt{(29)}}[2-(-1)]$$

$$= \frac{2(2)+3(6)+4(3)}{\sqrt{(29)}}=\frac{34}{\sqrt{(29)}}.$$ **Ans.**

THE PLANE AND THE STRAIGHT LINE

If a given line intersects a given plane to find the co-ordinates of the point of intersection and to deduce the conditions that (i) the line may be perpendicular to the plane, (ii) the line may be parallel to the plane and (iii) the line may by lying on the plane.

Let the equations of the given line and the given plane be

$$\frac{x-\alpha}{l}=\frac{y-\beta}{m}=\frac{z-\gamma}{n} = r \text{ (say)} \quad ...(i)$$

and $\qquad ax + by + cz + d = 0 \qquad$ ...(ii)

respectively.

Any point on the line (i) is $(\alpha + lr, \beta + mr, \gamma + nr)$.

If this point lies on (ii), then we have

$a(\alpha + lr) + b(\beta + mr) + c(\gamma + nr) + d = 0$

$\Rightarrow \quad r(al + bm + cn) = -(a\alpha + b\beta + c\gamma + d)$

$\Rightarrow \quad r = -(a\alpha + b\beta + c\gamma + d)/(al + bm + cn)$ ...(iii)

Therefore the point of intersection of (i) and (ii) is

$(\alpha + lr, \beta + mr, \gamma + nr)$,

where r is given by (iii).

(i) Condition of Perpendicularity

The direction cosines of the normal to the plane (ii) are proportional a, b, c

If the line (i) is perpendicular to the plane (ii), then it must be parallel to the normal to the plane and hence we have

$$\frac{l}{a}=\frac{m}{b}=\frac{n}{c},$$

which is the required condition.

(ii) Condition of Parallelism

If the line (i) is parallel to the plane (ii), then it must be perpendicular to the plane and hence we have

$$al + bm + cn = 0$$

Also (α, β, γ) should not be lie on the plane *i.e.*, $a\alpha + b\beta + c\gamma + d \neq 0$ as otherwise the line will not be parallel to the plane.

∴ the conditions for parallelism of the line (i) and the plane (ii) are

$$\mathbf{A}l + bm + cn = 0,\ a\alpha + b\beta + c\gamma + d \neq 0.$$

Aliter : If the line (i) is parallel to the plane (ii), then their point of intersection is at infinity, consequently the value of r given by (iii) should be infinite and the conditions for the same are

$$al + bm + cn = 0 \text{ and } a\alpha + b\beta + c\gamma + d \neq 0$$

(iii) Conditions for the Line to Lie on the Plane

If the line (i) lies on the plane (ii) then for all values of r, the point $(\alpha, + lr, \beta + mr, \gamma + nr)$ lies on the plane (ii) and so we have

$$a(\alpha + lr) + b(\beta + mr) + c(\gamma + nr) + d = 0$$

$$\Rightarrow \quad (al + bm + cn) + (a\alpha + b\beta + c\gamma + d) = 0.$$

If this is true for all values of r, then we must have.

$$al + bm + cn = 0 \quad \text{and} \quad a\alpha + b\beta + c\gamma + d = 0,$$

which are the required conditions.

Aliter : If the line (i) lies on the plane (ii), then it is perpendicular to the normal to the plane and therefore, we have

$$al + bm + cn = 0.$$

Also as the line (i) lies on the plane (ii), so the point (α, β, γ) through which the line (i) passes, lies on the plane (ii) and so we have

$$a\alpha + b\beta + c\gamma + d = 0.$$

Example 1:

Show that the equation of the plane, passing through the point (α, β, γ) and perpendicular to the line $x/l = y/m = z/n$ is

$$l(x-\alpha) + m(y-\beta) + n(z-\gamma) = 0$$

Solution:

The equation of any plane through (α, β, γ) is

$$A(x-\alpha) + B(y-\beta) + C(z-\gamma) = 0 \qquad ...(i)$$

The given line is $x/l = y/m = z/n$ ...(ii)

If the plane (i) is perpendicular to the line (ii), then its normal is parallel to the line (ii) and so we have $A/l = B/m = C/n = k$ (say).

Substituting these proportionate values of A, B, C in (i) we get the required equation as $l(x-\alpha) + m(y-\beta) + n(z-\gamma) = 0$. **Hence proved.**

Example 2:

Find the equation to the plane through (2, – 3, 4) normal to the line joining (3, 4, – 2) and (2, – 1, 6).

Solution:

The equation of any plane through (2, – 3, 4) is

$$A(x-2) + B(y+3) + C(z-4) = 0 \qquad ...(i)$$

Also the d.r.'s of the line joining (3, 4, – 2) and (2, – 1, 6) are 3 – 2, 4 – (– 1),(–2) – 6 *i.e.*, 1, 5, – 8.

If the plane (i) is at right angles to the line joining these points *i.e.*, the lines whose d.c.'s are 1, 5, –8, then the normal to the plane (i) whose d.r.'s are A, B, C is parallel to this line and consequently we have

$$A/1 = B/5 = C/(-8).$$

Substituting these proportionate values of A, B and C in (i) we have the required equation as

$$5(x+y) - 8(z-4) = 0 \quad \text{or} \quad x + 5y - 8z + 45 = 0. \qquad \textbf{Ans.}$$

Example 3:

Find the equation of the plane through the point (2, 1, 1), (1, –2, 3) and parallel to the x-axis.

Solution:

The equation of any plane through (2, 2, 1) is

$$A(x-2) + B(y-2) + C(z-1) = 0 \qquad ...(i)$$

If it passes through (1, –2, 3), then

$$A(1-2)+B(-2-2)+C(3-1)=0$$

$$\Rightarrow \quad -A-4B+2C=0 \text{ or } A+4B-2C=0 \quad ...(ii)$$

The equation of x-axis is $\frac{x}{1}=\frac{y}{0}=\frac{z}{0}$ ...(iii)

If the line given by (iii) is parallel to the plane given by (i), then this line is perpendicular to the normal to the plane given by (i) and so we have

$$A.1+B.0+C.0=0 \text{ or } A=0 \quad ...(iv)$$

Solving (ii) and (iv) we have $4B-2C=0$ or $C=2B$.

Substituting these values in (i) we get the required equation as

$B(y-2)+2B(z-1)=0$ or $y+2z-4=0$. **Ans.**

Example 4:

Find the angle between the line given by $y+z-5=0=x+y+z$ and the plane $x-y+z=0$.

Solution:

Let l, m, n be the d.c.'s of the given line, then as this is perpendicular to the normal to the planes $y+z-5=0$ and $x+y+z=0$, so we have

$$0.l+1.m+1.n=0$$

and $\quad 1.l+1.m+1.n=0$

Solving these, we get

$$\frac{l}{0}=\frac{m}{1}=\frac{n}{-1}=\frac{\sqrt{(l^2+m^2+n^2)}}{\sqrt{\left[0^2+1^2+(-1)^2\right]}}=\frac{1}{\sqrt{2}}$$

$$\therefore l=0,\ m=1/\sqrt{2},\ n=-1/\sqrt{2}$$

Also the direction ratios of the normal to the plane $x-y+z=0$ are 1, – 1 and 1

$\therefore$ The d.c.'s the normal to this plane are $\frac{1}{\sqrt{3}},-\frac{1}{\sqrt{3}},\frac{1}{\sqrt{3}}$

Hence if θ be required angle, then $\frac{1}{2}\pi-\theta$ will be the angle between the line and normal to the plane.

$$\therefore \quad \cos\left(\frac{1}{2}\pi-\theta\right)=\text{"}l_1l_2+m_1m_2+n_1n_2\text{"}$$

$$\Rightarrow \quad \sin\theta = 0.\frac{1}{\sqrt{3}}+\frac{1}{\sqrt{2}}.\left(-\frac{1}{\sqrt{2}}\right)+\left(-\frac{1}{\sqrt{2}}\right)\left(\frac{1}{\sqrt{3}}\right)=-\frac{2}{\sqrt{6}}$$

$$\Rightarrow \quad \theta = \sin^{-1}(-2/\sqrt{6}).$$ **Ans.**

Example 5:

Find the direction cosines of the line whose equations are x + y = 3, x + y + z = 0 and show that it makes an angle of 30° with the plane y − z + 2 = 0.

Solution:

Let l, m, n be the d.c.'s of the line. Then as this line is perpendicular to the normal to the planes $x + y - 3 = 0$ and $x + y + z = 0$, so we have

$$1.l + 1.m + 0.n = 0,$$

and $$1.l + 1.m + 1.n = 0$$

Solving these, we get

$$\frac{l}{1}=\frac{m}{-1}=\frac{n}{0}=\frac{\sqrt{(l^2+m^2+n^2)}}{\sqrt{[1^2+(-1)^2+0^2]}}=\frac{1}{\sqrt{2}}$$

$\therefore$ $l = 1/\sqrt{2}$, $m = -1/\sqrt{2}$ and $n = 0$.

Also the direction of the normal to the plane $y - z + 2 = 0$ are 0, 1 and − 1.

The d.c.'s of the normal to the plane are 0, $1/\sqrt{2}$, $-1/\sqrt{2}$.

Hence if θ be the required angle, then $\frac{1}{2}\pi-\theta$ will be the angle between the line and the normal to the plane.

$$\therefore \cos\left(\frac{1}{2}\pi-\theta\right) = \text{"}l_1l_2 + m_1m_2 + n_1n_2\text{"}$$

$$\Rightarrow \quad \sin\theta = 0.\frac{1}{\sqrt{2}}+\frac{1}{\sqrt{2}}\left(-\frac{1}{\sqrt{2}}\right)+\left(-\frac{1}{\sqrt{2}}\right).0=-\frac{1}{2}$$

$\Rightarrow \quad \theta = 150°$ or 30°, if the acute angle is taken.

TO FIND THE EQUATION OF A PLANE THROUGH A GIVEN LINE WHOSE EQUATIONS ARE GIVEN IN (I) GENERAL FORM, (II) SYMMETRIC FORM

(i) Let the equations of the line in the general form be

$$P_1 \equiv a_1x + b_1y + c_1z + d_1 = 0$$

and $P_2 \equiv a_2x + b_2y + c_2z + d_2=0$ ...(i)

then $P_1 + \lambda P_2 \equiv (a_1x + b_1y + c_1z + d_1) + \lambda(a_2x + b_2y + c_2z + d_2) = 0$...(ii)

being a first degree equation in x, y and z represents a plane, also it is satisfied by all those points which satisfy (i), hence $P_1 + \lambda P_2 = 0$, given by (ii) represents the plane through the line $P_1 = 0$ and $P_2 = 0$ given by (i).

(ii) Let the equations of the line in symmetric form be given as

$$\frac{x-\alpha}{l} = \frac{y-\beta}{m} = \frac{z-\gamma}{n} \quad \text{...(iii)}$$

As the required plane passes through this line, so it passes through (α, β, γ) which is a given point on this line.

The equation of any plane through (α, β, γ) is

$$A(x-\alpha) + B(y-\beta) + C(z-\gamma) = 0 \quad \text{...(iv)}$$

Since the plane passes through the line (ii) whose d.c.'s are l, m, n therefore the normal to this plane whose d.r.'s are A, B, C is perpendicular the normal to the line (iii), consequently we have

$$A.l + B.m + C.n = 0$$

Hence the required equation of the plane through the line (iii) is

$$A(x-\alpha) + B(y-\beta) + C(z-\gamma) = 0,$$

where $Al + Bm + Cn = 0$.

TO FIND THE EQUATION OF A PLANE THROUGH A GIVEN LINE AND PARALLEL TO ANOTHER

We have proved that the equation of the plane through the line

$$\frac{x-\alpha_1}{l_1} = \frac{y-\beta_1}{m_1} = \frac{z-\gamma_1}{n_1} \text{ is}$$

$$A(x-\alpha_1) + B(y-\beta_1) + C(z-\gamma_1) = 0 \quad \text{...(i)}$$

where $Al_1 + Bm_1 + Cn_1 = 0$ ...(ii)

Again if the plane given by (i) is parallel to another line then we have

$$\frac{x-\alpha_2}{l_2} = \frac{y-\beta_2}{m_2} = \frac{z-\gamma_2}{n_2} \quad \text{...(iii)}$$

then the normal to the plane given by (i) must be at right angles to the line given by (iii) whose d.c.'s are l_2, m_2, n_2.

$$\therefore \quad Al_2 + Bm_2 + Cn_2 = 0 \quad \text{...(iv)}$$

Eliminating A, B, C from (i), (ii) and (iv) we have the required equation as

$$\begin{vmatrix} x-\alpha_1 & y-\beta_1 & z-\gamma_1 \\ l_1 & m_1 & n_1 \\ l_2 & m_2 & n_2 \end{vmatrix} = 0$$

$\Rightarrow \quad \Sigma\,[(x - \alpha_1)(m_1n_2 - m_2n_1)] = 0.$

Example 1:

Prove that the line $\dfrac{x-3}{2} = \dfrac{y-4}{3} = \dfrac{z-5}{4}$ *is parallel to the plane* $4x + 4y - 5z = 0$.

Solution:

If the given line is parallel to the given plane, then the given line must be perpendicular to the normal to the given plane and here the direction ratios of the given the line are 2, 3, 4 whereas those of the normal to the given plane are 4, 4, – 5.

Also "$a_1a_2 + b_1b_2 + c_1c_2$" $= 2\,(4) + 3\,(4) - 4\,(-5) = 8 + 12 - 20 = 0$

Hence the given line is perpendicular to the normal to the given plane and thus parallel to the given plane. **Hence proved.**

Example 2:

Find the equation of the plane through the point (–1, 0, 1) and the lines

$$4x - 3y + 1 = 0 = y - 4z + 13;$$
$$2x - y - 2 = 0 = z - 5.$$

Solution:

The equation of the plane through the line

$$4x - 3y + 1 = 0 = y - 4z + 13$$

is $\quad (4x - 3y + 1) + \lambda\,(y - 4z + 13) = 0 \quad$ ...(i)

If its through (–1, 0, 1), then

$$[4\,(-1) - 3\,(0) + 1] + \lambda\,[0 - 4\,(1) + 13] = 0$$

or $\quad -3 + 9\lambda = 0$ or $\lambda = \dfrac{1}{3}$

From (i) the required equation is

$$(4x - 3y + 1) + \frac{1}{3}\,(y - 4z + 13) = 0.$$

$\Rightarrow \quad 12x - 9y + 3 + y - 4z + 13 = 0$

$\Rightarrow \quad 12x - 8y - 4z + 16 = 0$

$\Rightarrow \quad 3x - 2y - z + 4 = 0.$ **Ans.**

Similarly, we can find the equation of the plane through the point (–1, 0, 1) and the line $2x - y - 2 = 0 = z - 5$ as $2x - y - z + 3 = 0$. **Ans.**

Example 2:

Find the condition that the line $x/l = y/m = z/n$ may be (i) perpendicular, (ii) parallel to the plane $ax + by + cz = 0$.

Solution:

(i) If the given line is perpendicular to the given plane, then the given line must be parallel to the normal to the given plane and hence we have

$l/a = m/b = n/c.$ **Ans.**

(ii) If the given line is parallel to the given plane, then the given line must be perpendicular to the normal to the given plane and hence we have

$la + mb + nc = 0.$ **Ans.**

Example 3:

The condition that the line $(x - \alpha)/l = (y - \beta)/m = (z - \gamma)/n$ is perpendicular to the plane $ax + by + cz + d = 0$ is

(i) $a/l = b/m = c/n$;

(ii) $a/m = b/n = c/l$;

(iii) $l/a = m/b = n/c$;

(iv) $al = bm = cn$.

Solution:

Do yourself. **Ans.** (i), (iii)

Example 4:

Find the equation of a system of planes perpendicular to the line with direction ratios a, b, c.

Solution:

Let the required equation be $Ax + By + Cz + D = 0$ ...(i)

If this plane is perpendicular to the line with d. ratios a, b, c, then the normal to the plane (i) must be parallel to this line.

i.e., $$\frac{A}{a} = \frac{B}{b} = \frac{C}{c} = \lambda \text{ (say)}$$

i.e., $A = a\lambda$ $B = b\lambda$ and $C = c\lambda$

$\therefore$ From (i), we have $a\lambda x + b\lambda y + c\lambda z + D = 0$

$\Rightarrow$ $ax + by + cz + k = 0$, where $k = D/\lambda$, is the required equation of a system of planes where k is parameter *i.e.,* for different values of k we shall get different members of the above system of planes. **Ans.**

Example 5:

Find the equation of the plane containing the line $\frac{y}{b}+\frac{z}{c}=1$, $x = 0$ *and parallel to the line* $\frac{x}{a}-\frac{z}{c}=1$, $x = 0$.

Solution:

The equation of the line $x = 0$, $(y/b) + (z/c) = 1$ can be written in the symmetric forms

$$\frac{x}{0}=\frac{y-\frac{1}{2}b}{b}=\frac{z-\frac{1}{2}c}{-c} \qquad \text{...(i)}$$

And the equations of the line $(x/a) - (z/c) = 1$, $y = 0$ can be written in the symmetric form as

$$\frac{x-\frac{1}{2}a}{a}=\frac{z+\frac{1}{2}c}{c} \qquad \text{...(ii)}$$

where $A.0 + B.b + C(-c) = 0.$ ...(iv)

Also as the plane (iii) is parallel to the line (ii), so we have

$$A.a + B.0 + C.c = 0. \qquad \text{...(v)}$$

Solving (iv) and (v), we have $\frac{A}{bc}=\frac{B}{-ca}=\frac{C}{-ab}$

Substituting these proportionate values of A, B and C in (iii) we have the required equations as

$$bc\,(x) - ca\left(y-\frac{1}{2}b\right) - ab\left(z-\frac{1}{2}c\right) = 0$$

or $bcx - baz + abc = 0$

$\Rightarrow$ $(x/a) - (y/b) - (z/c) + 1 = 0.$ **Ans.**

Example 6:

Find the equation of the plane which pass through the z-axis and is perpendicular to the line

$$\frac{x-1}{\cos\theta}=\frac{y+2}{\sin\theta}=\frac{z-3}{0}$$

Solution:

The equations of z-axis are $\frac{x}{0}=\frac{y}{0}=\frac{z}{1}$

$\therefore$ The equation of any plane through z-axis is

$$A(x)+B(y)+C(z)=0 \qquad ...(i)$$

where $A.0 + B.0 + C.1 = 0$ *i.e.,* $C = 0$ ...(ii)

Also if the plane given by (i) is perpendicular to the given line, then the normal to the plane given by (i) must be parallel to the given line whose direction cosines are cos θ, sin θ, 0.

i.e.,
$$\frac{A}{\cos\theta}=\frac{B}{\sin\theta}=\frac{C}{0}. \qquad ...(iii)$$

Substituting the proportionate values of A and B from (iii) and the value of C from (ii) in (i) we get the required equation as

$$x\cos\theta + y\sin\theta = 0$$

or
$$x + y\tan\theta = 0.$$ **Ans.**

Example 7:

Find the equation of the plane which passes through the x-axis and is perpendicular to the line

$$(x-1)/\cos\theta = (y+2)/\sin\theta = (z-3)/0.$$

Solution:

The equations of x-axis in the symmetric form are

$$\frac{x-0}{1}=\frac{y-0}{0}=\frac{z-0}{0}$$ **(Note)**

$\therefore$ The equation of any plane through x-axis is

$$A(x-0)+B(y-0)+C(z-0)=0 \qquad ...(i)$$

where $A.1 + B.0 + C.0 = 0$ *i.e.,* $A = 0$ ...(ii)

$\therefore$ (i) reduces to $By + Cz = 0$ ...(iii)

$\therefore$ The direction ratios of its normal are 0, B, C.

If (iii) is perpendicular to the given line, then the normal to (iii) must be parallel to the given line

i.e.,
$$\frac{0}{\cos\theta}=\frac{B}{\sin\theta}=\frac{C}{0}=k \text{ (say)}$$

$\therefore$ B = k sin θ; C = 0 and k cos θ = 0.

If k cos θ = 0, then either k = 0 or cos θ = 0.

If k = 0, then B = 0 and from (iii), we do not find any plane, so

$$\cos\theta = 0 \text{ or } \theta = \pi/2.$$

$\therefore$ B = k sin ($\therefore$/2) = k and C = 0.

$\therefore$ From (iii) be required equation is ky = 0 or y = 0. **Ans.**

Example 8:

Find the equation of the plane through the line ax + by + cz + d = 0, a'x + b'y + c'z + d' = 0 and parallel to the line

$$x/l = y/m = z/n.$$

Solution:

The equation of the plane through the given line is

$$(ax + by + cz + d) + \lambda (a'x + b'y + c'z + d') = 0$$

$$\Rightarrow \quad (a + \lambda a')x + (b + \lambda b')y + (c + \lambda c')z + (d + \lambda d') = 0 \qquad ...(i)$$

If this plane is parallel to the line $x/l = y/m = z/n$, then the normal to (i) must be at right angles to the line $x/l = y/m = z/n$, whose d.c.'s are l, m, n.

$$\therefore (a + \lambda a')\lambda + (b + \lambda b')m + (c + \lambda c')n = 0$$

$$\Rightarrow \quad \lambda = -(al + bm + cn)/(a'l + b'm + c'n) \qquad ...(ii)$$

$\therefore$ The required plane from (i) is

$$(ax + by + cz + d) + \lambda . (a'x + b'y + c'z + d') = 0.$$

where λ is given by (ii).

Example 9:

Find the equation of the plane through the line

$$x + y - z = x - 2y + 3z - 5.$$

Solution:

The equation of the plane through the given line is

$$(x + y - z) + 1 (x - 2y + 3z - 5) = 0 \qquad \textbf{Ans.}$$

where λ is some constant which can be evaluated if some other condition be given.

Example 10:

Prove that the equation to the two planes inclined at an angle α to xy-plane and containing the line y = 0, z cos β = x sin β is $(x^2 + y^2)\tan^2\beta + z^2 - 2zx\tan\beta = y^2\tan^2\alpha$..

Solution:

The equation of the plane containing the line y = 0, x sin β = z cos β is $(x \sin\beta - z\cos\beta) + \lambda y = 0$

⇒ $x\sin\beta + \lambda . y - z\cos\beta = 0.$ ...(i)

The other plane is xy-plane *i.e.*, z = 0

i.e., $0.x + 0.y + 1.z = 0$ ...(ii) **(Note)**

It is given that the angle between plane (i) and (ii) is α, so we have

$$\cos\alpha = \frac{a_1a_2 + b_1b_2 + c_1c_2}{\sqrt{(a_1^2 + b_1^2 + c_1^2)}\,\sqrt{(a_2^2 + b_2^2 + c_2^2)}},$$

∴ angle between planes = angle between their normals.

⇒ $$\cos\alpha = \frac{0.\sin\beta + 0.\lambda + 1.(-\cos\beta)}{\sqrt{(\sin^2\beta + \lambda^2 + \cos^2\beta)}\,\sqrt{(0^2 + 0^2 + 1^2)}}$$

Squaring and cross-multiplying, we have

$(\lambda^2 + 1)\cos^2\alpha = \cos^2\beta$ or $\lambda^2 = (\cos^2\beta - \cos^2\alpha)/\cos^2\alpha$

⇒ $\lambda = \pm\left[\sqrt{(\cos^2\beta - \cos^2\alpha)}\right]/\cos\alpha = \pm\mu$ (say) ...(iii)

which gives two values of λ which are equal in magnitude but opposite in sign.

∴ From (i) combined equation of two required planes is

$[(x\sin\beta - z\cos\beta) + \mu y]\,[(x\sin\beta - z\cos\beta) - \mu y] = 0.$ **(Note)**

⇒ $(x\sin\beta - z\cos\beta)^2 - \mu^2 y^2 = 0$

⇒ $(x\sin\beta - z\cos\beta)^2\cos^2\alpha = (\cos^2\beta - \cos^2\alpha)\,y^2$, from (iii)

⇒ $(x^2\sin^2\beta + z^2\cos^2\beta - 2xz\sin\beta\cos\beta)$ χοσ2 $\alpha = (\cos^2\beta - \cos^2\alpha)y^2$

⇒ $x^2\tan^2\beta + z^2 - 2xz\tan\beta = (\sec^2\alpha - \sec^2\beta)\,y^2$,

dividing each term by $\cos^2\alpha\cos^2\beta$

⇒ $x^2\tan^2\beta + z^2 - 2xz\tan\beta = [(1 + \tan^2\alpha) - (1 + \tan^2\beta)]\,y^2$

⇒ $(x^2 + y^2)\tan^2\beta + z^2 - 2xz\tan\beta = y^2\tan^2\alpha.$ **Hence proved.**

Example 11:

Find the equation of the plane through the line of intersection of the planes of the planes 2x + y – z = 3 and 5x – 3y + 4z +9 = 0 and parallel to the line (x –1)/2 = (y – 3)/4 = (z – 5)/5.

Solution:

Equation of any plane through the intersection of the given planes is

$$(2x + y - z - 3) + \lambda\ (5x - 3y + 4z + 9) = 0$$

$$\Rightarrow \quad (2 + 5\lambda)\ x + (1 - 3\lambda)\ y + (4\lambda - 1)\ z + (9\lambda - 3) = 0 \quad ...(i)$$

If this plane is parallel to the given line, then the normal to the plane (i) must be perpendicular to the given line whose direction cosines are 2, 4, 5.

$$\therefore \quad 2\ (2 + 5\lambda) + 4\ (1 - 3\lambda) + 5\ (4\lambda - 1) = 0 \text{ or } \lambda = -\ 1/6$$

Substituting this value of λ in (i) and simplifying we get the required equation of the plane as $7x + 9y - 10z = 27$. **Ans.**

Example 12:

Find the equation of the plane through the line of intersection of the planes $ax + by + cz + d = 0$, $a'x + b'y + c'z + d' = 0$ and parallel to x-axis.

Solution:

We find that the equation of the plane through the line of intersection of the given planes is

$$(a + \lambda a')\ x + (b + \lambda b')\ y + (c + \lambda c')\ z + (d + \lambda d') = 0 \quad ...(i)$$

If this plane is parallel to x-axis whose d.c.'s are 1, 0, 0 then the normal to the plane (i) must be perpendicular to x-axis, so we have

$$(a + \lambda a').1 + (b + \lambda b').0 + (c + \lambda c').0 = 0$$

$$\Rightarrow \quad a + \lambda a' = 0 \quad \Rightarrow \lambda = -\ a/a'.$$

Substituting this value of λ in (i) and simplifying, we get the required equation as $(ba' - ab')\ y + (ca' - c'\ a)\ z + (da' - d'\ a) = 0$.

Example 13:

Find the equation of the plane through the line

$\frac{1}{2}(x-1) = \frac{1}{4}(y-6)\ \frac{1}{2}(z+1)$ *and parallel to*

the line $\frac{1}{2}(x-2) = -\frac{1}{3}(y-1)\ \frac{1}{5}(z+4)$.

Solution:

The equation of any plane through the line

$\frac{x-1}{3} = \frac{y+6}{4} = \frac{z+1}{2}$ is $A\ (x - 1) + B\ (y + 6) + C\ (z + 1) = 0$, ...(i) where $\quad A.3 + B.4 + C.2 = 0 \qquad ...(ii)$

Also if the plane (i) is parallel to the line $\frac{x-2}{2} = \frac{y-1}{-3} = \frac{z-4}{5}$, then the normal to (i) is at right angles to this line whose d.r.'s are 2, –3, 5

i.e., $\quad A.2 + B.(-3) + C.5 = 0 \qquad ...(iii)$

Eliminating A, B, C from (i), (ii) and (iii) we have required equation as

$$\begin{vmatrix} x-1 & y+6 & z+1 \\ 3 & 4 & 2 \\ 2 & -3 & 5 \end{vmatrix} = 0,$$

which on simplifying reduces to $26x - 11y - 17z - 109 = 0$. **Ans.**

Example 14:

Find the equation of the plane which contains the two parallel lines

$$x - 4 = -\frac{1}{4}(y-3) = \frac{1}{5}(z-2)$$

and $\quad x - 3 = -\frac{1}{4}(y+2) = \frac{1}{5}z.$

Solution:

The equation of any plane through $\frac{x-4}{1} = \frac{y-3}{-4} = \frac{z-2}{5}$

is $\quad A(x-4) + B(y-3) + C(z-2) = 0. \qquad ...(i)$

where $\quad A.1 + B.(-4) + C.5 = 0 \qquad ...(ii)$

If the plane (i) contains the parallel line $\frac{x-3}{1} = \frac{y+2}{-4} = \frac{z}{5}$ also then the point (3, – 2, 0) on this must lie on the plane (i) which gives

$$A(3-4) + B(-2-3) + C(0-2) = 0$$

$\Rightarrow \quad A + 5B + 2C = 0. \qquad ...(iii)$

Solving (ii) and (iii), we have

$$\frac{A}{-8-25} = \frac{B}{5-2} = \frac{C}{5+4} \Rightarrow \frac{A}{11} = \frac{B}{-1} = \frac{C}{-3} \qquad ...(iv)$$

Substituting the proportionate values of A, B, C from (iv) in (i) we have the required equation as

$$11(x-4) - (y-3) - 3(z-2) = 0$$

$$11x - y - 3z - 35 = 0.$$ **Ans.**

Example 15:

Find the equation of the plane through the lines

$$ax + by + cz = 0 = a'x + b'y + c'z$$

and $$\alpha x + \beta y + \gamma z = 0 = \alpha' x + \beta' y + \gamma' z.$$

Solution:

The equations of the lines (both of which pass through the origin) can be written in the symmetric form is

$$\frac{x}{bc'-b'c} = \frac{y}{ca'-c'a} = \frac{z}{ab'-a'b} \quad \text{...(i)}$$

and $$\frac{x}{\beta\gamma'-\beta'\gamma} = \frac{y}{\gamma\alpha'-\gamma'\alpha} = \frac{z}{\alpha\beta'-\alpha'\beta} \quad \text{...(ii)}$$

The equation of any plane through the line (i) is

$$Ax + By + Cz = 0, \quad \text{...(iii)}$$

where $$A(bc' - b'c) + B(ca' - c'a) + C(ab' - a'b) = 0 \quad \text{...(iv)}$$

Also if the line (ii) lies on the plane (iii), then the following conditions should be satisfied:

(a) The plane (iii) should pass through the origin [which is a point on the line (iii) and that is true.

And (b) the normal to the plane (iii) must be at right angles to the line (ii), the condition for the same is

$$A(\beta\gamma' - \beta'\gamma) + B(\gamma\alpha' - \gamma'\alpha) + C(\alpha\beta' - \alpha'\beta) = 0 \quad \text{...(v)}$$

Eliminating A, B, C from (iii), (iv) and (v), we have the required equation as

$$\begin{vmatrix} x & y & z \\ bc'-b'c & c'a-ca' & ab'-a'b \\ \beta\gamma'-\beta'\gamma & \gamma'\alpha-\gamma\alpha' & \alpha\beta'-\alpha'\beta \end{vmatrix} = 0.$$

VOLUME OF TETRAHEDRON IN TERMS OF THE CO-ORDINATES OF THE VERTICES

Let V be the volume of the tetrahedron ABCD, where the co-ordinates of the vertices, A, B, C and D are (x_1, y_1, z_1), (x_2, y_2, z_2), (x_3, y_3, z_3) and (x_4, y_4, z_4).

Let Δ be the area of ΔBCD and p be the length of perpendicular from the vertex A to opposite face BCD.

Then we have

Then $V = \frac{1}{3} p.\Delta$ ...(i)

Now if Δ_x, Δ_y and Δ_z, be the projections of the triangle BCD on the co-ordinates planes, then we have

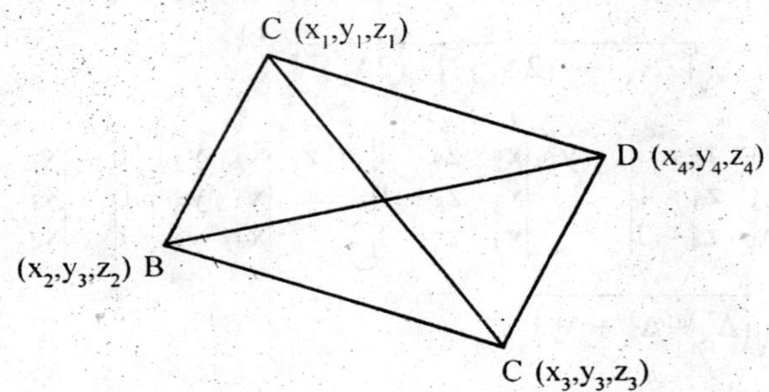

$$\Delta_x = \frac{1}{2} \begin{vmatrix} y_2 & z_2 & 1 \\ y_3 & z_3 & 1 \\ y_4 & z_4 & 1 \end{vmatrix};$$

$$\Delta_y = \frac{1}{2} \begin{vmatrix} x_2 & z_2 & 1 \\ x_3 & z_3 & 1 \\ x_4 & z_4 & 1 \end{vmatrix}$$

and $\Delta_z = \frac{1}{2} \begin{vmatrix} x_2 & y_2 & 1 \\ x_3 & y_3 & 1 \\ x_4 & y_4 & 1 \end{vmatrix}$ ...(ii)

Also $\Delta = \sqrt{(\Delta_x^2 + \Delta_y^2 + \Delta_z^2)}$ ...(iii)

Again the equation of the plane BCD is given as

$$\begin{vmatrix} x & y & z & 1 \\ x_2 & y_2 & z_2 & 1 \\ x_3 & y_3 & z_3 & 1 \\ x_4 & y_4 & z_4 & 1 \end{vmatrix} = 0$$

$$\Rightarrow x \begin{vmatrix} y_2 & z_2 & 1 \\ y_3 & z_3 & 1 \\ y_4 & z_4 & 1 \end{vmatrix} - y \begin{vmatrix} x_2 & z_2 & 1 \\ x_3 & z_3 & 1 \\ x_4 & z_4 & 1 \end{vmatrix} + z \begin{vmatrix} x_2 & y_2 & 1 \\ x_3 & y_3 & 1 \\ x_4 & y_4 & 1 \end{vmatrix} - \begin{vmatrix} x_2 & y_2 & z_2 \\ x_3 & y_3 & z_3 \\ x_4 & y_4 & z_4 \end{vmatrix} = 0$$

$$\Rightarrow x\,(2\Delta_x) - y\,(2\Delta_y) + z\,(2\Delta_z) - \begin{vmatrix} x_2 & y_2 & z_2 \\ x_3 & y_3 & z_3 \\ x_4 & y_4 & z_4 \end{vmatrix} = 0, \text{ from (ii)}$$

$\therefore$ p = length of perp. from A (x_1, y_1, z_1) to the plane BCD

$$= [x_1 (2\Delta_x) - y_1 (2\Delta_y) + z_1 (2\Delta_z) - \begin{vmatrix} x_2 & y_2 & z_2 \\ x_3 & y_3 & z_3 \\ x_4 & y_4 & z_4 \end{vmatrix}]$$

$$+\sqrt{[2\Delta_x)^2 + (2\Delta_y)^2] + (2\Delta_z)^2]}$$

$$= [x_1 \begin{vmatrix} y_2 & z_2 & 1 \\ y_3 & z_3 & 1 \\ y_4 & z_4 & 1 \end{vmatrix} - y_1 \begin{vmatrix} x_2 & z_2 & 1 \\ x_3 & z_3 & 1 \\ x_4 & z_4 & 1 \end{vmatrix} + z_1 \begin{vmatrix} x_2 & y_2 & 1 \\ x_3 & y_3 & 1 \\ x_4 & y_4 & 1 \end{vmatrix} - \begin{vmatrix} x_2 & y_2 & z_2 \\ x_3 & y_3 & z_3 \\ x_4 & y_4 & z_4 \end{vmatrix}]$$

$$\div 2 \sqrt{\left(\Delta_x^2 + \Delta_y^2 + \Delta_z^2\right)}$$

$$= \begin{vmatrix} x_1 & y_1 & z_1 & 1 \\ x_2 & y_2 & z_2 & 1 \\ x_3 & y_3 & z_3 & 1 \\ x_4 & y_4 & z_4 & 1 \end{vmatrix} . 2\Delta, \text{ from (iii)} \qquad \text{...(iv)}$$

$\therefore$ From (i), $V = \frac{1}{3} p.\Delta.$

$$\Rightarrow \quad V = (1/6) \begin{vmatrix} x_1 & y_1 & z_1 & 1 \\ x_2 & y_2 & z_2 & 1 \\ x_3 & y_3 & z_3 & 1 \\ x_4 & y_4 & z_4 & 1 \end{vmatrix}, \text{ substituting value of p from (iv)} \quad \text{...(v)}$$

Another Form

$$\begin{vmatrix} x_1 - x_2 & y_1 - y_2 & z_1 - z_2 & 0 \\ x_2 - x_3 & y_2 - y_3 & z_2 - z_3 & 0 \\ x_3 - x_4 & y_3 - y_4 & z_3 - z_4 & 0 \\ x_4 & y_4 & z_4 & 1 \end{vmatrix}$$

, substracting 2nd row from 1st, 3rd from 2nd and 4th from 3rd row respectively.

$$\begin{vmatrix} x_1 - x_2 & y_1 - y_2 & z_1 - z_2 \\ x_2 - x_3 & y_2 - y_3 & z_2 - z_3 \\ x_3 - x_4 & y_3 - y_4 & z_3 - z_4 \end{vmatrix} \qquad \text{...(vi)}$$

Particular case (One vertex at origin).

Let $x_4 = 0$ $y_4 = z_4$ i.e D is the origin

$$\text{Then from (v), we have } V = (1/6) \begin{vmatrix} x_1 & y_1 & z_1 \\ x_2 & y_2 & z_2 \\ x_3 & y_3 & z_3 \end{vmatrix} \qquad \text{...(vii)}$$

VOLUME OF TETRAHEDRON WHEN EQUATIONS OF FOUR FACES ARE GIVEN

Let the equations of four faces of the tetrahedron be

$$= \frac{1}{6D_1D_2D_3D_4}\Delta^{4-1},$$

since we know that if Δ' be the determinant formed by cofactors of a determinant Δ of nth order then $\Delta' = \Delta^{n-1}$

$\Rightarrow \quad V = \Delta^3/(6D_1D_2D_3D_4).$

VOLUME OF A TETRAHEDRON IN TERMS OF THREE CONTERMINUS EDGES AND THE ANGLES WHICH THEY MAKE WITH EACH OTHER

Let the lengths of the edges OA, OB and OC of the tetrahedron OABC be given as a, b, and c and the angles BOC, COA and OAB be λ μ ν respectively. Let the point O be taken as origin and any three mutually perpendicular lines through O be taken as co-ordinate axes.

Let the direction cosines of the lines OA, OB and OC be l_1, m_1, n_1; l_2, m_2, n_2 and l_3, m_3, n_3 respectively. Then the co-ordinates of A, B and C are (l_1a, m_1a, n_1a), (l_2b, m_2b, n_2b) and (l_3c, m_3c, n_3c) respectively. Then we have

$$\cos \angle BOC = \cos \lambda = l_2l_3 + m_2m_3 + n_2n_3, \quad \text{...(i)}$$

$$\cos \angle COB = \cos \mu = l_3l_1 + n_3n_1 + n_1n_2, \quad \text{...(ii)}$$

and $$\cos \angle AOB = \cos \nu$$

$$= l_1l_2 + m_1m_2 + n_1n_2, \quad \text{...(iii)}$$

Also $$l_1^2 + m_1^2 + n_1^2 = 1 = l_2^2 + m_2^2 + n_2^2 = l_3^2 + m_3^2 + n_3^2 \quad \text{...(iv)}$$

Therefore volume V of tetrahedron OABC is given by

$$V = \text{"}(1/6) \begin{vmatrix} x_1 & y_1 & z_1 \\ x_2 & y_2 & z_2 \\ x_3 & y_3 & z_3 \end{vmatrix}\text{"}$$

$$= (1/6) \begin{vmatrix} l_1a & m_1a & n_1a \\ l_2b & m_2b & n_2b \\ l_3c & m_1c & n_1c \end{vmatrix} \quad \therefore \text{ A is } (l_1a, m_1a, n_1a) \text{ etc.}$$

$$\Rightarrow \quad V = (1/6)\, abc \begin{vmatrix} l_1 & m_1 & n_1 \\ l_2 & m_2 & n_2 \\ l_3 & m_3 & n_3 \end{vmatrix} \quad \text{...(v)}$$

Now we have $\begin{vmatrix} l_1 & m_1 & n_1 \\ l_2 & m_2 & n_2 \\ l_3 & m_3 & n_3 \end{vmatrix}^2 = \begin{vmatrix} l_1 & m_1 & n_1 \\ l_2 & m_2 & n_2 \\ l_3 & m_3 & n_3 \end{vmatrix} \times \begin{vmatrix} l_1 & m_1 & n_1 \\ l_2 & m_2 & n_2 \\ l_3 & m_3 & n_3 \end{vmatrix}$

$$= \begin{vmatrix} l_1^2 + m_1^2 + n_1^2 & l_1 l_2 + m_1 m_2 + n_1 n_2 & l_1 l_3 + m_1 m_3 + n_1 n_3 \\ l_2 l_1 + m_2 m_1 + n_2 n_1 & l_2^2 + m_2^2 + n_2^2 & l_2 l_3 + m_2 m_3 + n_2 n_3 \\ l_3 l_1 + m_3 m_1 + n_3 n_1 & l_3 l_2 + m_3 m_2 + n_3 n_2 & l_3^2 + m_3^2 + n_3^2 \end{vmatrix}$$

$$a_1 x + b_1 y + c_1 z + d_1 = 0 \quad \text{...(i)}$$

$$a_2 x + b_2 y + c_2 z + d_2 = 0 \quad \text{...(ii)}$$

$$a_3 x + b_3 y + c_3 z + d_3 = 0 \quad \text{...(iii)}$$

$$a_4 x + b_4 y + c_4 z + d_4 = 0 \quad \text{...(iv)}$$

Solving (ii), (iii) and (iv), we have

$$\frac{x}{\begin{vmatrix} b_2 & c_2 & d_2 \\ b_3 & c_3 & d_3 \\ b_4 & c_4 & d_4 \end{vmatrix}} = \frac{-y}{\begin{vmatrix} a_2 & c_2 & d_2 \\ a_3 & c_3 & d_3 \\ a_4 & c_4 & d_4 \end{vmatrix}} = \frac{z}{\begin{vmatrix} a_2 & b_2 & d_2 \\ a_3 & b_3 & d_3 \\ a_4 & b_4 & d_4 \end{vmatrix}} = \frac{-1}{\begin{vmatrix} a_2 & b_2 & d_2 \\ a_3 & b_3 & d_3 \\ a_4 & b_4 & d_4 \end{vmatrix}}$$

$$\Rightarrow \quad \frac{x}{A_1} = \frac{-y}{-B_1} = \frac{z}{C_1} = \frac{-1}{-D_1},$$

where A_1, B_1, C_1, D_1 are cofactors of a_1, b_1, c_1 and d_1 respectively in the determinant

$$\Delta = \begin{vmatrix} a_1 & b_1 & c_1 & d_1 \\ a_2 & b_2 & c_2 & d_2 \\ a_3 & b_3 & c_3 & d_3 \\ a_4 & b_4 & c_4 & d_4 \end{vmatrix}$$

$\therefore$ The point of intersection of the planes (ii), (iii) and (iv) above is

$$\left(\frac{A_1}{D_1}, \frac{B_1}{D_1}, \frac{C_1}{D_1}\right)$$

Similarly solving other sets of three planes from the given planes (i), (ii), (iii) and (iv) the other points of intersection *i.e.,* other three vertices of the tetrahedron are

$$\left(\frac{A_2}{D_2}, \frac{B_2}{D_2}, \frac{C_2}{D_2}\right); \left(\frac{A_3}{D_3}, \frac{B_3}{D_3}, \frac{C_3}{D_3}\right) \text{ and } \left(\frac{A_4}{D_4}, \frac{B_4}{D_4}, \frac{C_4}{D_4}\right)$$

$\therefore$ The required volume of tetrahedron is given by

$$= "(1/6) \begin{vmatrix} x_1 & y_1 & z_1 & 1 \\ x_2 & y_2 & z_2 & 1 \\ x_3 & y_3 & z_3 & 1 \\ x_4 & y_4 & z_4 & 1 \end{vmatrix}$$

$$\begin{vmatrix} A_1/D_1 & B_1/D_1 & C_1/D_1 & 1 \\ A_2/D_2 & B_2/D_2 & C_2/D_2 & 1 \\ A_3/D_3 & B_3/D_3 & C_3/D_3 & 1 \\ A_4/D_4 & B_4/D_4 & C_4/D_4 & 1 \end{vmatrix}$$

$$= \frac{1}{6D_1D_2D_3D_4} \begin{vmatrix} A_1 & B_1 & C_1 & D_1 \\ A_2 & B_2 & C_2 & D_2 \\ A_3 & B_3 & C_3 & D_3 \\ A_4 & B_4 & C_4 & D_4 \end{vmatrix}$$

$$= \begin{vmatrix} 1 & \cos\nu & \cos\mu \\ \cos\nu & 1 & \cos\lambda \\ \cos\mu & \cos\lambda & 1 \end{vmatrix}, \text{ from (i), (ii), (iii), and (iv)}$$

$\therefore$ From (v), we have $V = \pm (1/6)\, abc \begin{vmatrix} 1 & \cos\nu & \cos\mu \\ \cos\nu & 1 & \cos\lambda \\ \cos\mu & \cos\lambda & 1 \end{vmatrix}^{1/2}$

Here the negative sign may be neglected while calculating magnitude of the volume.

Example 1:

Prove that the volume of a tetrahedron of which a pair of opposite edges is formed by lengths r and r' on the straight lines whose equations are

$$\frac{x-a}{l} = \frac{y-b}{m} = \frac{z-c}{m}, \frac{x-a'}{l'} = \frac{y-b'}{m'} = \frac{z-c'}{n'}$$

$$\begin{vmatrix} a-a' & b-b' & c-c' \\ l & m & n \\ l' & m' & n' \end{vmatrix}.$$

Solution:

If a vertex of the tetrahedron be (a, b, c) on the line

$$\frac{x-a}{l} = \frac{y-b}{m} = \frac{z-c}{n},$$

then the other vertex on this line at a distance r from the vertex (a, b, c) is (a + *l*r, b + mr, c + nr).

Similarly if (a', b', c') be a vertex of the tetrahedron on the other given line, then the remaining vertex of the tetrahedron at a distance r' from (a', b', c') on this line is (a' + l', b' + m' r', c' + n'r').

∴ Required volume of the tetrahedron

$$= (1/6)\begin{vmatrix} a & b & c & 1 \\ a+lr & b+mr & c+nr & 1 \\ a' & b' & c' & 1 \\ a'+l'r' & b'+m'r' & c'+n'r' & 1 \end{vmatrix}$$

$$= (1/6)\, rr' \begin{vmatrix} a & b & c & 1 \\ lr & mr & nr & 0 \\ a' & b' & c' & 1 \\ l'r' & m'r' & n'r' & 0 \end{vmatrix}$$, subtracting lst row from 2nd and 3rd from 4th row

$$= (1/6)\, rr' \begin{vmatrix} a-a' & b-b' & c-c' & 0 \\ l & m & n & 0 \\ a' & b' & c' & 1 \\ l' & m' & n' & 0 \end{vmatrix}$$, subtracting 3rd row from lst and taking r, r' common from 2nd and 4th rows respectively.

$$= (1/6)\, rr' \begin{vmatrix} a-a' & b-b' & c-c' \\ l & m & n \\ l' & m' & n' \end{vmatrix}$$, expanding with respect to 4th column.

SKEW LINES AND CHANGE OF AXES SKEW LINES

Equations of Two Skew Lines

To show that by proper choice of axes equations of two skew lines can be written in a simple form as

$y = x \tan \alpha,\ z = c$

and $y = -x \tan \alpha,\ z = -c$.

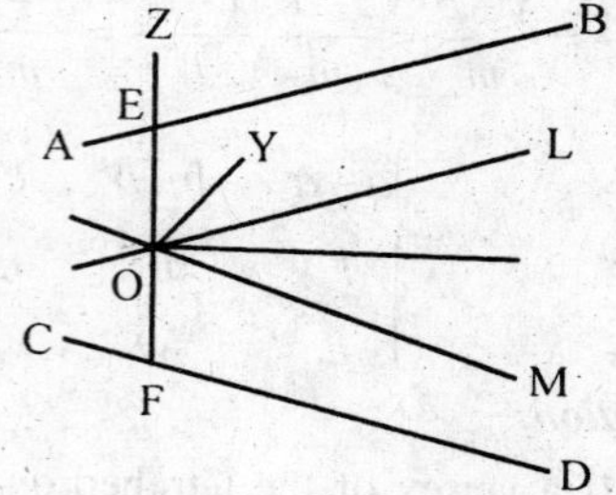

Let AB and CD be the skew lines and EF be the shortest distance them, where EF = 2c (say).

Let O be the mid-point of EF and through O draw OL and OM parallel to AB and CD respectively.

Choose O as origin and the internal and external bisectors of the angle LOM be taken as x and y axes respectively. Let OE be taken as z-axis. If ∠ LOM = 2α, then the equations of

OL and OM are $y = x \tan \alpha,\ z = 0;\ y = -x \tan \alpha,\ z = 0$.

Now AB and CD are lines parallel to OL and OM at distance c above and below xy-plane respectively.

∴ The equations of AB and CD are given by

$$y = x \tan \alpha, \; z = c$$

and $$y = -x \tan \alpha, \; z = -c.$$

LOCUS OF A LINE INTERSECTING THREE LINES

To find the locus of a line which intersects three lines.

$$u_1 = 0 = v_1 \quad \text{...(i)}$$

$$u_2 = 0 = v_2 \quad \text{(ii)}$$

and $$u_3 = 0 = v_3. \quad \text{...(iii)}$$

Equations of any line intersecting the lines (i) and (ii)

are $$\lambda_1 v_1 = 0, \; u_2 - \lambda_2 v_2 = 0 \quad \text{...(iv)}$$

If the lines (iv) intersects the line (iii) also then at the point of intersection, the equations (iii) and (iv) have common values of x, y, z.

∴ Eliminating x, y, z, from (iii) and (iv) we get a relation in λ_1, λ_2 which can be written as $f(\lambda_1, \lambda_2) = 0$ ...(v)

The required locus is obtained by eliminating λ_1, λ_2 from (iv) and (v) and is

$$\left(\frac{u_1}{v_1}, \frac{u_2}{v_2}\right) = 0$$

Note: If we are given two lines and a curve (instead of third line), the method of procedure is the same.

Example 1:

A straight line is drawn through a variable point on the ellipse $(x^2/a^2) + (y^2/b^2) = 1$, $z = 0$ to meet two fixed lines $y = mx$, $z = c$ and $y = -mx$, $z = -c$. Find the locus of the straight line.

Solution:

Given fixed lines are $y - mx = 0, \; z - c = 0;$ ...(i)

$$y + mx = 0, \; z + c = 0 \quad \text{...(ii)}$$

and the ellipse is $(x^2/a^2) + (y^2/b^2) = 1, \; z = 0$ ...(iii)

And line intersecting lines (i) and (ii) is

$$(y - mx) + k_1 (z - c) = 0,$$

$$(y + mx) + k_2 (z + c) = 0. \quad \text{...(iv)}$$

If it meets the ellipse (iii), then we are to eliminate k_1, k_2 from (iii) and (iv).

Putting z = 0 in (iv), we get $y - mx - k_1c = 0$, $y + mx + k_2c = 0$

$\Rightarrow$ $$mx - y + k_1c = 0,\ mx + y + k_2c = 0$$

Adding and subtracting these, we get

$$x = -\frac{(k_1 + k_2)\,c}{2m},\ y = \frac{(k_1 - k_2)\,c}{2}$$

Substituting these values of x and y in (iii), we get

$$\frac{(k_1 + k_2)^2\,c^2}{4a^2m^2} + \frac{(k_1 - k_2)^2\,c^2}{4b^2} = 1$$

$$\Rightarrow \quad (k_1 + k_2)^2\,c^2b^2 + (k_1 - k_2)^2\,c^2\,a^2\,m^2 = 4a^2b^2m^2$$

$$\Rightarrow \quad \left\{\left(\frac{mx - y}{z - c}\right) + \left(-\frac{mx + y}{z + c}\right)\right\}^2 c^2\,b^2 + \left\{\frac{mx - y}{z - c} + \frac{mx + y}{z + c}\right\}^2 c^2a^2m^2$$

$$= 4a^2b^2m^2m \text{ from (iv)}$$

$$\Rightarrow \quad \{(mx - y)(z + c) - (mx + y)(z - c)\}^2\,c^2b^2$$

$$+\{(mx - y)(z+c) + (mx+y)(z - c)\}^2\,c^2a^2m^2 = 4a^2b^2m^2\,(z^2 - c^2)^2$$

$$\Rightarrow \quad (cmx - yz)^2\,c^2b^2 + (mxz - cy)^2\,c^2a^2m^2 = a^2b^2m^2\,(z^2 - c^2)^2,$$

which is the required locus. **Ans.**

Example 2:

A point moves so that the line joining the feet of the perpendiculars from it to two given straight lines subtends a right angle at mid-point of their S.D. Prove its locus is hyperbolic cylinder.

Solution:

Let O the mid-point of the S.D. between given lines AB and CD, be taken as origin and let the equations of AB and CD,

$$\frac{x}{l} = \frac{y}{m} = \frac{z - c}{0} = r \qquad \text{...(i)}$$

and $$\frac{x}{1} = \frac{y}{-m} = \frac{z + c}{0} = r \qquad \text{...(ii)}$$

respectively.

Let the moving point P be (x_1, y_1, z_1) and feet of perpendicular from P to AB and CD be Q and Q'.

Then from (i) and (ii) Q is (r, mr, c) and Q' is (r', mr', –c).

$\therefore$ The direction ratios of PQ and PQ'' are respectively

$$x_1 - r,\ y_1 - mr,\ z_1 - c$$

and $\quad x_1 - r',\ y_1 - mr,\ z_1 + c.$

Now according to the problem PQ is perpendicular to AB and PQ' is perpendicular to CD, so we have

$$1.(x_1 - r) + m(y_1 - mr) + 0.(z_1 - c) = 0$$

and $\quad 1.(x_1 - r') + m(y_1 + mr') = 0\ (z_1 + c) = 0$

i.e., $\quad r = (x_1 + my_1)/(1 + m^2)$

and $\quad r' = (x_1 - my_1)/(1 + m^2)$ ...(iii)

Also QQ' subtends a right angle at O *i.e.,* OQ is perpendicular to OQ

$\therefore$ r.r' + mr (–mr') + c (– c) = 0, since the direction ratios of OQ are r, mr, c etc.

$\Rightarrow \quad rr'(1 - m^2) = c^2$

$\Rightarrow \quad \left(\dfrac{x_1 + my_1}{1 + m^2}\right)\left(\dfrac{x_1 - my_1}{1 + m_2}\right)(1 - m^2) = c^2$, from (iii)

$\Rightarrow \quad x_1^2 - m^2y_1^2 = [c^2(1 + m^2)^2/(1 - m^2)]$

$\therefore$ The required locus of P (x_1, y_1, z_1) is

$x^2 - m^2y^2 = [c^2(1 + m^2)^2/(1 - m^2)]$, which represents a hyperbolic cylinder.

Example 3:

A variable lines intersects the x-axis and curve $x = y$, $y^2 = cz$ and is parallel to the plane $x = 0$. Prove that it generates the paraboloid $xy = cz$.

Solution:

The plane parallel to the plane x = 0 is x = λ. ...(i)

And the plane containing x-axis *i.e.,* y = 0, z = 0 is y = μz ...(ii)

$\therefore$ The line which intersects the x-axis and is parallel to the plane x =0 is the line of intersection of (i) and (ii).

If this meets the curve x = y, $y^2 = cz$, then with the help of (i) and (ii)

we have $x = \lambda = y$ and $\mu = \dfrac{y}{z} = \dfrac{cy}{cz} = \dfrac{cy}{y^2}$ from $y^2 = cz$

$\Rightarrow \qquad \mu = \frac{c}{y} = \frac{c}{\lambda}$ or $\lambda\mu = c$ ...(iii) **(Note)**

This required is obtained by eliminating λ and μ from (i), (ii) and (iii) and is $c = \lambda\mu = x.\ (y/z)$ or $xy = cz$. **Hence proved.**

TO CHANGE THE ORIGIN OF CO-ORDINATES WITHOUT CHANGING THE DIRECTION OF THE AXES

Let (x, y, z) be the co-ordinates of a point P referred to original axes of co-ordinates OX, OY and OZ O as origin.

Let the co-ordinates of a point O' be (α, β, γ) referred to OX, OY and OZ.

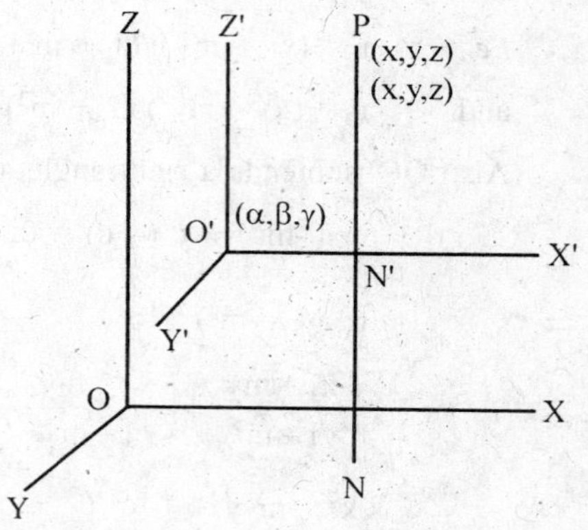

Through O' draw lines O' X', O' Y' and O'Z' parallel to OX, OY and OZ respectively.

Let co-ordinates of P be (x' y', z') referred to these parallel axes through O'.

From P draw PN perpendicular to XOY plane meeting the plane X'O'Y' in N'. Then PN = z and PN' = z'.

Also N'N = perpendicular distance between the planes XOY and X'O'Y'

= perpendicular distance of O' from XOY plane = γ.

Now from the figure it is evident that PN = PN' + N'N

i.e., $\qquad z = z' + \gamma$.

Similarly we can show that $x = x' + \alpha$ and $y = y' + \beta$.

TO CHANGE THE DIRECTION OF CO-ORDINATE AXES WITHOUT CHANGING THE ORIGIN

Let OX, OY, OZ and OX', OY', OZ' be two sets of co-ordinate axes through the common origin O.

Let the direction of cosines of OX', OY' and OZ' be $l_1, m_1, n_1, l_2, m_2, n_2$ and l_3, m_3, n_3 respectively referred to OX, OY and OZ.

The direction cosines of OX, OY, OZ referred to OX', OY', OZ' are evidently l_1, l_2, l_3; $m_1\ m_2, m_3$ and n_1, n_2, n_3 respectively. **(Note)**

Let the co-ordinates of P be (x, y, z) and (x', y', z') referred to the original axes OX, OY, OZ and the new axes OX', OY', OZ' respectively.

From P draw PN perpendicular to OX.

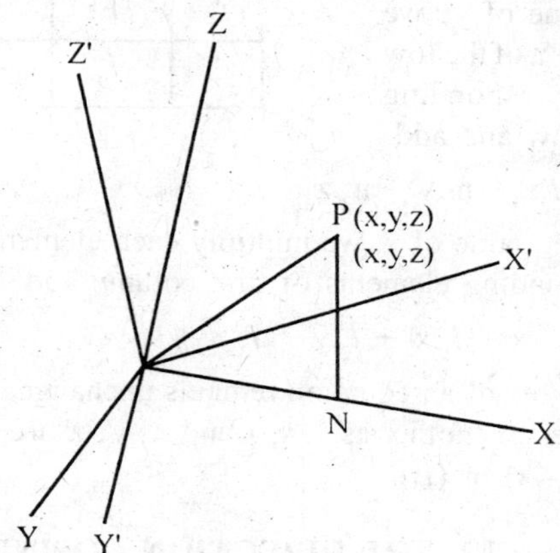

Then x = ON = Projection of OP on OX. ...(i)

Now d.c.'s of OX referred to the new axes are l_1, l_2, l_3 and the co-ordinates of P referred to the new axes are (x', y', z').

∴ From (i), we have

$$x = l_1 (x' - 0) + l_2 (y' - 0) + l_3 (z' - 0).$$

$$\Rightarrow \qquad \mathbf{x = l_1\ x + l_2 y' + l_3 z'}$$

Similarly $\qquad \mathbf{y = m_1 x' + m_2 y' + m_3 z'}$ ...(A)

and $\qquad \mathbf{z = n_1 x' + n_2 y' + n_3 z'}$

Multiplying these relations by l_1, m_1, n_1 respectively and adding we get

$$l_1 x + m_1 y + n_1 z = x' \left(\Sigma l_1^2\right) + y' (\Sigma l_1 l_2) + z' (\Sigma l_1 l_3)$$

$$x';\ \Sigma l^2{}_1 = 1\ \Sigma l_1 l_2 = 0 = \Sigma l_1 l_3$$

$$\Rightarrow \qquad \mathbf{x' = l_1 x + m_1 y + n_1 z}$$

Similarly $\qquad \mathbf{y' = l_2 x + m_2 y + n_2 z}$ ...(B)

and $\qquad \mathbf{z' = l_3 x + m_3 y + n_3 z}$

The relations (A) express the old co-ordinates x, y, z in terms of the new co-ordinates x', y', z' and the relations (B) express x', y', z', in terms x, y, z. These relations are also written conveniently with the help of the adjoining table. In this table direction cosines of mutually perpendicular axes.

How to write the relation (A) or (B) with the help of the above table?

	x	y	z
x'	l_1	m_1	n_1
y'	l_2	m_2	n_2
z'	l_3	m_3	m_3

To get the value of y', we multiply each element of the row of y' with the corresponding elements of first row and add

i.e., $y' = l_2 x + m_2 y + n_2 . z$

Similarly to get value of x, we multiply each element of the column of x with the corresponding elements of first column add

i.e., $x = l_1 . x' + l_2 . y' + l_3 . z'$

Cor. : The degree of an equation remains unchanged by transformation from one set of axes to another as x, y, z and x', y', z' are connected by inter relations given by (A) or (B).

RELATION BETWEEN THE DIRECTION COSINES OF THREE MUTUALLY PERPENDICULAR LINES

Let l_1, m_1, n_1; l_2, m_2, n_2 and l_3, m_3, n_3 be the direction cosines of any three mutually perpendicular lines OX', OY', OZ', referred to rectangular axes OX, OY, OZ.

Hence we have following two sets of relations:

$$\left.\begin{aligned} l_1^2 + m_1^2 + n_1^2 = 1, \\ l_2^2 + m_2^2 + n_2^2 = 1, \\ l_3^2 + m_3^2 + n_3^2 = 1, \end{aligned}\right\} \quad \text{...(i)}$$

$$\left.\begin{aligned} l_1 l_2 + m_1 m_2 + n_1 n_2 = 0, \\ l_2 l_3 + m_2 m_3 + n_2 n_3 = 0, \\ l_3 l_1 + m_3 m_1 + n_3 n_1 = 0, \end{aligned}\right\} \quad \text{...(ii)}$$

We know that the direction cosines of the lines OX, OY, OZ are l_1, l_2, l_3; m_1, m_2, m_3 and n_1, n_2, n_3 referred to OX', OY', OZ'. Hence as above we have

$$\left.\begin{aligned} l_1^2 + l_2^2 + l_3^2 = 1, \\ m_1^2 + m_2^2 + m_3^2 = 1, \\ n_1^2 + n_2^2 + n_3^2 = 1 \end{aligned}\right\} \quad \text{...(iii)}$$

$$\left.\begin{aligned} l_1 m_1 + l_2 m_2 + l_3 m_3 = 0, \\ m_1 n_1 + m_2 n_2 + m_3 n_3 = 0, \\ n_1 l_1 + n_2 l_2 + n_3 l_3 = 0 \end{aligned}\right\} \quad \text{...(iv)}$$

Now two cases arise:

Case I : *If only is shifted to (x_1, y_1, z_1) axes remaining parallel.*

In this case the given expression transforms into

$$a(x + x_1)^2 + b(y + y_1)^2 + c(z + z_1)^2 + 2f(y + y_1)(z + z_1)$$
$$+ 2g(z + z_1)(x + x_1) + 2h(x + x_1)(y + y_1)$$

$$\Rightarrow [ax^2 + by^2 + cz^2 + 2fyz + 2gzx + 2hxy]$$
$$+ 2x(ax_1 + hy_1 + gz_1) + 2y(hx_1 + by_1 + fz_1)$$
$$+ 2z(gx_1 + fy_1 + cz_1) + \ldots$$

The second degree terms remain unaltered after the transformation and as $a + b + c$, $A + B + C$ and Δ contain a, b, c, f, g, h *i.e.*, the coefficients of second degree terms, so these are invariants *i.e.*, remain unchanged.

Case II : *If only the (rectangular) axes are rotated, origin remaining same.*

In this case let the given expression $ax^2 + by^2 + cz^2 + 2fyz + 2gzx + 2hxy$ transform into $a_1x^2 + b_1y^2 + c_1z^2 + 2f_1yz + 2g_1zx + 2h_1xy$

Also as origin remains unchanged, so the distance of any point $P(x, y, z)$ from the origin remains unchanged and so $x^2 + y^2 + z^2$ remains the same.

$\therefore$ The expression

$$ax^2 + by^2 + cz^2 + 2fyz + 2gzx + 2hxy + \lambda(x^2 + y^2 + z^2) \quad \ldots(i)$$

transforms into

$$a_1x^2 + b_1y^2 + c_1z^2 + 2f_1yz + 2g_1zx + 2h_1xy + \lambda(x^2 + y^2 + z^2) \ldots(ii)$$

If for some value of λ, the expression (i) is the product of two linear factors, then for the same value of λ, the expression (ii) can also be written as the product of two linear factors.

Now (i) can be written as

$$(a + \lambda)x^2 + (b + \lambda)y^2 + (c + \lambda)z^2 + 2fyz + 2gxz + 2hxy$$

and this can be written as the product of two linear factors if

$$\text{“}abc + afgh - af^2 - bg^2 - ch^2 = 0\text{”}$$

i.e., if $(a + \lambda)(b + \lambda)(c + \lambda) + 2fgh - (a + \lambda)f^2 - (b + \lambda)g^2 - (c + \lambda)h^2 = 0$

i.e., if $\lambda^3 + (a + b + c)^2\lambda^3 + (ab + bc + ca - f^2 - g^2 - h^2)\lambda + (abc + 2fgh - af^2 - bg^2 - ch^2) = 0$

i.e if $\lambda^3 + (a + b + c)\lambda^2 + (a + B + C)\lambda + \Delta = 0,$ ...(iii)

where the symbols have their usual meanings.

Similarly (ii) can be written as the product of two linear factors if

$$\lambda^3 + (a_1 + b_1 + c_1)\lambda^2 + (A_1 + B_1 + C_1)\lambda + \Delta_1 = 0, \qquad ...(iv)$$

where

$$A = b_1c_1 - f_1^2 \text{ etc. and } \Delta_1 = a_1b_1c_1 + 2f_1g_1h_1 - a_1f_1^2 - b_1g_1^2 - c_1h_1^2$$

∴ The equations (iii) and (iv) have the same roots, so comparing the coefficients, we get

$$\frac{1}{1} = \frac{a+b+c}{a_1+b_1+c_1} = \frac{A+B+C}{A_1+B_1+C_1} = \frac{\Delta}{\Delta_1}$$

$\Rightarrow$ $a + b + c = a_1 + b_1 + c_1,$

$A + B + C = A_1 + B_1 + C_1$ and $\Delta = \Delta_1$

$\Rightarrow$ a + b + c, A + B + C and D remain unchanged.

These two sets of relations can also be deduced algebraically from the two sets (i) and (ii).

Again from the relations (ii), we have

$$l_1l_2 + m_1m_2 + n_1n_2 = 0,\ l_1l_3 + m_1m_3 + n_1n_3 = 0$$

Solving these simultaneously, we get

$$\frac{l_1}{m_2n_3 - m_3n_2} = \frac{m_1}{n_2l_3 - n_3l_2} = \frac{n_1}{l_2m_3 - l_3m_2}$$

$$= \frac{\sqrt{\left[\left(l_1^2 + m_1^2 + n_1^2\right)\right]}}{\sqrt{\left[\Sigma\left(m_2n_3 - m_3n_2\right)^2\right]}} = \frac{\pm\sqrt{(1)}}{\sqrt{(\sin 90^\circ)}} \qquad \textbf{(Note)}$$

$$= \pm 1. \qquad \textbf{(Note)}$$

These express the d.c.'s of one line in terms of those of the other two.

$$\text{Also let } \Delta = \begin{vmatrix} l_1 & m_1 & n_1 \\ l_2 & m_2 & n_2 \\ l_3 & m_3 & n_3 \end{vmatrix}$$

$$\text{Then} \quad \Delta^2 = \begin{vmatrix} l_1 & m_1 & n_1 \\ l_2 & m_2 & n_2 \\ l_3 & m_3 & n_3 \end{vmatrix} \times \begin{vmatrix} l_1 & m_1 & n_1 \\ l_2 & m_2 & n_2 \\ l_3 & m_3 & n_3 \end{vmatrix}$$

$$= \begin{vmatrix} \Sigma l_1^2 & \Sigma l_1 l_2 & \Sigma l_1 l_2 \\ \Sigma l_1 l_2 & \Sigma l_2^2 & \Sigma l_2 l_3 \\ \Sigma l_3 l_1 & \Sigma l_3 l_2 & \Sigma l_3^2 \end{vmatrix} = \begin{vmatrix} 1 & 0 & 0 \\ 0 & 1 & 0 \\ 0 & 0 & 1 \end{vmatrix},$$

from relations (i) and (ii)

$$\Rightarrow \Delta^2 = 1 \text{ or } \Delta = \pm 1 \text{ i.e., } \begin{vmatrix} l_1 & m_1 & n_1 \\ l_2 & m_2 & n_2 \\ l_3 & m_3 & n_3 \end{vmatrix} = \pm 1.$$

INVARIANTS

Show that if by any change of rectangular axes, the expression $ax^2 + by^2 + cz^2 + 2fyz + 2gzx + 2hxy$ be transformed, then the expressions $a + b + c$, $A + B + C$ and Δ are invariants i.e., remain changed in value and A, B, C are the cofactors of a, b, c in the determinant

$$\Delta = \begin{vmatrix} a & h & g \\ h & b & f \\ g & f & c \end{vmatrix}$$

Proof:

Here $A = bc - f^2$,

$B = ca - g^2$, $C = ab - h^2$ and the value of

$\Delta = abc + 2fgh - af^2 - bg^2 - ch^2$

$\therefore \quad A + B + C = bc + ca + ab - f^2 - g^2 - h^2.$

Example 1:

Two systems of rectangular axes have the same origin. If a plane cuts them at distances a, b, c and a', b,' c from the origin then show that $a^{-2} + b^{-2} + c^{-2} = (a')^{-2} + (b')^{-2} + (c')^{-2}$.

Solution:

Referred to the first system of axes, let the equation of the plane in the intercept form be $\frac{x}{a} + \frac{y}{b} + \frac{z}{c} = 1$. ...(i)

Replacing x by $l_1x + l_2y + l_2z$, y by $m_1x + m_2y + m_3z$ and z by $n_1x + n_2y + n_3z$ the equation of the plane referred to the second system is

$$\frac{l_1x + l_2y + l_3z}{a} + \frac{m_1x + m_2y + m_3z}{b} + \frac{n_1x + n_2y + n_3z}{c} = 1 \qquad ...(ii)$$

The intercept on new x-axis by this plane is a' (given), so putting $x = a'$, $y = 0 = z$ in (ii) we get

$$\left(\frac{l_1}{a}+\frac{m_1}{b}+\frac{n_1}{c}\right)a' = 1$$

or $$\frac{1}{a'} = \left(\frac{l_1}{a}+\frac{m_1}{b}+\frac{n_1}{c}\right)$$

Similarly $$\frac{l}{b'} = \frac{l_2}{a}+\frac{m_2}{b}+\frac{n_2}{c}; \frac{l}{c'} = \frac{l_3}{a}+\frac{m_3}{b}+\frac{n_3}{c}$$

$$\therefore \frac{1}{a'^2}+\frac{1}{b'^2}+\frac{1}{c'^2} = \left(\frac{l_1}{a}+\frac{m_1}{b}+\frac{n_1}{c}\right)^2+\left(\frac{l_2}{a}+\frac{m_2}{b}+\frac{n_2}{c}\right)^2+\left(\frac{l_3}{a}+\frac{m_3}{b}+\frac{n_3}{c}\right)^2$$

$$= \frac{1}{a^2}\left(l_1^2+l_2^2+l_3^2\right)+ \ldots + \ldots + \frac{2}{ab}(l_1m_1 + l_2m_2 + l_3m_3) + \ldots + \ldots$$

$$= \frac{1}{a^2}(1)+\frac{1}{b^2}(1)+\frac{1}{c^2}(1)+\frac{2}{ab}(0)+\frac{2}{bc}(0)+\frac{2}{ca}(0) \quad \Sigma\, l_1^2 = 1, \Sigma\, l_1m_1 = 0.$$

Hence $(a')^{-2} + (b')^{-2} + (c')^{-2} = a^{-2} + b^{-2} + c^{-2}$. **Hence proved.**

Example 2:

If $ax^2 + by^2 + cz^2 + 2fyz + 2hgz + 2hxy + 2ux + 2vy + 2wz + d$ be transformed by change of coordinates from are one set of rectangular axes to another with the same origin, the expression $a + b + c$, $u^2 + v^2 + w^2$ are invariants.

Solution:

Replacing x by $l_1x + l_2y + l_3z$ etc. The given expression can be transformed as

$$\Sigma\, a\,(l_1x + l_2y + l_3z)^2 + \Sigma\, 2f\,(m_1x + m_2y + m_3z)\,(n_1x + n_2y + n_3z) + \Sigma\, 2u\,(l_1x + l_2y + l_3z) + d$$

$$= a'x^2 + b'y^2 + c'z^2 + 2f'yz + 2g'zx + 2h'xy + 2u'x + 2v'y + 2w'z + d' \quad \ldots(ii)$$

Comparing coefficients of x^2, y^2, z^2, x, y, z on both sides we get

$$a' = al_1^2 + bm_1^2 + cn_1^2 + 2fm_1n_1 + 2gl_1n_1 + 2hl_1m_1$$

$$b' = al_2^2 + bm_2^2 + cn_2^2 + 2fm_2n_2 + 2gl_2n_2 + 2hl_2m_2$$

$$c' = al_3^2 + bm_3^2 + cn_3^2 + 2fm_3n_3 + 2gl_3n_3 + 2hl_3m_3$$

$u' = = ul_1 + vm_1 + wn_1$, $v' = ul_2 + vm_2 + wn_2$, $w' = ul_3 + vm_3 + wn_3$

Now $a' + b' + c' = a\Sigma\, l_1^2 + b\Sigma m_1^2 + c\Sigma n_1^2 + 2f\Sigma m_1n_1 + 2g\,\Sigma n_1l_1 + 2h\Sigma\, l_1m_1$

$= a(1) + (b)(1) + c(1) + 2f(0) + 2g(0) + 2h(0)$, from result (iii) and (iv).

i.e., $a' + b' + c' = a + b + c$. **Hence proved.**

Also $(u')^2 + (v')^2 + (w')^2 = (ul_1 + vm_1 + wm_1)^2 + (ul_2 + vm_2 + wn_2)^2 + (ul_3 + vm_3 + wn_3)^2$

$= u^2 \Sigma l_1^2 + v^2 \Sigma m_1^2 + w^2 \Sigma n_1^2 + 2uv \Sigma l_1 m + 2vw \Sigma m_1 n_1 + 2uw \Sigma l_1 n_1$

$= u^2(1) + v^2 + (w^2)(1) + 2uw(0) + 2vw(0) + 2uw(0)$, as before

$= u^2 + v^2 + w^2$. **Hence proved.**

SOME SOLVED EXAMPLES

Example 1:

Find the incentre of the tetrahedron formed by the planes $x = 0$, $y = 0$, $z = 0$ and $x + y + z = a$.

Solution:

Evidently the planes $x = 0$, $y = 0$ and $z = 0$ meet in (0, 0, 0). Hence the incentre lies on the perpendicular from (0, 0, 0) to the plane $x + y + z = a$ and divides it in the ratio 3: 1 [3 from the vertex (0, 0, 0) and 1 from the plane $x + y + z = a$].

The equations of the perpendicular from (0, 0, 0) to the plane $x + y + z = a$ is

$$\frac{x}{1} = \frac{y}{1} = \frac{z}{1} = r \text{ (say)}$$

Any point on this perpendicular is (r, r, r). If it lies on the plane $x + y + z = a$, then we have $r + r + r = a$ or $r = a/3$.

$\therefore$ The perpendicular from (0, 0, 0) meets the plane $x + y + z = a$ in (r, r, r) *i.e.,* $\left(\frac{1}{3}a, \frac{1}{3}a, \frac{1}{3}a\right)$. Also the incentre divides the join of (0, 0, 0) and $\left(\frac{1}{3}a, \frac{1}{3}a, \frac{1}{3}a\right)$ in the ratio 3: 1, therefore if (x_1, y_1, z_1) be the required incentre,

Then we have $x_1 = \dfrac{3.\frac{1}{3}a + 1.0}{3+1} = \dfrac{1}{4}a$.

Similarly $y_1 = \dfrac{1}{4}a = z$.

$\therefore$ The required incentre is $\left(\frac{1}{4}a, \frac{1}{4}a, \frac{1}{4}a\right)$. **Ans.**

Example 2:

Find the angle between the lines

$3x + 2y + z = 0 = x + y - 2z$ *and*

$2x - y - z = 0 = 7x + 10y - 8z.$

Solution:

Let l_1, m_1, n_1 be the d.c.'s of the first line, then we have

$$3l_1 + 2m_1 + n_1 = 0 \quad \text{...(i)}$$

$$l_1 + m_1 - 2n_1 = 0 \quad \text{...(ii)}$$

Solving (i) and (ii) we get

$$\frac{l_1}{-4-1} = \frac{m_1}{1+6} = \frac{n_1}{3-2}$$

$$\Rightarrow \frac{l_1}{5} = \frac{m_1}{7} = \frac{n_1}{1} = \frac{\sqrt{(l_1^2 + m_1^2 + n_1^2)}}{\sqrt{(5^2 + 7^2 + 1^2)}} = \frac{1}{\sqrt{(75)}} = \frac{1}{5\sqrt{3}} \quad \textbf{(Note)}$$

$\therefore l_1 = -5/5\sqrt{3}$, $m_1 = 7/5\sqrt{3}$, $n_1 = 1/5\sqrt{3}$.

Similarly, if l_2, m_2, n_2 be the d.c.'s of the second line, then we have

$2l_2 - m_2 - n_2 = 0$, $7l_2 + 10m_2 - 8n_2 = 0$ (iv) Solving (iii) and (iv)

We get $$\frac{l_2}{8+10} = \frac{m_2}{-7+16} = \frac{n_2}{20+7}$$

$$\Rightarrow \frac{l_2}{2} = \frac{m_2}{1} = \frac{n_2}{3} = \frac{\sqrt{(l_2^2 + m_2^2 + n_2^2)}}{\sqrt{(2^2 + 1^2 + 3^2)}} = \frac{1}{\sqrt{(14)}}$$

$\therefore l_2 = 2/\sqrt{(14)}$, $m_2 = 1/\sqrt{(14)}$, $n_2 = 3/\sqrt{(14)}$.

= If θ be the required angle then we have

$\cos\theta = 'l_1l_2 + m_1m_2 + n_1n_2'$

$$= \frac{-5}{5\sqrt{3}}.\frac{2}{\sqrt{(14)}} + \frac{7}{5\sqrt{3}}.\frac{1}{\sqrt{(14)}} + \frac{1}{5\sqrt{3}}.\frac{3}{\sqrt{(14)}} = 0$$

$\Rightarrow \quad \theta = \pi/2.$ **Ans.**

Example 3:

Find the symmetric equations of the line

$x - y = 0;\ 3x - z + 8 = 0.$

Solution:

The equations of the given planes can be written as

$$1.x - 1 \cdot y + 0 \cdot z = 0,\ 3x + 0 \cdot y - 1 \cdot z + 8 = 0 \qquad ...(i)$$

Let l, m, n be the d.c.'s of the line. Then as their lines lies on both the planes given by (i), so it is perpendicular to the normals of the planes given by (i). Then we have

$$\therefore l \cdot 1 + m(-1) + n \cdot 0 = 0 \qquad ...(ii)$$

and $\quad l \cdot 3 + m(0) + n(-1) = 0. \qquad ...(iii)$

Solving (ii) and (iii) we get $l/1 = m/1 = n/3$

Putting $x = 0$ in (i) we have $y = 0$, $z = 8$.

$\therefore$ The line (i) meets the plane $x = 0$ in (0, 0, 8) $\qquad$ ...(iv)

Hence the required form of the equations of the given line from (ii) and (iii) is $\quad (x - 0)/1 = (y - 0)/1 = (z - 8)/3.$ **Ans.**

Example 4:

Find the angle between the lines whose equations are $x + 2y - 2z = 0$, $x - 2y + z = 7$ and $x - 1 = -\frac{1}{2}(y + 2) = \frac{1}{3}z$.

Solution:

Let l_1, m_1, n_1 be the d.c.'s of the first line, then we have

$$l_1 + 2m_1 - 2n_1 = 0, \qquad ...(i)$$

$$l_1 - 2m_1 + m_1 = 0. \qquad ...(ii)$$

Solving (i) and (ii)

We get $\quad \dfrac{l_1}{2-4} = \dfrac{m_1}{-2-1} = \dfrac{n_1}{-2-2}$

$$\Rightarrow \quad \frac{l_1}{2} = \frac{m_1}{3} = \frac{n_1}{4} = \frac{\sqrt{(l_1^2 + m_1^2 + n_1^2)}}{\sqrt{(2^2 + 3^2 + 4^2)}} = \frac{1}{\sqrt{(29)}}$$

$$\therefore l_1 = 2/\sqrt{(29)},\ m_1 = 3/\sqrt{(29)},\ n_1 = 4/\sqrt{(29)}.$$

Now from the equation of second line it is obvious that its d.r.'s are 1, – 2, 3 and so its d.c.'s are $1/\sqrt{(14)}$, $-2\ 2/\sqrt{(14)}$, $3/\sqrt{(14)}$.

$\therefore$ If θ be the required angle, then we know that

$$\cos\theta = 'l_1 l_2 + m_1 m_2 + n_1 n_2'$$

$$= \frac{2}{\sqrt{(29)}}\cdot\frac{1}{\sqrt{(14)}}+\frac{3}{\sqrt{(29)}}\cdot\frac{-2}{\sqrt{(14)}}+\frac{4}{\sqrt{(29)}}\cdot\frac{3}{\sqrt{(14)}}$$

$$= 8/[\sqrt{(29)}\ \sqrt{(14)}]=8/\sqrt{(406)}$$

$$\theta = \cos^{-1}\left[8/\sqrt{(406)}\right].$$ **Ans.**

Example 5:

Find the symmetric form the equations of the line given by

$$x = ay + b,\ z = cy + d.$$

Solution:

The equations of the given planes can be written as

$$1 \,.\, x - a \,.\, y + 0 \,.\, z - b = 0$$

and $$0 \,.\, x + c \,.\, y - 1 \,.\, z + d = 0 \quad \text{...(i)}$$

Let l, m, n be the d.c.'s of the line. Then as this line lies on both the planes given by (i), so it is perpendicular to the normals to the planes given by (i)

There we have $l \,.\, 1 + m \,.\, (-a) + n \,.\, 0 = 0$ ...(i)

and $1 \,.\, 0 + m \,.\, c + n \,.\, (-1) = 0$ ...(ii)

Solving (i) and (ii) we get $l/a = m/1 = n/c$

Putting $y = 0$ in (i), we have $x = b$ and $z = d$

∴ The line (i) meets the plane $y = 0$ at $(b, 0, d)$.

Hence the required equations are $\frac{x-b}{a}=\frac{y-0}{1}=\frac{z-d}{c}$. **Ans.**

Example 6:

Find the equations of the line through (α, β, γ) *and parallel to the line*

$$a_1x + b_1y + c_1z + d_1 = 0,$$
$$a_2x + b_2y + c_2z + d_2 = 0.$$

Solution:

Let l, m, n be the direction cosines of the required line. Then as this line is parallel to the line given by the planes

$$a_1x + b_1y + c_1z + d_1 = 0 \quad \text{...(i)}$$

and $$a_2x + b_2y + c_2z + d_2 = 0 \quad \text{...(ii)}$$

$$\therefore\ a_1l + b_1m + c_1n = 0,$$

$$a_2l + b_2m + c_2n = 0$$

Solving (i) and (ii) we get $\frac{l}{b_1c_2 - b_2c_1} = \frac{m}{c_1a_2 - c_2a_1} = \frac{n}{a_1b_2 - a_2b_1}$

$\therefore$ The required equations are $\frac{x-\alpha}{b_1c_2 - b_2c_1} = \frac{y-\beta}{c_1a_2 - c_2a_1} = \frac{z-\gamma}{a_1b_2 - a_2b_1}$.

Example 7:

Prove that the lines $x = ay + b$, $z = cy + d$ and $x = a'y + b'$, $z = c'y + d'$ are perpendicular if $aa' + cc' = -1$.

Solution:

If l_1, m_1, n_2 be the d.c.'s of the first line, then

$l_1 . 1 + m_1 (-a) + m_1 (-a) + n_1 . 0 = 0$; $l_1.0 + m_1c + n_1 (-1) = 0$.

Solving these we get $\frac{l_1}{a} = \frac{m_1}{1} = \frac{n_1}{c}$ ...(i)

Similarly if l_2, m_2, n_2 be the d.c.'s of the second line then we can get

$$\frac{l_2}{a'} = \frac{m_2}{1} = \frac{n_2}{c'} \quad ...(ii)$$

If the given lines are perpendicular, then we have

$$'l_1l_2 + m_1m_2 + n_1n_2 = 0'$$

$\Rightarrow$ $aa' + 1.1 + cc' = 0$, from (i) and (ii)

$\Rightarrow$ $aa' + cc' = -1$. **Hence proved.**

Example 8:

Find the symmetric form of the line

$$3x + 2y + z = 5,\ x + y - 2z = 3$$

Solution:

Let l, m, n be the d.c.'s of the line. Then as this line lies on both the given planes, so it is perpendicular to the normal to these planes.

$\therefore$ We have $3l + 2m + n = 0$ ...(i)

$l + m - 2n = 0$ (ii)

Solving (i) and (ii) we get $\frac{l}{2(-2) - 1.1} = \frac{m}{1.1 - 3.(-2)} = \frac{n}{3.1 - 2.1}$

$\Rightarrow$ $l/(-5) = m/7 = n/1$ ...(iii)

Putting z = 0 in the given equation we get $3x + 2y = 5$, $x + y = 3$

Solving these we get $x = -1$, $y = 4$

$\therefore$ The required line the plane z = 0 in (−1, 4, 0) ...(iv)

∴ From (i) and (ii) the required equations are

$$\frac{x+1}{-5}=\frac{y-4}{7}=\frac{z}{1}.$$ **Ans.**

Example 9:

Find the equations of the line through the point (–2, 3, 4) and parallel to the planes 2x + 3y + 4z = 5 and 3x + 4y + 5z = 6.

Solution:

Let l, m, n be the d.c.'s of the line. We have

$$2l + 3m + 4n = 0, \quad \text{...(i)}$$

$$3l + 4m + 5n = 0 \quad \text{...(ii)}$$

Solving (i) and (ii) we get

$$\frac{l}{3.5-4.4}=\frac{m}{4.3-2.5}=\frac{n}{2.4-3.3} \text{ or } \frac{l}{-1}=\frac{m}{2}=\frac{n}{-1}$$

∴ The required equations of the line through (–2, 3, 4) are

$$\frac{x-(-2)}{-1}=\frac{y-3}{2}=\frac{z-4}{-1}$$

or $$\frac{x+2}{1}=\frac{y-3}{-2}=\frac{z-4}{1}.$$ **Ans.**

Example 10:

Find the equations of the line through the origin parallel to the line

$$x + y + z + 2 = 0 = 4x + 3y + 2z + 1.$$

Solution:

Let l, m, n be the direction cosines of the given line. Then as this line lies on both the given planes which constitute the given line, so it is perpendicular to the normals of these planes and thus we have

$$l \cdot 1 + m \cdot 1 + n \cdot 1 = 0$$

and $$4l + 3m + 2n = 0$$

Solving these we get $\frac{l}{1}=\frac{m}{-2}=\frac{n}{1}$

∴ The required equations of the line through (0, 0, 1) and parallel to the line whose d.c.'s are proportional to 1, –2, 1 are

$$\frac{x-0}{1}=\frac{y-0}{2}=\frac{z-0}{1}$$

or $$x = -\frac{1}{2}y = z.$$ **Ans.**

Example 11:

Prove that the lines $x + y - z = 5$, $9x - 5y + z = 4$ and $6x - 8y + 4z = 3$, $x + 8y - 6z + 7 = 0$ are parallel.

Solution:

If l_1, m_1, n_1 be the d.c.'s of the first line, then we have

$$l_1 + m_1 - n_1 = 0 \text{ (i) and } 9l_1 - 5m_1 + n_1 = 0 \qquad \text{...(ii)}$$

Solving (i) and (ii) we get $\dfrac{l_1}{1.1-1.5} = \dfrac{m_1}{-1.9-1.1} = \dfrac{n_1}{-1.5-1.9}$

$$\Rightarrow \quad \frac{l_1}{-4} = \frac{m_1}{-10} = \frac{n_1}{-14} \text{ or } \frac{l_1}{2} = \frac{m_1}{5} = \frac{n_1}{7} \qquad \text{(A)}$$

Similarly if l_2, m_2, n_2 be the d.c.'s of the second line, then we have

$$6l_2 - 8m_2 + 4n_2 = 0' \qquad \text{...(iii)}$$

$$l_2 + 8m_2 - 6n_2 = 0 \qquad \text{...(iv)}$$

Solving (iii) and (iv) we get $\dfrac{l_2}{8.6-4.8} = \dfrac{m_2}{4.1+6.6} = \dfrac{n_2}{6.8-8.1}$

$$\Rightarrow \quad \frac{l_2}{8.6} = \frac{m_2}{40} = \frac{n_2}{-56} \text{ or } \frac{l_2}{2} = \frac{m_2}{5} = \frac{n_2}{7} \qquad \text{(B)}$$

From (A) and (B) we find that $\dfrac{l_1}{l_2} = \dfrac{m_1}{m_2} = \dfrac{n_1}{n_2}$

Hence the given lines are parallel.

Example 12:

Find the equations of the line joining the points (4, –5, –2) and (–1, 5, 3) and show that it meets the surface $2x^2 + 3y^2 - 4z^2 = 1$ in coincident points.

Solution:

The equations of the line joining (4, – 5, – 2) and (– 1, 5, 3) are

$$\frac{x-4}{-1-4} = \frac{y-(-5)}{5-(-5)} = \frac{z-(-2)}{3-(-2)} \text{ or } \frac{x-4}{-1} = \frac{y+5}{2} = \frac{z+2}{1} = r \text{, say}$$

$\therefore$ Any point on it is $(4 - r, -5 + 2r, -2 + r)$

If it lies on the surface $2x^2 + 3y^2 - 4z^2 = 1$, then

$$2(4 - r)^2 + 3(-5 + 2r)^2 - 4(-2 + r)^2 = 1$$

$$\Rightarrow 2(16 - 8r + r^2) + 3(25 - 20r + 4r^2) - 4(r^2 - 4r + 4) = 1$$

$$\Rightarrow 10r^2 - 60r + 90 = 0 \text{ or } r^2 - 6r + 9 = 0 \text{ or } (r - 3)^2 = 0 \text{ or } r = 3, 3.$$

These two values of r being coincident the line (i) meets the given surface in coincident points.

Example 13:

The plane $lx + my = 0$ is rotated about the line of intersection with the plane $z = 0$ through an angle α. Prove that the equation to the plane in its new position is

$$lx + my \pm z\sqrt{(l^2 + m^2)}\ \tan\alpha = 0.$$

Solution:

The equation of any plane through the line of intersection of the planes $lx + my = 0$ and $z = 0$ is given by $(lx + my) = \lambda z = 0$ ...(i)

If (i) represents the planes obtained by rotating the given plane

$$lx + my = 0 \qquad \text{...(ii)}$$

through an angle α about the line of intersection of (ii) and the plane $z = 0$, then the angle between the planes (i) and (ii) is α and thus we have

$$\cos\alpha = \frac{l.l + m.m + \lambda.0}{\sqrt{(l^2 + m^2 + \lambda^2)}.\sqrt{(l^2 + m^2 + 0^2)}}$$

$\Rightarrow$ $(l^2 + m^2 + \lambda^2)(l^2 + m^2)\cos^2\alpha = (l^2 + m^2)^2$, squaring and cross multiplying

$\Rightarrow$ $(l^2 + m^2 + \lambda^2)\cos^2\alpha = (l^2 + m^2)$ or $l^2 + m^2 + \lambda^2 = (l^2 + m^2)\sec^2\alpha$

$\Rightarrow$ $\lambda^2 = (l^2 + m^2)\sec^2\alpha - (l^2 + m^2) = (l^2 + m^2)\tan^2\alpha$

$\Rightarrow$ $\lambda = \pm\sqrt{(l^2 + m^2)}\ \tan\alpha$

$\therefore$ From (i) the required equation is

$$lx + my \pm z\sqrt{(l^2 + m^2)}\ \tan\alpha = 0.$$

Ans.

Example 14:

Prove that the line $\frac{1}{2}(x - 3) = \frac{1}{3}(y - 4) = \frac{1}{4}(z - 5)$ lies on the plane $4x + 4y - 5z - 3 = 0$.

Solution:

From the equations of the line it is evident that the given line passes through (3, 4, 5) and its direction ratios are 2, 3, 4.

Now we find that the point (3, 4, 5) lies on the given plane

$$4x + 4y - 5z - 3 = 0 \qquad \text{...(i)}$$

as the co-ordinates of this point satisfy (i).

Also we find (i) that the d. ratios of the normal to the given plane (i) are 4, 4, –5.

$\because$ 2.4 + 3.4 + 4.(–5) = 0, so the normal to the plane (i) is perpendicular to the given line.

Hence we find that (a) the given the line passes through (3, 4, 5) and this point lies on the plane (i) and (b) the given line is perpendicular to the normal to the given plane (i).

$\therefore$ We conclude that the given line lies on the given plane.

Example 15:

Find the equation of the plane through the line

$$\frac{1}{2}(x-2)=\frac{1}{3}(y-3)=\frac{1}{5}(z-4) \text{ and parallel to x-axis.}$$

Solution:

The given line is $\dfrac{x-2}{2}=\dfrac{y-3}{3}=\dfrac{z-4}{5}$

So (2, 3, 4) is a point on this line and is d.r.'s are 2, 3, 5.

Equation of an plane A (x – 2) + B (y – 3) + C (z – 4) = 0, ...(i)

where A.2 + B.3 + C.5 – 0. ...(ii)

If the plane (ii) is parallel to x-axis whose d.c.'s are 1, 0, 0 then the normal to (ii) is at right angles to x-axis.

i.e., A.*l* + B.0 = C.0 = 0 ...(iii)

Eliminating A, B, C from (i), (ii) and (iii) we get the required equation as

$$\begin{vmatrix} x-2 & y-3 & z-4 \\ 2 & 3 & 5 \\ 1 & 0 & 0 \end{vmatrix} = 0$$

which on simplifying reduces to 5y – 3z – 3 = 0. **Ans.**

Example 16:

Find the equation of the plane through the point (α', β', γ') and the line

$$\frac{x-\alpha}{l}=\frac{y-\beta}{m}=\frac{z-\gamma}{n}.$$

Solution:

The equation of the plane through the given line is given as

$$A(x-\alpha)+B(y-\beta)+C(z-\gamma)=0, \quad \text{...(i)}$$

where we have A.*l* + B.m + C.n = 0. ...(ii)

If this plane (i) passes through (α, β, γ) also, then we have

$$A(\alpha' - \alpha) + B(\beta' - \beta) + C(\gamma' - \gamma) = 0 \qquad \text{...(iii)}$$

Eliminating A, B, C from (i), (ii) and (iii), we have the required plane as

$$\begin{vmatrix} x-\alpha & y-\beta & z-\gamma \\ l & m & n \\ \alpha'-\alpha & \beta'-\beta & \gamma'-\gamma \end{vmatrix} = 0. \qquad \textbf{Ans.}$$

Example 17:

Find the equation of the plane passing through the line of intersection of the planes $x + y + z = 6$ and $2x + 3y + 4z + 5 = 0$ and perpendicular to the plane $4x + 5y - 3z = 8$.

Solution:

The equation of any plane which passes through the line of intersection of the given planes $x + y + z = 6$ and $2x + 3y + 4z + 5 = 0$ is

$$(x + y + z - 6) + \lambda(2x + 3y + 4z + 5) = 0 \qquad \text{...(i)}$$

$$\Rightarrow \quad (1 + 2\lambda)x (1 + 3\lambda)y + (1 + 4\lambda)z + (5\lambda - 6) = 0 \qquad \text{...(ii)}$$

If this plane (ii) is perpendicular to the plane $4x + 5y - 3z = 8$,

then we have $\quad 4(1 + 2\lambda) + 5(1 + 3\lambda) - 3(1 + 4\lambda) = 0$

$$\Rightarrow \quad 11\lambda + 6 = 0 \text{ or } \quad \lambda = -6/11$$

∴ From (i), the required equation is given as

$$(x + y + z - 6) - (6/11)(2x + 3y + 4z + 5) = 0$$

$$\Rightarrow \quad x + 7y + 13z + 96 = 0. \qquad \textbf{Ans.}$$

Example 18:

Find the equation of the plane which is perpendicular to the plane $5x + 3y + 6z + 8 = 0$ and which contains the line of intersection of the planes $x + 2y + 3z - 4 = 0$ and $2x + y - z + 5 = 0$.

Solution:

Do yourself. **Ans.** $51x + 15y - 50z + 173 = 0$

Example 19:

Find the equation of the plane which contains the line $(x - 1)/2 = (y + 1)/(-1) = (z - 3)/4$ and is perpendicular to the plane $x + 2y + z = 12$.

Solution:

The equation of any plane through the given line is given by

$$A\,(x-1)+B\,(y+1)+C\,(z-3)=0 \quad \text{...(i)}$$

where $A.2+B.(-1)+C.4=0$ ...(ii)

Also if the plane (i) is perpendicular to the plane $x+2y+z=12$, we have $A.1+B.2+C.2=0$ ...(iii)

Solving (ii) and (iii), we have

$$\frac{A}{-2-8}=\frac{B}{4-4}=\frac{C}{4+1} \text{ or } \frac{A}{-2}=\frac{B}{0}=\frac{C}{1}$$

Substituting the proportionate values of A, B, C in (i), we have the required plane as

$-2(x-1)+0(y+1)+(z-3)=0$ or $2x-z+1=0$. **Ans.**

Example 20:

Find the equation of the plane which passes through the point (1, 2, −1) and which contains the line

$$\frac{1}{2}(x+1)=\frac{1}{3}(y-1)=-(z+2).$$

Solution:

The equation of the plane through the given line is

$$A\,(x+1)+B\,(y-1)+C\,(z+2)=0 \quad \text{...(i)}$$

where $A.2+B.3+C.(-1)=0$ ...(ii)

If this plane (i) passes through the point (1, 2, −1) also, then we have

$A\,(1+1)+B\,(2-1)+C\,(-1+2)=0$ or $2A+B+C=0$ ...(iii)

Eliminating A, B, C from (i), (ii) and (iii) we get

$$\begin{vmatrix} x+1 & y-1 & z+2 \\ 2 & 3 & -1 \\ 2 & 1 & 1 \end{vmatrix}=0$$

$\Rightarrow 4\,(x+1)-4\,(y-1)-4\,(z+2)=0$, on expanding the determinant

$\Rightarrow \quad x-y-z=0.$ **Ans.**

Example 21:

Find the equation of the plane through the point (2, −1, 1) and the line $4x-3y+5=0=y-2z-5$.

Solution:

The equation of any plane through the line

$$4x-3y+5=0,\; y-2z-5=0.$$

is $(4x - 3y + 5) + \lambda (y - 2z - 5) = 0$

If it passes through (2, – 1, 1), then from (i) we have

$$[4.2 - 3(-1) + 5] + \lambda [(-1) - 2(1) - 5] = 0 \text{ or } \lambda = 2$$

Substituting this value in (i) the required equation is

$(4x - 3y + 5) + 2(y - 2z - 5) = 0$ or $4x - y - 4z - 5 = 0$. **Ans.**

Example 22:

Prove that the plane through the point (α, β, γ) and the line $x = py + q = rz + s$ is given by

$$\begin{vmatrix} x & py+q & rz+s \\ \alpha & p\beta+q & r\gamma+s \\ 1 & 1 & 1 \end{vmatrix} = 0.$$

Solution:

The equations of the line $x = py + q = rz + s$ can be written in the symmetric form as $\frac{x}{1} = \frac{y+(q/p)}{1/p} = \frac{z+(s/r)}{(1/r)}$ ...(i) **(Note)**

∴ The equation of the plane through the line (i) is given as

$$A(x) + B\left(y + \frac{q}{p}\right) + C\left(z + \frac{s}{r}\right) = 0 \quad \text{...(ii)}$$

where $A.1 + B.(1/p) + C(1/r) = 0$. ...(iii)

If this plane (ii) passes through the point (α, β, γ), then

We have

$$A.\alpha + B\left(\beta + \frac{q}{p}\right) + C\left(\gamma + \frac{s}{r}\right) = 0 \quad \text{...(iv)}$$

Eliminating A, B and C from (ii), (iii) and (iv), we have

$$\begin{vmatrix} x & y+\frac{q}{p} & z+\frac{s}{r} \\ 1 & \frac{1}{p} & \frac{1}{r} \\ \alpha & \beta+\frac{q}{p} & \gamma+\frac{s}{r} \end{vmatrix} = 0.$$

Multiplying C_2 and C_3 by p and r respectively and interchanging R_2 and R_3 this reduces to

$$\begin{vmatrix} x & py+q & rz+s \\ \alpha & p\beta+q & r\gamma+s \\ 1 & 1 & 1 \end{vmatrix} = 0.$$ **Hence proved.**

Example 23:

Find the equations to the line through the plane (1, 2, 4) and perpendicular to the line 3x + 2y – z – 4 = 0 = x – 2y – 2z – 5.

Solution:

The equation of any plane through the given line is

$(3x + 2y - z - 4) + \lambda (x - 2y - 2z - 5) = 0$ ...(i)

If it passes through the point (1, 2, 4), then we have

$(3.1 + 2.2. - 4 - 4) + \lambda (1 - 2.2 - 2.4 - 5) = 0$ or $\lambda = -(1/16)$.

∴ From (i) the equation of the plane through (1, 2, 4) and the given line is given by

$$16 (3x + 2y - z - 4) - (x - 2y - 2z - 5) = 0$$

⇒ $47x + 34y - 14z - 59 = 0$ ...(ii)

Also if λ, m, n be the direction cosines of the given line, then we have

$3l + 2m - n = 0$ and $l - 2m - 2n = 0$

Solving these equations we have

$$\frac{l}{-4-2} = \frac{m}{-1+6} = \frac{n}{-6-2}$$

or $\frac{l}{-6} = \frac{m}{5} = \frac{n}{-8}$ ...(iii)

Also any plane through (1, 2, 4) is

$A (x - 1) + B (y - 2) + C (z - 4) = 0$ ...(iv)

If this plane is perpendicular to the given line then the normal to this plane must be parallel to the given line. So from (iii) we have

$A/l = B/m = C/n$ or $A/(-6) = B/5 = C/(-8)$

Hence from (iv) the equation of the plane through (1, 2, 4) and perpendicular to the given line is given as

$-6 (x - 1) + 5 (y - 2) - 8 (z - 4) = 0$ or $6x - 5y + 8z = 28$ ...(v)

The line of intersection of the planes (ii) and (v) is the perpendicular from (1, 2, 4) to the given line, hence the required general equations of the perpendicular are

$47x + 34y - 14z = 59;\ 6x - 5y + 8z = 28$ ...(vi)

Now if we want to express the equation of this line in the symmetric form, then let l_1, m_1, n_1 be d.c.'s of this perpendicular and so we have

$$47l_1 + 34m_1 - 14n_1 = 0 \text{ and } 6l_1 - 5m_1 + 8n_1 = 0$$

Solving these, we obtain

$$\frac{l_1}{272-70} = \frac{m}{-84-376} = \frac{n_1}{-235-204}$$

or
$$\frac{l_1}{202} = \frac{m_1}{-460} = \frac{n_1}{-439}$$

∴ The equations (vi) of perpendicular from (1, 2, 4) to the given line can be written in the symmetric form as

$$\frac{x-1}{202} = \frac{y-2}{-460} = \frac{z-4}{-439}.$$ **Ans.**

Example 24:

Find k so that the lines given by the following equations may be perpendicular to each other.

$$\frac{x-1}{-3} = \frac{y-2}{2k} = \frac{z-3}{2}$$

and
$$\frac{x-1}{3k} = \frac{y-5}{1} = \frac{z-6}{-5}.$$

Solution:

The direction ratios of the given lines are – 3, 2k, 2 and 3k, 1, – 5

∴ If these lines are perpendicular to each other, then we have

$$'a_1a_2 + b_1b_2 + c_1c_2 = 0'$$

i.e., $(-3)(3k) + (2k)(1) + 2(-5) = 0$

i.e., $-9k + 2k - 10 = 0$ or $7k + 10 = 0$ or $k = -10/7$. **Ans.**

Example 25:

Find the equations of the line through (α, β, γ) at right angles to the lines $\frac{x}{l_1} = \frac{y}{m_1} = \frac{z}{n_1}$ *and* $\frac{x}{l_2} = \frac{y}{m_2} = \frac{z}{n_2}$.

Solution:

Let the required line through (α, β, γ) be given as

$$\frac{x-\alpha}{l} = \frac{y-\beta}{m} = \frac{z-\gamma}{n}. \qquad ...(i)$$

If this line is perpendicular to the given lines, then we have

$$ll_1 + mm_1 + nn_1 = 0 \qquad ..(ii)$$

and $$ll_2 + mm_2 + nn_2 = 0 \qquad ...(iii)$$

Solving (ii) and (iii)

We obtain $\dfrac{l}{m_1n_2 - m_2n_1} = \dfrac{m}{n_1l_2 - n_2l_1} = \dfrac{n}{l_1m_2 - l_2m_1}$

Substituting these proportionate values of l, m, n in (i) we have the required equations as $\dfrac{x-\alpha}{m_1n_2 - m_2n_1} = \dfrac{y-\beta}{n_1l_2 - n_2l_1} = \dfrac{z-\gamma}{l_1m_2 - l_2m_1}$. **Ans.**

Example 26:

Does the lines $\dfrac{x-1}{5} = \dfrac{y+2}{6} = \dfrac{z-3}{4}$ *intersect any one of the planes*

(i) $2x + 3y - 7z + 29 = 0$

(ii) $2x + 3y - 7z + 25 = 0,$

(iii) $2x + 3y - 6z + 20 = 0$

If it does, find the point of intersection.

Solution:

Any point on the given line is $(5r + 1, 6r - 2, 4r + 3)$ ...(A)

If the given line intersects the plane (i), then we have

$$2(5r + 1) + 3(6r - 2) - 7(4r + 3) + 29 = 0$$

$$\Rightarrow \quad r(10 + 18 - 28) + (2 - 6 - 21 + 29) = 0$$

from which r cannot be determined. Hence the given line does not meet the plane (i).

Similarly if the given line intersects the plane (ii), then proceeding as above we find that r can not be determined, hence the given line does not intersect the plane (ii).

Finally if the given line intersects the plane (iii), then

$$2(5r + 1) + 3(6r - 2) - 6(4r + 3) + 20 = 0$$

$$\Rightarrow r(10 + 18 - 24) + (2 - 6 - 18 + 20) = 0 \text{ or } 4r - 2 = 0 \text{ or } r = 1/2$$

Hence the given line intersects the plane (iii) and the required point of intersection from (A) is

$[5(1/2) + 1, 6(1/2) - 2, 4(1/2) + 3]$ *i.e.,* $(7/2, 1, 5)$. **Ans.**

Example 27:

Find the ratio in which the join of (2, 3, 1) and (–2, 1, – 3) is cut by the plane x – 2y + 3z + 4 = 0. Find also the coordinates of the point of intersection.

Solution:

The direction ratios of the line joining the points (2, 3, 1) and (–2, 1, –3) are – 2 – 2, 1 – 3, – 3 – 1 or – 4, – 2, – 4 or 2, 1, 2.

∴ The equations of the line through (2, 3, 1) and joining the given points

are $$\frac{x-2}{2}=\frac{y-3}{1}=\frac{z-1}{2}=r \text{ (say)} \qquad ...(i)$$

Any point on this line is (2 + 2r, 3 + r, 1 + 2r) ...(ii)

If this point lies on the plane x – 2y + 3z + 4 = 0, then

We have (2 + 2r) – 2(3 + r) + 3 (1 + 2r) + 4 = 0

$$\Rightarrow 6r = -3 \text{ or } = -\frac{1}{2}$$

Substituting this value of r in the co-ordinates of the point given by (ii), we have the required point of intersection as

$$\left(2-1, 3-\frac{1}{2}, 1-1\right) \text{ i.e., } \left(1, \frac{5}{2}, 0\right) \qquad \textbf{Ans.}$$

Let the point $\left(1, \frac{5}{2}, 0\right)$ divide the join of (2, 3, 1) and (– 2, 1, – 3) in the ratio m : n, then

$$1=\frac{m(-2)+n(2)}{m+n}, \frac{5}{2}=\frac{m(1)+n(3)}{m+n}; 0=\frac{m(-3)+n(1)}{m+n}$$

which give 3m = n or m : n = 1 : 3. **Ans.**

Example 28:

Find the image of the point P (3, 5, 7) in the plane

2x + y + z = 6.

Solution:

The image of the point (3, 5, 7) on the given plane is the point in which the line through (3, 5, 7) perpendicular to the given plane meets it.

Now the direction ratio of the normal to the given plane are 2, 1, 1.

∴ Equations of the line through P (3, 5, 7) perpendicular to the given plane are

$$\frac{x-3}{2}=\frac{y-5}{1}=\frac{z-7}{1} \quad ...(i)$$

Any point on this line (i) is (3 + 2r, 5 + r, 7 + r)

If it lies on the given plane 2x + y + z = 6, then we have

$$2(3+2r)+(5+r)+(7+r)=6 \text{ or } 6r=-12 \text{ or } r=-2$$

∴ Required point is (3 + 2r, 5 + r, 7 + r), where r = –2

i.e., (3 – 4, 5 – 2, 7 – 2) *i.e.,* (– 1, 3, 5). **Ans.**

Example 29:

Find the image of the point (1, 3, 4) in the plane 2x – y + z + 3 = 0.

Solution:

The d.r.'s of the normal to the given plane are 2, – 1, 1, so the equations of the line through (1, 3, 4) perpendicular to the given plane are

$$\frac{x-1}{2}=\frac{y-3}{-1}=\frac{z-4}{1}=r \text{ (say)} \quad ...(i)$$

∴ Any point on this line (i) is (2r + 1, – r + 3, r + 4)

If it is lies on the given plane, then we have

$$2(2r+1)-(-r+3)+(r+4)+3=0 \text{ or } r=-1$$

Substituting this value of r in (i), the required image is

(– 2 + 1, 1 + 3, – 1 + 4) *i.e.,* (– 1, 4, 3). **Ans.**

Example 30:

Find the coordinates of the foot of the perpendicular drawn from the origin to the plane 2x + 3y – 4z + 1 = 0.

Solution:

The equations of the perpendicular from origin to the plane

$$2x+3y-4z+1=0 \quad ...(i)$$

is $\frac{x-0}{2}=\frac{y-0}{3}=\frac{z-0}{-4}$ **(Note)**

Any point on it is (2r, 3r, – 4r). If this point lies on the plane (i), then

$$2(2r)+3(3r)-4(-4r)+1=0 \text{ or } 29r=1 \text{ or } r=1/29$$

∴ The coordinates of the required point are (2r, 3r,–4r),

where r = 1/29

i.e., (2/29, 3/29, – 4/29). **Ans.**

Example 31:

Prove that the distance of the point of intersection of the line $x - 3 = \frac{1}{2}(y-4) = \frac{1}{2}(z-5)$ *and the planes* $x + y + z = 17$ *from the point (3, 4, 5) is 3.*

Solution:

Any point on the given line is

(3 + 1, r, 4 + 2r, 5 + 2r)

If it lies on the plane x + y + z = 17, then we have

(3 + r) + (4 + 2r) + (5 + 2r) = 17 or 5r = 5 or r = 1.

Substituting this value of r in the coordinates of the point above, we have the point of intersection of the given line and plane as (4, 6, 7).

∴ The required distance

= the distance between (4, 6, 7) and (3, 4, 5)

$= \sqrt{[(4-3)^2 + (6-4)^2 + (7-5)^2]} = \sqrt{(1^2 + 2^2 + 2^2)}$

= 3. **Hence proved.**

Example 32:

A variable plane makes intercepts on the co-ordinate axes the sum of whose squares is constant and equal to k^2*. Show that locus of the foot of the perpendicular from the origin to the plane is*

$$(x^{-2} + y^{-2} + z^{-2})(x^2 + y^2 + z^2)^2 = k^2.$$

Solution:

Let the variable plane be (x/a) + (y/b) + (z/c) = 1, ...(i)

where a, b and c are its intercepts on the co-ordinate axes

∴ according to the given condition $a^2 + b^2 + c^2 = k^2$. ...(ii)

Now the equations of the like through origin and perpendicular to the plane (i) [or parallel to the normal to (i) whose direction ratios are 1/a, 1/b and 1/c] are given by $\frac{x-0}{1/a} = \frac{y-0}{(1/b)} = \frac{z-0}{(1/c)} = r$ (say).

Any point on this line is (r/a, r/b, r/c). ...(iii)

If this point lies on the plane (i) then

$$\frac{(r/a)}{a}+\frac{(r/b)}{b}+\frac{(r/c)}{c}=1 \text{ or } r\,(a^{-2}+b^{-2}+c^{-2})=1.$$

Substituting this value of r in (iii) we find that the co-ordinates of the foot of the perpendicular from the origin to the plane (i) are given by

$$x=\frac{1}{a(a^{-2}+b^{-2}+c^{-2})},\ y=\frac{1}{b(a^{-2}+b^{-2}+c^{-2})},\ z=\frac{1}{c(a^{-2}+b^{-2}+c^{-2})}$$

$$\therefore\ x^2+y^2+z^2=\frac{1}{(a^{-2}+b^{-2}+c^{-2})^2}\left[\frac{1}{a^2}+\frac{1}{b^2}+\frac{1}{c^2}\right]$$

$$=\frac{(a^{-2}+b^{-2}+c^{-2})}{(a^{-2}+b^{-2}+c^{-2})^2}=\frac{1}{(a^{-2}+b^{-2}+c^{-2})} \quad \text{...(iv)}$$

And $\frac{1}{x^2}+\frac{1}{y^2}+\frac{1}{z^2}=(a^{-2}+b^{-2}+c^{-2})^2\,[a^2+b^2+c^2]$

$\Rightarrow \quad x^{-2}+y^{-2}+z^{-2}=(a^{-2}+b^{-2}+c^{-2})\,k^2$, from (ii)

$= k^2/(x^2+y^2+z^2)^2$, from (iv)

$\Rightarrow \quad (x^2+y^2+z^2)^2\,(x^{-2}+y^{-2}+z^{-2})=k^2.$ **Hence proved.**

Example 33:

Find the distance of the point (1, 5, 10) from the point of intersection of the line $\frac{x-2}{3}=\frac{y+1}{4}=\frac{z-2}{10}$ *and plane* $x-y+z=16$. *(Kanpur 90)*

Solution:

Any point on the line is $(2+3r, -1+4r, 2+10r)$

If this point lies on the give plane $x-y+z=16$

then $(2+3r)-(-1+4r)+(2+10r)=16 \Rightarrow r=11/9$.

Substituting this value of r in the coordinates of the point of intersection B of the given line and plane is given as

$[2+3\,(11/9), -1+4\,(11/9), 2+10\,(11/9)]$ or $\left(\frac{17}{3},\frac{35}{9},\frac{128}{9}\right)$

$\therefore$ Required distance

= distance between (1, 5, 10) and (17/9, 35/9, 128/9)

$=\sqrt{[\{1-(17/3)\}^2+\{5-(35/9)\}^2+\{10-(128/9)\}^2]}$

$=\sqrt{[3308/81]}=\sqrt{(3308)}/9$. **Ans.**

Example 34:

Find the distance of the point (3, –4, 6) from the plane 2x + 5y– 6z =16 measured along a line with direction cosines proportional to (2, 1, – 2).

Solution:

Do yourself. **Ans.** (22/7) $\sqrt{(14)}$

Example 35:

Find the equations of the straight line through (a, b, c) which are (i) parallel to z-axis and (ii) perpendicular to z-axis.

Solution:

The equations of any line through (a, b, c) are

$$\frac{x-a}{l}=\frac{y-b}{m}=\frac{z-c}{n}, \quad \text{...(A)}$$

where l, m, n are the d.c.'s of the line

(i) If this line is parallel to z-axis, then its d.c.'s are proportional to (0, 0, 1) the d.c.'s of the z-axis.

Hence from (A) above the required equations are

$$\frac{x-a}{0}=\frac{y-b}{0}=\frac{z-c}{1}.$$ **Ans.**

(ii) If line given by (A) is perpendicular to z-axis *i.e.*, parallel to xy-plane, then we have $l.0 + m.0 + n.1 = 0$ or $n = 0$.

From (A) above the required equations are

$$\frac{x-a}{l}=\frac{y-b}{m}=\frac{z-c}{0}.$$ **Ans.**

Example 36:

P is a point on the plane lx + my + nz = p. A point Q is taken on the line OP such that OP.OQ = P^2, prove that the locus of Q is

$$p\,(lx + my + nz) = x^2 + y^2 + z^2.$$

Solution:

Let Q be the point (α, β, γ) and OQ = R. Then the direction ratios of the line OQ are α – 0, β – 0, γ – 0, *i.e.*, α, β, γ.

∴ The direction cosines of OQ are α/R, β/R, γ/R, where

$$R = OQ = \sqrt{(\alpha^2+\beta^2+\gamma^2)} \quad \text{...(i)}$$

∴ The equations of the line OQ are $\frac{x-0}{\alpha/R}=\frac{y-0}{\beta/R}=\frac{z-0}{\gamma/R}=r$ (say),

where r is the distance of any point from (0, 0, 0)

Let OP = r, then the co-ordinates of P are $\left(\frac{\alpha r}{R}, \frac{\beta r}{R}, \frac{\gamma r}{R}\right)$

But it is given that P is a point on the plane $lx + my + nz = p$. Then we have

$$\therefore \quad l\frac{\alpha r}{R}+m\frac{\beta r}{R}+n\frac{\gamma r}{R}=P$$

or $$\frac{r}{R}(l\alpha + m\beta + n\gamma) = p, \qquad \text{...(ii)}$$

It is given that OP.OQ = p^2.

⇒ $r \cdot R = p^2$, ∴ OP = r, OQ = R

⇒ $r = p^2/R$.

∴ from (ii) we get $(p^2/R^2)(l\alpha + m\beta + n\gamma) = p$

⇒ $p(l\alpha + m\beta + n\gamma) = R^2 = \alpha^2 + \beta^2 + \gamma^2$, from (i).

∴ The locus of Q (α, β, γ) is $p(lx + my + nz) = x^2 + y^2 + z^2$.

Hence proved.

Example 37:

Show that if the axes are rectangular, the equations to the perpendicular from the point (α, β, γ) to the plane

$$ax + by + cz + d = 0 \quad \textit{are} \quad \frac{x-\alpha}{a}=\frac{y-\beta}{b}=\frac{z-\gamma}{c}$$

and deduce the perpendicular distance of the point (α, β, γ) from the plane.

Solution:

The perpendicular from (α, β, γ) to the given plane is parallel to the normal to the given plane viz. $ax + by + cz + d = 0$.

∴ The d.c.'s of this perpendicular are proportional to a, b, c.

Hence the equations of this perpendicular line are given by

$$\frac{x-\alpha}{a}=\frac{y-\beta}{b}=\frac{z-\gamma}{c}=r \text{ (say).} \qquad \text{...(i)}$$

Any point on this line given by (i) is $(\alpha + ar, \beta + br, \gamma + cr)$. If this point lies on the plane $ax + by + cz + d = 0$ we have

$$a(\alpha + ar) + b(\beta + br) + c(\gamma + cr) + d = 0$$

$\Rightarrow$ $r(a^2 + b^2 c^2) = -(a\alpha + b\beta + c\gamma + d)$

$\Rightarrow$ $r = -(a\alpha + b\beta + c\gamma + d)/(a^2 + b^2 + c^2)$ ...(ii)

Now the required distance = distance between the points (α, β, γ) and $(\alpha + ar, \beta + br, \gamma + cr)$

$$= \sqrt{\left[(\alpha + ar - \alpha)^2 + (\beta + br - \beta)^2 + (\gamma + cr - \gamma)^2\right]}$$

$$= \sqrt{(a^2r^2 + b^2r + c^2r^2)} = r\sqrt{(a^2 + b^2 + c^2)}$$

$$= \frac{-(a\alpha + b\beta + c\gamma + d)\sqrt{(a^2 + b^2 + c^2)}}{(a^2 + b^2 + c^2)}, \text{ from (ii)}$$

$= (a\alpha + b\beta + c\gamma + d)/\sqrt{(a^2 + b^2 + c^2)}$, numerically.

The foot of the perpendicular from the point (α, β, γ) on the plane $ax + by + cz + d = 0$ is the point $(\alpha + ar, \beta + br, \gamma + cr)$.

Substituting value of r from (ii), we can get the coordinates of the foot of the above perpendicular.

Example 38:

Find the distance of the point (1, –2, 3) from the plane $x - y + z = 5$ measured parallel to the line $\frac{1}{2}x = \frac{1}{3}y = -\frac{1}{6}z$.

Solution:

We have to find the distance of this point from the given plane measured parallel to the given line whose direction ratios are 2, 3, – 6.

Now the equations of the line through (1, – 2, 3) and parallel to the line whose d.c.'s are 2, 3, – 6 are $\frac{x-1}{2} = \frac{y+2}{3} = \frac{z-3}{-6}$ ...(i)

Any point on this line (i) is $(1 + 2r, -2 + 3r, 3 - 6r)$

If it is lies on the given plane $x - y + z = 5$, then we have

$$(1 + 2r) - (-2 + 3r) + (3 - 6r) = 5$$

$\Rightarrow$ $1 - 7r = 0$ or $r = (1/7)$. Substituting this value of r in (i), the point is $[(9/7), (-11/7), (15/7)]$ and therefore the required distance of this point from the given point (1, – 2, 3)

$$= \left[\left(\frac{9}{7} - 1\right)^2 + \left(-\frac{11}{7} + 2\right)^2 + \left(\frac{15}{7} - 3\right)^2\right] = \sqrt{\left[\frac{4}{49} + \frac{9}{49} + \frac{36}{49}\right]} = 1$$ **Ans.**

Example 39:

Find the locus of a point which moves so that its distance from the line $x = y = -z$ is twice its distance from the plane

$$x + y + z = 0.$$

Solution:

Let P (x_1, y_1, z_1) be the point whose locus is required under the given condition.

Its distance from the given plane are

$$= \frac{x_1 + y_1 + z_1}{\sqrt{(1^2 + 1^2 + 1^2)}} = \frac{x_1 + y_1 + z_1}{\sqrt{3}} = p \quad \text{(say)}, \qquad \text{...(i)}$$

Also any point on the given line $\frac{x}{1} = \frac{y}{1} = \frac{z}{-1}$ is A (0, 0, 0).

Let the foot of the perpendicular from P to this line be N and let it be at distance r from A (0, 0, 0), then the coordinates of N are (r, r, –r) and the d.r.'s of PN are $x_1 - r$, $v_1 - r$, $z_1 + r$.

Also the d.r.'s of the given line AB (say) are 1, 1, –1.

Since PN is perpendicular to AB, thus we have

$$1.(x_1 - r) + 1.(y_1 - r) - 1.(z_1 + r) = 0 \text{ or } r = \frac{1}{3}(x_1 + y_1 - z_1)$$

∴ The coordinates of N are (r, r, –r), where $r = \frac{1}{3}(x_1 + y_1 - z_1)$

$$\therefore PN^2 = (x_1 - r)^2 + (y_1 - r)^2 + (z_1 + r)^2$$

$$= x_1^2 + y_1^2 + z_1^2 + 3r^2 - 2r(x_1 + y_1 - z_1)$$

$$= x_1^2 + y_1^2 + z_1^2 + 3.(1/9)(x_1 + y_1 - z_1)^2 - \frac{2}{3}.(x_1 + y_1 - z_1)^2$$

$$= x_1^2 + y_1^2 + z_1^2 - \frac{1}{3}(x_1 + y_1 - z_1)^2$$

$$= (2/3)\,[x_1^2 + y_1^2 + z_1^2 - x_1y_1 + y_1z_1 + z_1x_1] \qquad \text{...(ii)}$$

Also according to the problem PN = 2p or $PN^2 = 4p^2$

$$\Rightarrow \quad (2/3)\,[x_1^2 + y_1^2 + z_1^2 - x_1y_1 + y_1z_1 + z_1x_1] = (4/3)(x_1 + y_1 + z_1)^2.$$

from (i) and (ii)

$$\Rightarrow \quad x_1^2 + y_1^2 + z_1^2 + 5x_1y_1 + 3y_1z_1 + 3z_1x_1 = 0, \text{ on simplifying.}$$

The required locus of P (x_1, y_1, z_1) is

$$x^2 + y^2 + z^2 + 5xy + 3yz + 32x = 0.$$ **Ans.**

Example 40:

Find the equation of the two planes through the origin which are parallel to the line $\frac{x-1}{2} = \frac{y+3}{-1} = \frac{z-1}{-2}$ *and distance* $\frac{5}{3}$ *from it.*

Solution:

The equation of any plane through origin is

$$Ax + By + Cz = 0. \quad ...(i)$$

If this plane is parallel to the given line, whose direction ratios are 2, –1, –2, then the normal to this plane (i) must be perpendicular to the given line *i.e.,* $A.2 + B.(-1) + C(-2) = 0$ or $2A - B - 2C = 0$ ...(ii)

Also the plane (i) is at a distance (5/3) from the given line *i.e.,* at a distance 5/3 from the point (1, – 3, – 1) on this line.

$$\therefore \quad \frac{A(1) + B(-3) + C(-1)}{\sqrt{(A^2 + B^2 + C^2)}} = \frac{5}{3}$$

$$\Rightarrow \quad 9(A - 3B - C)^2 = 25(A^2 + B^2 + C^2)$$

$$\Rightarrow 9(A^2 + 9B^2 + C^2 - 6AB + 6BC - 2AC) = 25(A^2 + B^2 + C^2)$$

$$\Rightarrow \quad 8A^2 - 28B^2 + 8C^2 + 27AB - 27BC + 9CA = 0 \quad ...(iii)$$

From (ii) we have B = 2 (A – C). Substituting this in (iii) we get

$$8A^2 - 28\{4(A - C)^2\} + 8C^2 + 54A(A - C) - 54C(A - C) + 9CA = 0$$

$$\Rightarrow -50A^2 - 50C^2 + 125AC = 0 \quad \text{or} \quad 2A^2 + 2C^2 - 5AC = 0$$

$$\Rightarrow \quad (2A - C)(A - 2C) = 0 \text{ or } A = \frac{1}{2}C, 2C$$

$$\therefore \text{ From (ii) we have } B = 2(A - C) = 2\left(\frac{1}{2}C - C\right), \text{ if } A = \frac{1}{2}C$$

$$\Rightarrow \quad 2(2C - C) \text{ if } A = 2C.$$

i.e., $B = -C, 2C$

Thus we have two cases $A = \frac{1}{2}C$, $B = -C$ and A $A = 2C = B$

∴ From (i) the required equations are

$$1/2\, Cx - Cy + Cz = 0 \text{ and } 2Cx + 2Cy + Cz = 0$$

$$x - 2y + 2z = 0 \text{ and } 2x + 2y + z = 0$$ **Ans.**

Example 41:

From the point P (1, 2, 3), PN is drawn perpendicular to the straight line (x – 2)/3 = (y – 3)/4 = (z – 4)/5. Find the distance PN, the equations to PN coordinates of N.

Solution:

The equations of the given line AB (say) are given as

$$\frac{x-2}{3} = \frac{y-3}{4} = \frac{z-4}{5} \qquad ...(i)$$

where A is (2, 3, 4) say.

Let N, the foot of the perpendicular from P to AB be at a distance r from A. Then from (i) are coordinates of N are (2 + 3r, 3 + 4r, 4 + 5r). ...(ii)

∴ The direction ratios of the line PN are given by

$(2 + 3r) - 1, (3 + 4r) - 2, (4 + 5r) - 3$ *i.e.,* $1 + 3r, 1 + 4r, 1 + 5r$

Also direction ratios of the line AB given by (i) are 3, 4, 5.

As PN is perpendicular to AB, so we have

$$(1 + 3r).3 + (1 + 4r).4 + (1 + 5r).5 = 0$$

or $\quad 50r + 12 = 0$ or $r - 6/25$.

∴ From (ii) the coordinates of N are $\left(\frac{32}{25}, \frac{51}{25}, \frac{70}{25}\right)$ **Ans.**

∴ The distance PN = distance between P and N is given by

$$= \sqrt{\left[\left(\frac{32}{25}-1\right)^2 + \left(\frac{51}{25}-2\right)^2 + \left(\frac{70}{25}-3\right)^2\right]} \qquad \textbf{(Note)}$$

$$= \frac{1}{2}\sqrt{[(7)^2 + (1)^2 + (-5)^2]} = \frac{5\sqrt{3}}{25} = \frac{\sqrt{3}}{5}$$ **Ans.**

And the direction ratios of PN are 1 + 3r, 1 + 4r, 1 + 5r

where we have r = – 6/25

i.e., d ratios of PN are 7/25, 1/25, – 5/25 *i.e.,* 7, 1, – 5.

∴ The equations of the perpendicular PN which passes through (1, 2, 3) and whose d. ratios are 7, 1, –5 are

$$\frac{x-1}{7}=\frac{y-2}{1}=\frac{z-3}{-5}.$$ **Ans.**

Example 42:

Prove that the equations of the perpendicular from the point (1, 6, 3) to the line $x=\frac{y-1}{2}=\frac{z-2}{3}$ *are* $\frac{x-1}{0}=\frac{y-6}{-3}=\frac{z-3}{2}$ *and the coordinates of the foot of the perpendicular are (1, 3, 5).*

Solution:

Here the point P is (1, 6, 3) and the equations of line AB are given by

$\frac{x}{1}=\frac{y-1}{2}=\frac{z-2}{3}$ ∴ The d.r.'s of the line AB are 1, 2, 3.

Let N be the foot of the perpendicular from P to AB be given as

$$(r, 1+2r, 2+3r) \quad \text{...(i)}$$

The d.r.'s of the line PN are given as

$$r-1, (1+2r)-6, (2+3r)-3 \textit{ i.e., } r-1, 2r-5, 3r-1 \quad \text{...(ii)}$$

As PN is perpendicular to AB whose d.r.'s are 1, 2, 3 so we get

$$1.(r-1)+2(2r-5)+3(3r-1)=0 \text{ or } r=1$$

∴ From (i), the coordinates of N are (1, 3, 5). **Hence proved.**

And from (ii), the d.r.'s of the line PN are 0, – 3, 2. Also perpendicular PN passes through P (1, 6, 3). Hence the equations of the perpendicular PN are

$$\frac{x-1}{0}=\frac{y-6}{-3}=\frac{z-3}{2}$$

Example 43:

Find the length and equations of the perpendicular from the origin to the line $x+2y+3z+4=0=2x+3y+4z+5$. *Also find the coordinates of the foot of the perpendicular.*

Solution:

The equations of the given line AB (say) in the symmetric form can be found as

$$\frac{x-2}{1}=\frac{y+3}{-2}=\frac{z-0}{1}. \quad \text{...(i)}$$

A (2, –3, 0) is any point on this line. Also here P is (0, 0, 0) and let N, the foot of the perpendicular from P to the line (i), be at a distance r from A.

Then the coordinates of N are (2 + r, – 3 – 2r, r) ...(ii)

∴ The d.r.'s of the line PN are (2 + r, – 3 – 2r, r) ...(iii)

Also the d.r.'s of the line AB from (i) are 1, – 2, 1.

∴ As PN is perpendicular to AB, so we have

$1.(2 + r) - 2(-3 - 2r) + 1.r = 0$ or $6r + 8 = 0$ or $r = -(4/3)$

∴ From (ii) coordinates of N are $\left(\frac{2}{3}, -\frac{1}{3}, -\frac{4}{3}\right)$. **Ans.**

and from (iii) the d.r.'s of the line PN are

$(2/3), -(1/3), -(4/3)$ or $2, -1, -4,$ ...(iv)

The required length of the perpendicular

$$= PN = \sqrt{\left[(2/3)^2 + (-1/3)^2 + (-4/3)^2\right]} = \frac{1}{3}\sqrt{(21)}$$ **Ans.**

Also the equations of the line PN passing through P (0, 0, 0) and d.r.'s 2, – 1, – 4 [See results (iv) above] are

$$\frac{x-0}{2} = \frac{y-0}{-1} = \frac{z-0}{-4} \text{ or } \frac{x}{2} = \frac{y}{-1} = \frac{z}{-4}.$$ **Ans.**

Example 44:

How for is the point (4, 1, 1,) from the line of intersection of x + y + z – 4 = 0 = x – 2y – z – 4?

Solution:

The equation of the line AB are

$$(x - 4) + y + z = 0 \text{ and } (x - 4) - 2y - z = 0$$

The equation of the line in symmetric form are

$$\frac{x-4}{1} = \frac{y}{2} = \frac{z}{-3}$$ ...(i)

A (4, 0, 0) is any point on this line given by (i)

Here P is (4, 1, 1) and let N, the foot of the perpendicular from P on the line (i), be at a distance r from A.

Then the coordinates of N are (4 + r, 2r, – 3r) ...(ii)

∴ The d.r.'s of the line PN are given by

$(4 + r) - 4, 2r - 1, -3r - 1$

or $r, 2r - 1, -3r - 1$

Also from (i) we find that the d.c.'s the line AB are 1, 2, –3

Since PN is perpendicular to AB, so we have

$1.r + 2(2r - 1) - 3(-3r - 1) = 0$ or $14r + 1 = 0$ or $r = -1/14$

$\therefore$ From (ii) the coordinates of N are $\left(\frac{55}{14}, -\frac{1}{7}, \frac{3}{14}\right)$

$\therefore$ The required distance = PN is given as

$$= \sqrt{\left[\left\{4 - \frac{55}{14}\right\}^2 + \left\{1 + \frac{1}{7}\right\}^2 + \left\{1 - \frac{3}{14}\right\}^2\right]}$$

$$= \sqrt{\left[\left(\frac{1}{14}\right)^2 + \left(\frac{16}{14}\right)^2 + \left(\frac{11}{14}\right)^2\right]} = \frac{1}{14}\sqrt{[1 + 256 + 121]}$$

$= (1/14)\sqrt{(378)} = (3/14)\sqrt{(42)}$. **Ans.**

Example 45:

Show that the distance of the line $ny - mx = x$ and $mx - ny = v$ from the origin is $\sqrt{\left[\left(\lambda^2 + \mu^2 + v^2\right)\right] / \left(l^2 + m^2 + n^2\right)}$.

Solution:

If l_1, m_1, n_1 be the d.c.'s of the common line, then this line being perpendicular the given three planes, we get

$0.l_1 + n.m_1 - m.n_1 = 0$, $-n.l_1 + 0.m_1 + l.n_1 = 0$

and $m.l_1 - n.m_1 + 0.n_1 = 0$

Solving the above three equations in pairs, we obtain

$$\frac{l_1}{nl} = \frac{m_1}{mn} = \frac{n_1}{n^2}$$

or $$\frac{l_1}{l} = \frac{m_1}{m} = \frac{n_1}{n} \quad ...(i)$$

$$\frac{l_1}{nl} = \frac{m_1}{lm} = \frac{n_1}{n^2} \quad ...(ii)$$

and $$\frac{l_1}{-mn} = \frac{m_1}{-m^2} = \frac{n_1}{-nm}$$

or $$\frac{l_1}{n} = \frac{m_1}{m} = \frac{n_1}{n} \quad ...(iii)$$

Also we have $l = n$ as a condition that the plane intersects in a line. Substituting n for l, all the three equations (i), (ii) and (iii) reduce to

$$\frac{l_1}{n} = \frac{m_1}{m} = \frac{n_1}{n}$$

or $$\frac{l_1}{l} = \frac{m_1}{m} = \frac{n_1}{n} \qquad ...(iv)$$

Also putting $x = 0$ in the equations of the given planes

$$ny - mz = \lambda,\ lz = \mu - ny = \nu$$

These give $x = 0,\ y = \dfrac{\nu}{n},\ z = \dfrac{\nu}{l}$

and $$l\nu + m\mu + l\lambda = 0$$

or $$l\lambda + m\mu + n\nu = 0, \quad \therefore l = n \qquad ...(v)$$

which is true.

$\therefore$ Any point on the line common to three given planes is $(0, -n/n, m/l)$

$\therefore$ The equations of this common line are

$$\frac{x-0}{l/\sqrt{(\Sigma l^2)}} = \frac{y+(\nu/n)}{m/(\Sigma l^2)} = \frac{z-(\mu/l)}{n/\sqrt{(\Sigma l^2)}} \qquad ...(v)$$

$\therefore$ Square of the length of the perpendicular from the origin (0, 0, 0) to the line (v)

$$= \begin{vmatrix} m/\left(\sqrt{\Sigma l^2}\right) & n/\left(\sqrt{\Sigma l^2}\right) \\ 0+(\nu/n) & 0-(\mu/l) \end{vmatrix}^2 + \begin{vmatrix} n/\left(\sqrt{\Sigma l^2}\right) & l/\left(\sqrt{\Sigma l^2}\right) \\ 0+(\mu/l) & 0-0 \end{vmatrix}^2$$

$$+ \begin{vmatrix} l/\left(\sqrt{\Sigma l^2}\right) & m/\left(\sqrt{\Sigma l^2}\right) \\ 0-0 & 0+(\nu/n) \end{vmatrix}^2$$

$$= \frac{1}{\Sigma l^2}\left[\begin{vmatrix} m & n \\ \nu/n & -\mu/l \end{vmatrix}^2 + \begin{vmatrix} n & l \\ -\mu/l & 0 \end{vmatrix}^2 + \begin{vmatrix} l & m \\ 0 & \nu/n \end{vmatrix}^2\right]$$

$$= \frac{1}{\Sigma l^2}\left[\left(\frac{-m\mu}{l} - \nu\right)^2 = \mu^2 + \left(\frac{l\nu}{n}\right)^2\right]$$

$$= \frac{1}{\Sigma l^2}\left[\left(\frac{-m\mu - l\nu}{l}\right)^2 + \mu^2 + \left(\frac{l\nu}{n}\right)^2\right],$$

$$= \frac{1}{\Sigma l^2}\left[\left(\frac{m\mu - n\nu}{l}\right)^2 + \mu^2 + \left(\frac{n\nu}{n}\right)^2\right], \quad \therefore l = n$$

$$= \frac{1}{\Sigma l^2}\left[\left(\frac{-l\mu}{l}\right) + \mu^2 + \nu^2\right], \quad \therefore l\lambda + m\mu + n\nu = 0$$

$$= \frac{1}{\Sigma l^2}\left[\lambda^2 + \mu^2 + \nu^2\right] = \frac{\lambda^2 + \mu^2 + \nu^2}{(l^2 + m^2 + n^2)}$$

$\therefore$ Required perpendicular distance

$$= \sqrt{\left[\left(\lambda^2 + \mu^2 + \nu^2\right)/\left(l^2 + m^2 + n^2\right)\right]}.$$ **Hence proved.**

Example 46:

Find the equations of the perpendicular from (1, 3, 7) on the line x = 3 – 5t, y = 2 + 5t, z = – 7 + 2t.

Solution:

The equations of the given line AB (say) are given as

$$x = 3 - 5t,\ y = 2 + 5t,\ z = -7 + 2t.$$

$$\Rightarrow \quad \frac{x-3}{-5} = \frac{y-2}{5} = \frac{z+7}{2} = t \qquad ...(i)$$

Let the given point (1, 3, 7) be P and N be the foot of the perpendicular from P (1, 3, 7) on the line AB given by (i). Let A be (3, 2, – 7), which evidently lies on (i). Let AN = t then from (i) the coordinates of N are given

$$(3 - 5t,\ 2 + 5t,\ -7 + 2t). \qquad (ii)$$

$\therefore$ d ratios of the line PN are

(3 – 5t) – 1, (2 + 5t) – 3, (– 7 + 2t) – 7 *i.e.*, 2 – 5t, – 1 + 5t, – 14+2t

Also d ratios of the line AB given by (i) are – 5, 5, 2.

As PN is perpendicular to AB, so we have

$$(2 - 5t).(-5) + (-1 + 5t)\ .\ 5 + (-14 + 2t)\ .\ 2 = 0$$

$$\Rightarrow \quad 54t - 43 = 0 \text{ or } t = 43/54.$$

$\therefore$ From (ii) the coordinates of N are $\left(-\frac{53}{54}, \frac{323}{54}, -\frac{292}{54}\right)$

Also d. ratios of PN are

2 – 5 (43/54), – 1 + 5 (43/54), – 14 + 2 (43/54)

i.e., $2 - \frac{215}{54}, -1 + \frac{215}{54}, -14 + \frac{86}{54}$ *i.e.*, **– 107, 161, – 670**

$\therefore$ The equations of PN are $\frac{x-1}{-107} = \frac{y-3}{161} = \frac{z-7}{-670}$. **Ans.**

Example 47:

Find the equation of the right circular cylinder of radius 2 whose axis passes through (1, 2, 3) and has direction cosines proportional to (2, 3, 6).

Solution:

The equations of the axis of the cylinder are given by

$$\frac{x-1}{2} = \frac{y-2}{-3} = \frac{z-3}{6}.$$

If P (x, y, z) by any point on the cylinder, then the length of the perpendicular from P to the line (i) is equal to radius 2 of the cylinder.

$$i.e.,\ (2)^2 = \frac{1}{2^2+3^2+6^2}\left[\begin{vmatrix} x-1 & y-2 \\ 2 & -3 \end{vmatrix}^2 + \begin{vmatrix} y-2 & z-3 \\ -3 & 6 \end{vmatrix}^2 + \begin{vmatrix} z-3 & x-1 \\ 6 & 2 \end{vmatrix}^2\right]$$

$\Rightarrow 4 = (1/49)\ [(-3x - 2y + 7)^2 + (6y - 3z - 21)^2 + (2z - 6x)^2]$

$\Rightarrow 196 = (3x + 2y - 7)^2 + 9\ (2y - z - 7)^2 + 4\ (3x - z)^2.$ **Ans.**

Example 48:

Find the locus of a point whose distance from x-axis is twice its distance from the yz-plane.

Solution:

Let P (α, β, γ) be the point whose locus is to be found according to the given problem.

Distance of P from yz-plane = x-coordinate of the point P = α **(Note)**

Again square of distance of P (α, β, γ) from x-axis *i.e.*,

$$\frac{x-0}{1} = \frac{y-0}{0} = \frac{z-0}{0} \text{ is}$$

$$\begin{vmatrix} \alpha-0 & \beta-0 \\ 1 & 0 \end{vmatrix}^2 + \begin{vmatrix} \beta-0 & \gamma-0 \\ 0 & 0 \end{vmatrix}^2 + \begin{vmatrix} \gamma-0 & \alpha-0 \\ 0 & 1 \end{vmatrix}^2$$

i.e., $(-\beta)^2 + (0)^2 + (\gamma)^2$ *i.e.*, $\beta^2 + \gamma^2$

$\therefore$ According to the problem we have

$$\sqrt{(\beta^2+\gamma^2)} = 2\alpha \text{ or } 4\alpha^2 + \beta^2 + \gamma^2$$

$\therefore$ Required locus of P (α, β, γ) is $4x^2 = y^2 + z^2$. **Ans.**

Example 49:

Show that the S.D. between any two opposite edges of the tetrahedron formed by the planes $y + z = 0$, $z + x = 0$, $x + y = 0$, $x + y + z = a$ is $2a/\sqrt{6}$ and the three lines of S.D. intersect at the point $x = y = z = -a$.

Solution:

The equations of the line (or edge) of intersection of the planes

$$x + z = 0$$

and

$$z + x = 0$$

are given as $\dfrac{x}{1} = \dfrac{y}{1} = \dfrac{z}{-1}$. ...(i)

Similarly the equations of the edge of intersection of the planes

$x + y = 0$ and $x + y + z = a$ are given as $\dfrac{x}{1} = \dfrac{y}{-1} = \dfrac{z-a}{0}$. ...(ii)

Let l, m, n be the d.c.'s of the S.D. between the lines (i) and (ii). Then as it is perpendicular to both (i) and (ii), so we have

$$1.l + 1.m - 1.n = 0 \text{ and } 1.l - 1.m + 0.n = 0.$$

Solving above equations we have

$$\frac{l}{-1} = \frac{m}{-1} = \frac{n}{-2} = \frac{\sqrt{(l^2 + m^2 + n^2)}}{\sqrt{\left[(-1)^2 + (-1)^2 + (-2)^2\right]}} = \frac{1}{\sqrt{6}}$$

$$\therefore l = -\frac{1}{\sqrt{6}},\ m = -\frac{1}{\sqrt{6}},\ n = -\frac{1}{\sqrt{6}} \quad ...(iii)$$

From (i) and (ii) it is evident that A (0, 0, 0) is a point on line (i) and B (0, 0, a) is a point on the line (ii). Also required S.D. is the projection of AB on the line whose d.c.'s are given by (iii).

$\therefore$ Required S.D. $= l\,(0 - 0) + m\,(0 - 0) + n\,(0 - a)$,

where l, m, n are given by (iii)

$= -\left(2/\sqrt{6}\right)(-a) = \left(2/\sqrt{6}\right)a$. **Hence proved.**

Now the equation of the plane through (i) and the S.D. is given as

$$\begin{vmatrix} x & y & z \\ 1 & 1 & -1 \\ -1 & -1 & -2 \end{vmatrix} = 0 \text{ or } \begin{vmatrix} x & y & z \\ 1 & 1 & -1 \\ 0 & 0 & -3 \end{vmatrix} = 0, \text{ adding 2nd row to 3rd.}$$

$x - y = 0$ ...(iv)

Similarly the plane through the line (ii) and S.D. is

$$\begin{vmatrix} x & y & z-a \\ 1 & -1 & 0 \\ -1 & -1 & -2 \end{vmatrix} = 0 \text{ or } \begin{vmatrix} x & y+x & z-a \\ 1 & 0 & 0 \\ -1 & -2 & -2 \end{vmatrix} = 0, \text{ adding first column to 2nd}$$

$\Rightarrow (y + x) - (z - a) = 0$ or $x + y + z + a = 0$ ...(v)

$\therefore$ The equations of S.D. of (i) and (ii) are given by (iv) and (v)

i.e., $x - y = 0$ and $x + y - z + a = 0$.

Both these equations are satisfied by $x = y = z = - a$.

Similarly we can show that other S.D.'s between other pairs of opposite edges of the tetrahedron are also satisfied by $x = y = z - a$.

Hence these S.D.'s meet in the point $x = y = z = -a$. **Hence proved.**

Example 50:

Find the equations of the straight line through the origin and cutting each of the lines

$$\frac{x-\alpha_1}{l_1} = \frac{y-\beta_1}{m_1} = \frac{z-\gamma_1}{n_1} \text{ and } \frac{x-\alpha_2}{l_2} = \frac{y-\beta_2}{m_2} = \frac{z-\gamma_2}{n_2}.$$

Solution:

The equation of the plane containing the first line is given by

$$A(x - \alpha_1) + B(y - \beta_1) + C(z - \gamma_1) = 0 \quad ...(i)$$

where we have $Al_1 + Bm_1 + Cm_1 = 0$ ...(ii)

If (i) passes through the origin, then we have

$$A(0 - \alpha_1) + B(0 - \beta_1) + C(0 - \gamma_1) = 0$$

$$\Rightarrow \quad A\alpha_1 + B\beta_1 + C\gamma_1 = 0 \quad ...(iii)$$

Solving (ii) and (iii) we obtain

$$\frac{A}{m_1\gamma_1 - n_1\beta_1} = \frac{B}{n_1\alpha_1 - l_1\gamma_1} = \frac{C}{l_1\beta_1 - m_1\alpha_1} \quad ...(iv)$$

$\therefore$ From (i) the equation of the plane through the origin and the first line is $(m_1\gamma_1 - n_1\beta_1)(x - \alpha_1) + (n_1\alpha_1 - l_1\gamma_1)(y - \beta_1) + (l_1\beta_1 - m_1\alpha_1)(z - \gamma_1) = 0$

$\Rightarrow (m_1\gamma_1 - n_1\beta_1)x + (n_1\alpha_1 - l_1\gamma_1)y + (l_1\beta_1 - m_1\alpha_1)z = 0$ ...(v)

Similarly the equations of the plane through the origin and the second line is $(m_2\gamma_2 - n_2\beta_2)x + (n_2\alpha_2 - l_2\gamma_2)y + (l_2\gamma_2 - m_2\alpha_2)z = 0$ ...(vi)

The equations (v) and (vi) together the required line.

Example 51:

Find the length and equation of the line of S.D. between the two lines

$$\frac{1}{2}(x-1)+\frac{1}{4}(y-3)=z+2 \text{ and}$$

$$3x-y-2z+4=0=2x+y+z+1$$

Solution:

Any plane through the second line is given by

$$(3x-y-2z+4)+\lambda(2x+y+z+1)=0 \quad \text{...(i)}$$

$$\Rightarrow (3+2\lambda)x+(\lambda-1)y+(\lambda-2)z+(4+\lambda)=0$$

If this plane is parallel to the first line then its normal must be at right angles to first line and as such, we have

$$(3+2\lambda)2+(\lambda-1).4 \ (\lambda-2).1=0 \quad \text{or } \lambda=0.$$

∴ From (i) the plane through the second line and parallel to the first line is given by $3x-y-2z+4=0$ ...(ii)

Now the required S.D. between the given lines

= perpendicular distance of a point (1, 3, – 2) on the first line to the plane (ii)

$$=\frac{3.1-1.3-2.(-2)+4}{\sqrt{[3^2+(-1)^2+(-2)^2]}}=\frac{8}{\sqrt{(14)}} \quad \textbf{Ans.}$$

Now the S.D. is the line of intersection of the plane through the given lines and perpendicular to the plane (ii) found above.

The equation of the plane through the first line is given by

$$A(x-1)+B(y-3)+C(z+2)=0 \quad \text{...(iii)}$$

where we have $A.2+B.4+C.1=0$ ...(iv)

Also as the plane (ii) is perpendicular to plane (ii), so we get

$$A.3+B.(-1)+C(2)=0 \quad \text{...(v)}$$

Eliminating A, B and C between (iii), (iv) and (v) we get the equation of the plane through the first line and perpendicular to plane (ii) is given as

$$\begin{vmatrix} x-1 & y-3 & z+2 \\ 2 & 4 & 1 \\ 3 & -1 & -2 \end{vmatrix}=0$$

$$\Rightarrow \quad x-y+2z+6=0 \quad \text{...(vi)}$$

Also as in (i) the equation of any plane through the second line is

$$(3 + 2\lambda)\, x + (\lambda - 1)\, y + (\lambda - 2)\, z + (4 + \lambda) = 0 \qquad \text{...(vii)}$$

If it is perpendicular to the plane (ii), then

$$(3 + 2\lambda).3 + (\lambda - 1)(-1) + (\lambda - 2)(-2) = 0 \text{ or } \lambda = -14/3.$$

∴ From (vii), the equation of the plane through the second line and perpendicular to the plane (ii) is $19x + 17y + 20z + 2 = 0$ ...(viii)

The required equations of S.D. between the given line are (vi) and (viii)

Ans.

Example 52:

Prove that if the straight line $\dfrac{x-\alpha}{l} = \dfrac{y-\beta}{m} = \dfrac{z-\gamma}{n}$ *intersect the curve* $ax^2 + by^2 = 1,\ z = 0$ *then* $a(\alpha n - \gamma l)^2 + b(\beta n - \gamma m)^2 = n^2$.

Solution:

Any point on the given line is $(\alpha + lr,\ \beta + mr,\ \gamma + nr)$.

If this point lies on the curve $ax^2 + by^2 = 1,\ z = 0$, then we have

$$a(\alpha + lr)^2 + b(\beta + mr)^2 = 1 \qquad \text{...(i)}$$

and $\gamma + nr = 0$ or $r = -\gamma/n$ ...(ii)

Substituting the value of r from (ii) in (i) we have the required condition as

$$a\left[\alpha + l\left(-\frac{\gamma}{n}\right)^2\right] + b\left[\beta + m\left(-\frac{\gamma}{n}\right)\right]^2 = 1$$

$\Rightarrow$ $a(\alpha n - l\gamma)^2 + b(\beta n - \gamma m)^2 = n^2$ **Hence proved.**

Example 53:

Show that the shortest distance between an edge of cube and a diagonal which does not meet it is the joining of their mid-points.

Solution:

Here a = b – c and so we can find that S.D. between AE and

$$OB = \frac{a.a}{\sqrt{(a^2 + a^2)}}$$

i.e., S.D. between AE and OB $= a/\sqrt{2}$

Also here A, B and E are the points (a, 0, 0), (0, a, 0) and (0, a, a).

∴ The mid-points of AE and OB are $\left(\frac{1}{2}a, \frac{1}{2}a, \frac{1}{2}a\right)$ and $\left(0, \frac{1}{2}a, 0\right)$

So the length of their join

$$= \sqrt{\left[\left(\frac{1}{2}a-0\right)^2+\left(\frac{1}{2}a-\frac{1}{2}a\right)^2+\left(\frac{1}{2}a-0\right)^2\right]} = \sqrt{\left(\frac{1}{2}a^2\right)} = a/\sqrt{2}.$$

Hence proved

Example 54:

Show that the equation of the plane containing the line $\frac{y}{b}+\frac{z}{c}=1, x=0$ *and parallel to the line* $\frac{x}{a}-\frac{z}{c}=1,\ y=0$ *is* $\frac{x}{a}-\frac{y}{b}-\frac{z}{c}+1=0$ *and if 2d is the S.D. show that* $d^{-2}=a^{-2}+b^{-2}+c^{-2}$.

Solution:

The equation of the plane containing the line is given as

$$\frac{y}{b}+\frac{z}{c}=1,\ x=0 \text{ is } \left(\frac{y}{b}+\frac{z}{c}-1\right)+\lambda x=0$$

$\lambda x + (1/b)\,y + (1/c)\,z - 1 = 0$ ...(i)

If it is parallel to the line $\frac{x}{a}-\frac{z}{c}=1,\ y=0$ *i.e.,* $\frac{x-a}{a}=\frac{y}{0}=\frac{z}{c}$, then the normal to the plane (i) must be perpendicular to the line and so we have

$$\lambda.a + (1/b).0 + (1/c).c = 0 \text{ or } \lambda = -1/a.$$

∴ From (i), the equation of the required plane is given by

$$\left(\frac{y}{b}+\frac{z}{c}-1\right)-\frac{1}{a}x=0 \text{ or } \frac{x}{a}-\frac{y}{b}-\frac{z}{c}+1=0 \quad \text{...(ii)}$$

Hence proved.

Now any point on the line $\frac{x-a}{a}=\frac{y}{0}=\frac{z}{c}$ is (a, 0, 0) therefore

2d = S.D. between the given lines.

⇒2d = perpendicular distance of the point (a, 0, 0) from the plane (ii)

$$= \frac{a.(1/a)-0.(1/b)-0.(1.c)+1}{\sqrt{[(1/a)^2+(-1/b)^2+(-1/c)^2]}} = \frac{2}{\sqrt{[a^{-2}+b^{-2}+c^{-2}]}}$$

⇒ $d^{-2} = a^{-2} + b^{-2} + c^{-2}$. **Hence proved.**

Example 55:

Find the length and equations of the S.D. between

$$3x - 9y + 5z = 0 = x + y - z \quad ...(i)$$

and $$6x + 8y + 3z - 13 = 0 = x + 2y + z - 3 \quad ...(ii)$$

Solution:

The equations of the plane through the given lines are given as

$$(3x - 9y + 5z) + \lambda (x + y - z) = 0$$

and $$(6x + 8y + 3z - 13) + \mu (x + 2y + z - 3) = 0$$

i.e., $$(3 + \lambda) x + (\lambda - 9) y + (5 - \lambda) z = 0 \quad ...(iii)$$

and $$(6 + \mu) x + (8 + 2\mu) y + (3 + \mu) z - (13 + 3\mu) = 0 \quad ...(iv)$$

If the plane (iii) and (iv) are parallel, then we have

$$\frac{3+\lambda}{6+\mu} = \frac{\lambda-9}{8+2\mu} = \frac{5-\lambda}{3+\mu}, \text{ comparing coeff. of x, y, z.}$$

From $\frac{3+\lambda}{6+\mu} = \frac{\lambda-9}{8+2\mu}$ we obtain $(3 + \lambda)(8 + 2\mu) = (\lambda - 9)(6 + \mu)$

$\Rightarrow \quad 24 + 6\mu + 8\lambda + 2\lambda\mu = 6\lambda + \lambda\mu - 54 - 9\mu$

$\Rightarrow \quad \lambda\mu + 2\lambda + 15\mu + 78 = 0 \quad ...(v)$

From $\frac{3+\lambda}{6+\mu} = \frac{5-\lambda}{3+\mu}$ we get $(3 + \lambda)(3 + \mu) = (5 - \lambda)(6 + \lambda)$

$\Rightarrow \quad 9 + 3\lambda + 3\mu + \lambda\mu = 30 - 6\lambda + 5\mu - \lambda\mu$

$\Rightarrow \quad 2\lambda\mu + 9\lambda - 2\mu - 21 = 0 \quad ...(vi)$

From (v) and (vi) on eliminating $\lambda\mu$, we have

$$5\lambda - 32\mu - 177 = 0 \quad \text{or } 5\lambda = 32\mu + 177 \quad ...(vii)$$

From (v) we get $5\lambda\mu + 10\lambda + 75\mu + 390 = 0$

$\Rightarrow \quad \mu(32\mu + 177) + 2(32\mu + 177) + 75\mu + 390 = 0$, from (vii)

$\Rightarrow \quad 32\mu^2 + 316\mu + 744 = 0$

$\Rightarrow \quad 8\mu^2 + 79\mu + 186 = 0$

$\Rightarrow \quad (8\mu + 31)(\mu + 6) = 0$

$\Rightarrow \quad \mu = -6, -31/8.$

$\therefore$ From (vii), when $\mu = -6$, $\lambda = -3$ and when

$$\mu = -\frac{31}{8}, \ \lambda = \frac{53}{5}$$

Putting values $\lambda = \frac{53}{5}, \ \mu = -\frac{31}{8}$ in (iii) and (iv), we get the parallel planes as 17x + 2y – 7z = 0, 17x + 2y – 7z – 11 = 0

[The other pair of values of λ and μ does not give parallel planes].

Any point on the plane 17x + 2y – 7z = 0 is (0, 0, 0)

∴ The required length of S.D.

= length of the perpendicular from (0, 0, 0) to – 17x – 2y + 7z +11=0

$$= \frac{11}{\sqrt{[(-17)^2 + (-2)^2 + (7)^2]}} = \frac{11}{\sqrt{(342)}} = \frac{11}{3\sqrt{(38)}}$$ **Ans.**

Also we know the equation of any plane through (i) is

(3 + λ) x + (λ – 9) y + (5 – λ) z = 0.

If it is perpendicular to the plane 17x + 2y – 7z = 0,

then we have (3 + λ).17 + (λ – 9).2 + (5 – λ) (– 7) = 0 or λ = 1/(13).

∴ The equation of the planes through (i) and perpendicular to the plane

$$17x + 2y - 7z = 0 \text{ is } \left(3+\frac{1}{13}\right)x + \left(\frac{1}{13} - 9\right)y + \left(5 - \frac{1}{13}\right)z = 0$$

or 40x – 116y + 64z = 0 or 10x – 29y + 16z = 0 ...(viii)

And the equation of any plane through the line (ii) is given as

(6 + μ) x + (8 + 2μ) y + (3 + μ) z – (13 + 3μ) = 0

If it is perpendicular to the plane 17x _ 2y – 72 = 0

then we have (6 + μ) 17 + (8 + 2μ) (2) + (3 + μ) (–7) = 0

⇒ 14μ + 97 = 0 or μ = – 97/14.

∴ The equation of the plane through the line (ii) and perpendicular to the plane 17x + 2y – 7z = 0 is given as

[6 – (97/14)] x + [8 – (97/7)] y + [3 – (97/14)] z – [13 – (291/14)] =0

⇒ – (13/14) x – (41/7) y – (55/14) z + (109/14) = 0

⇒ 13x + 82y + 55z – 109 = 0. ...(ix)

The required equations of the S.D. between the given lines are (viii) and (ix). **Ans.**

Example 56:

Prove that the S.D. between the diagonals of rectangular parallelopiped and the edges not meeting it are

$$\frac{bc}{\sqrt{(b^2+c^2)}}, \frac{ca}{\sqrt{(c^2+a^2)}}, \frac{ab}{\sqrt{(a^2+b^2)}},$$

where a, b, c are the lengths of the edges.

Solution:

In the parallelopiped (as shown in the adjoining Fig.) let the sides OA = a, OB = b and OC = c.

Let the edges OA, OB and OC be taken as co-ordinates axes as shown in adjoining fig.

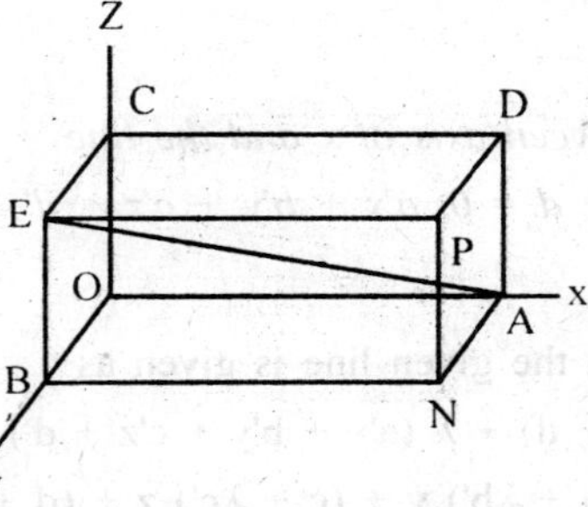

Consider the diagonal AE and the edge OB not intersecting this diagonal.

From the figure it is evident that the co-ordinates of A, B and E are (a, 0, 0), (0, b, 0) and (0, b, c) respectively.

The direction ratios of AE are a – 0, 0 – b, 0 – c.

i.e., a, – b, – c respectively.

$\therefore$ The equations of AE. and OB are $\frac{x-a}{a}=\frac{y-0}{-b}=\frac{z-0}{-c}$ ...(i)

and $\frac{x}{0}=\frac{y}{b}=\frac{z}{0}$. ...(ii)

If l, m, n be the d.c.'s of the S.D. between AE and OB, then as the line of S.D. is perpendicular to both AE and OB, therefore, we have

$$l.a - m.b - n.c = 0 \text{ and } l.0 + m.b + n.b = 0.$$

Solving these we get $\frac{l}{bc}=\frac{m}{0}=\frac{n}{ab}=\frac{\sqrt{(l^2+m^2+n^2)}}{\sqrt{(b^2c^2+a^2b^2)}}$

$$\Rightarrow \quad \frac{l}{bc} = \frac{m}{0} = \frac{b}{ab} = \frac{1}{b\sqrt{(c^2 + a^2)}}$$

$$\Rightarrow \quad l = \frac{c}{\sqrt{(c^2 + a^2)}},\ m = 0,\ n = \frac{a}{\sqrt{(c^2 + a^2)}} \qquad ...(iii)$$

Now the S.D. between AE and OB.

= the projection of the join A (a, 0, 0) and O (0, 0, 0) on the line whose d.c.'s are given by (iii).

= l (a – 0) + m (0 – 0) + n (0 – 0), where l, m, n are given by (iii)

$$= ac / \sqrt{(a^2 + c^2)} \qquad ...(iv)$$

Similarly we can find other S.D.'s.

Example 57:

Find the S.D. between axis of x and the line

$$ax + by + cz + d = 0,\ a'x + b'y + c'z + d' = 0.$$

Solution:

The plane through the given line is given as

$$(ax + by + cz + d) + \lambda\,(a'x + b'y + c'z + d') = 0 \qquad ...(i)$$

$$\Rightarrow (a + \lambda a')\,x + (b + \lambda b')\,y + (c + \lambda c')\,z + (d + \lambda d') = 0$$

If this plane is parallel to x-axis, whose d.c.'s are 1, 0, 0 then the normal to the plane is perpendicular to x-axis and so we have

$$(a + \lambda a').1 + (b + \lambda b').0 + (c + \lambda c').0 = 0 \text{ or } \lambda = -\,a/a'$$

∴ From (i), the equation of the plane through the given line and parallel to x-axis is (ax + by + cz + d) – (a/a') (a'x + b'y + c'z + d') = 0

$$\Rightarrow \quad (ba' - ab')\,y + (ca' - c'a)\,z + (da' + d'a) = 0 \qquad ...(ii)$$

Also any point on x-axis can be taken as origin (0, 0, 0)

∴ Required S.D. = length of perpendicular from (0, 0, 0) to the plane (ii)

$$= \frac{(da' - d'a)}{\sqrt{\left[(ba' - b'a)^2 + (ca' - c'a)^2\right]}} \qquad \textbf{Ans.}$$

Example 58:

Find the S.D. between the z-axis and the line

$$x + y + 2z = 3,\ 2x + 3y + 4z = 4.$$

Solution:

The plane through the given line is

$(x + y + 2z - 3) + \lambda (2x + 3y + 4z - 4) = 0$ ...(i)

$\Rightarrow \quad (1 + 2\lambda) x + (1 + 3\lambda) y + (2 + 4\lambda) z - (3 + 4\lambda) = 0$

If this plane is parallel to z-axis whose d.c.'s are 0, 0, 1, then the normal to this plane must be perpendicular to z-axis and so we have

$(1 + 2\lambda).0 + (1 + 3\lambda).0 + (2 + 4\lambda).1 = 0$ or $2 + 4\lambda = 0$ or $\lambda = -\frac{1}{2}$.

$\therefore$ From (i), the equation of the plane through the given line and parallel to z-axis is $(x + y + 2z - 3) - (1/2)(2x + 3y + 4z - 4) = 0$

$\Rightarrow \quad (2x + 2y + 4z - 6) - (2x + 3y + 4z - 4) = 0$ or $y + 2 = 0$ ...(ii)

Also any point on z-axis is (0, 0, 0).

$\therefore$ Required S.D. = length of perpendicular from (0, 0, 0) to the line (ii)

Example 59:

Find the S.D. between the z-axis and the line

$$ax + by + cz + d = 0 = a'x + b'y + c'z + d'.$$

Show that it meets the z-axis at a point whose distance from the origin is $\dfrac{(bc'-b'c)(db'-d'b)+(ca'-c'a)(ad'-a'd)}{(bc'-b'c)^2+(ca'-c'a)^2}$.

Solution:

The plane through the line is given by

$(ax + by + cz + d) + \lambda (a'x + b'y + c'z + d') = 0$

$\Rightarrow (a + \lambda a') x + (b + \lambda b') y + (c + \lambda c') z + (d + \lambda d') = 0$ ...(i)

If this plane is parallel to z-axis, whose d.c.'s are 0, 0, 1 then the normal to the plane (i) is perpendicular to z-axis and so we get

$(a + \lambda a').0 + (b + \lambda b').0 + (c + \lambda c').1 = 0$ or $\lambda = -c/c'$.

From (i) the equation of the plane through the given line and parallel to z-axis is $(ax + by + cz + d) - (c/c')(a'x + b'y + c'z + d') = 0$

$\Rightarrow \quad (c'a - ca') x + (c'b - b'c) y + (d'c - c'd) = 0$ ...(ii)

Also any point on the z-axis can be taken as origin *i.e.,* (0, 0, 0).

$\therefore$ Required S.D. = length of perpendicular from (0, 0, 0) to the plane (ii)

$$= \frac{(dc'-cd')}{\sqrt{[(c'a-ca')^2+(c'b-b'c)^2]}}$$ **Ans.**

Let the S.D. meet the z-axis and the given line at A and B and let A be $(0, 0, z_1)$ and B (x_2, y_2, z_2). Then the direction ratios of the line AB are $x_2, y_2, z_2 - z_1$ respectively.

Also the direction ratios of the given line are

$$bc' - b'c, ca' - c'a, ab' - a'b. \quad ...(iii)$$

Now AB is perpendicular to z-axis, whose d.c.'s are 0, 0, 1, so we have

$$0.x_2 + 0.y_2 + 1.(z_2 - z_1) = 0 \text{ or } z_1 = z_2 \quad ...(iv)$$

And as AB is perpendicular to the given line, whose d.c.'s are given by (iii) above, so we have

$$x_2 (bc' - b'c) + y_2 (ca' - c'a) + (z_2 - z_1) (ab' - a'b) = 0$$

$$\Rightarrow \quad x_2 (bc' - b'c) + y_2 (ca' - c'a) = 0 \quad ...(v)$$

with the help of (iv).

Also B (x_2, y_2, z_2) being a point on the given line we have

$$ax_2 + by_2 + (cz_2 + d) = 0 \quad ...(vi)$$

and $$a'x_2 + b' y_2 + (c' z_2 + d') = 0 \quad \textbf{(Note)} \quad ...(vii)$$

Eliminating x_2, y_2 between (v), (vi) and (vii) we get

$$\begin{vmatrix} bc'-b'c & ca'-c'a & 0 \\ a & b & cz_2+d \\ a' & b' & c'z_2+d' \end{vmatrix} = 0$$

$$\Rightarrow z_2 \begin{vmatrix} bc'-b'c & ca'-c'a & 0 \\ a & b & c \\ a' & b' & c' \end{vmatrix} + \begin{vmatrix} bc'-b'c & ca'-c'a & 0 \\ a & b & d \\ a' & b' & d' \end{vmatrix} = 0$$

$$\Rightarrow \quad z_2 [(bc' - b'c)^2 + (ca' - c'a)^2] + [(bc' - b'c) (bd' - b'd) + (ca' - c'a) (da' - c'a) (da' - d'a)] = 0,$$

expanding the determinants

$$\Rightarrow \quad z_2 = \frac{[(bc'-b'c)(bd'-b'd) + (ca'-c'a)(da'-d'a)]}{(bc'-b'c)^2 (ca'-c'a)^2} \quad ...(viii)$$

$\therefore$ The required distance $= z_1 = z_2$ (iv) and the value of z_2 is given by (viii) above. **Hence proved.**

Example 60:

Find the S.D. between lines

$$\frac{x-3}{3}=\frac{y-8}{-1}=\frac{z-3}{1} \quad ...(i)$$

and
$$\frac{x+3}{-3}=\frac{y+7}{2}=\frac{z-6}{4} \quad ...(ii)$$

Find also its equations and the points in which it meets the given lines.

Solution:

Any point on the line (i) is (3 – 3r, 8 – r, 3 + r), say point P

And any point on the line (ii) is (– 3 – 3r', – 7 + 2r', 6 + 4r') say point Q

Then the direction ratios of the line PQ are given as

(3 + 3r) – (– 3 – 3r'), (8 – r) – (– 7 + 2r'), (3 + r) – (6 + 4r')

⇒ 3r + 3r' + 6, – r – 2r' + 15, r – 4r' – 3 ...(iii)

Now if PQ is the S.D. between the given lines then PQ is perpendicular to both (i) and (ii), the conditions for the same are

3 (3r + 3r' + 6) –1 (– r – 2r' + 15) + 1. (r – 4r' – 3) = 0

and – 3 (3r + 3r' + 6) + 2 (– r –2r' + 15) + 4 (r – 4r' – 3) = 0

⇒ 11r + 7r' = 0 and 7r + 29r' = 0

Solving these we find r = 0 and r' = 0

Substituting these values and r', find that the coordinates of P and Q are (3, 8, 3) and (– 3, – 7, 6) respectively. **Ans.**

And the d.r.'s of the line PQ from (iii) are 6, 15, – 3 or 2, 5 – 1

Now the required S.D.

$= PQ = \sqrt{[\{3-(-3)\}^2+\{8-(-7)^2\}+(3-6)^2]}$

$= \sqrt{(36+225+9)} = 3\sqrt{(30)}$. **Ans.**

Also PQ is a line through P (3, 8, 3) and of direction ratios 2, 5, –1 so its equations are $\frac{x-3}{2}=\frac{y-8}{5}=\frac{z-3}{1}$. **Ans.**

Example 61:

Prove that a line which passes through (α, β, γ) and intersects the parabola y = 0, z^2 = 4ax lies on the surface

$(\beta z - y\gamma)^2 = 4a\,(\beta - \gamma)\,(\beta x - \alpha y).$

Solution:

Let the equation of the line is given as

$$\frac{x-\alpha}{l}=\frac{y-\beta}{m}=\frac{z-\gamma}{n} \qquad ...(i)$$

$\therefore$ Any point on this line is $(\alpha + lr, \beta + mr, \gamma + nr)$

If this point lies on the parabola $y = 0, z^2 = 4ax$.

Then we have $\beta + mr = 0$ or $r = \beta/m$ ...(ii)

and $(\gamma + nr)^2 + 4a(\alpha + lr)$ ...(iii)

Eliminating r between (ii) and (iii) we get

$$\left(\gamma-\frac{n\beta}{m}\right)^2=4a\left(\alpha-\frac{l\beta}{m}\right)$$

$$\Rightarrow \quad (m\gamma - n\beta)^2 = 4am(\alpha m - l\beta) \qquad ...(iv)$$

Eliminating l, m, n between (i) and (iv) we get the required locus as

$$[(y-\beta)\gamma-(z-\gamma)\beta]^2=4a(y-\beta)[\alpha(y-\beta)-(x-\alpha)\beta]$$

$$\Rightarrow \quad (y\gamma - z\beta)^2 = 4a(y-\beta)(\alpha y - \beta x)$$

$$\Rightarrow \quad (z\beta - \gamma y)^2 = 4a(\beta - y)(\beta x - \alpha y).$$ **Hence proved.**

Example 62:

Two straight lines $\frac{x-\alpha_1}{l_1}=\frac{y-\beta_1}{m_1}=\frac{z-\gamma_1}{n_1}, \frac{x-\alpha_2}{l_2}=\frac{y-\beta_2}{m_2}=\frac{z-\gamma_2}{n_2}$

are cut by a third line whose d.c.'s are λ, μ, ν. *Show that d the length intercepted on the third line is given by*

$$d\begin{vmatrix} l_1 & m_1 & n_1 \\ l_2 & m_2 & n_2 \\ \lambda & \mu & \nu \end{vmatrix}=\begin{vmatrix} \alpha_1-\alpha_2 & \beta_1-\beta_2 & \gamma_1-\gamma_2 \\ l_1 & m_1 & n_1 \\ l_2 & m_2 & n_2 \end{vmatrix}$$

Deduce the length of S.D. between the first two lines.

Solution:

Let the third line with d.c.'s λ, μ, ν meet the first line

$$\frac{x-\alpha_1}{l_1}=\frac{y-\beta_1}{m_1}=\frac{z-\gamma_1}{n_1}, \qquad ...(i)$$

at P $(\alpha_1 + l_1r_1, \beta_1 m_1r_1, \gamma_1 + n_1r_1)$.

Then the equations of third line can be written as

$$\frac{x-(\alpha_1+l_1r_1)}{\lambda}=\frac{y-(\beta_1+m_1r_1)}{\mu}=\frac{z-(\gamma_1+a_1r_1)}{\nu}=d \text{ (say)} \quad ...(ii)$$

$\therefore$ The coordinates of any point Q at a distance d from the point P on the line (ii) are $\alpha_1 + l_1r_1 + d\lambda, \beta_1 + m_1r_1 + d\mu, \gamma_1 + n_1r_1 + d\nu$. ...(iii)

According to the given problem this point Q given by (iii) is a point on the second line $\frac{x-\alpha_2}{l_2}=\frac{y-\beta_2}{m_2}=\frac{z-\gamma_2}{n_2}$ at a distance d from P.

Hence the point (iii) is the same as $(\alpha_2 + l_2r_2, \beta_2 + m_2r_2, \gamma_2 + n_2r_2)$...(iv)

Comparing (iii) and (iv) we get $d\lambda + \alpha_1 + \lambda_1r_1 = \alpha_2 + l_2r_2$ etc.

$\Rightarrow \quad d\lambda + (\alpha_1 - \alpha_2) + l_1r_1 - l_2r_2 = 0;$

Similarly $d\mu + (\beta_1 - \beta_2) + m_1r_1 = m_2r_2 = 0.$

$d\nu + (\gamma_1 - \gamma_2) + n_1r_1 - n_2r_2 = 0.$

Eliminating r_1 and r_2 from these we get

$$\begin{vmatrix} d\lambda+(\alpha_1-\alpha_2) & l_1 & l_2 \\ d\mu+(\beta_1-\beta_2) & m_1 & m_2 \\ d\nu+(\gamma_1-\gamma_2) & n_1 & n_2 \end{vmatrix}=0$$

$$\Rightarrow \quad d\begin{vmatrix} \lambda & l_1 & l_2 \\ \mu & m_1 & m_2 \\ \nu & n_1 & n_2 \end{vmatrix}+\begin{vmatrix} \alpha_1-\alpha_2 & l_1 & l_2 \\ \beta_1-\beta_2 & m_1 & m_2 \\ \gamma_1-\gamma_2 & n_1 & n_2 \end{vmatrix}=0$$

$$\Rightarrow \quad d\begin{vmatrix} l_1 & m_1 & n_1 \\ l_2 & m_2 & n_2 \\ \lambda & \mu & \nu \end{vmatrix}=-\begin{vmatrix} \alpha_1-\alpha_2 & \beta_1-\beta_2 & \gamma_1-\gamma_2 \\ l_1 & m_1 & n_1 \\ l_2 & m_2 & n_2 \end{vmatrix} \quad ...(v)$$

d being the distance between two points, neglecting the negative sign we have the required result.

If d stands for the S.D. between the given lines, then the lines with d.c.'s $\lambda, \mu\ \nu$ is perpendicular to both the given lines and as such we have $\lambda.l_1 + \mu.m_1 + \nu.n_1 = 0$ and $\lambda.l_1 + \mu.m_2 + \nu.n_2 = 0$.

Solving these we get

$$\frac{\lambda}{m_1n_2-m_2n_1}=\frac{\mu}{n_1l_2-n_2l_1}=\frac{\nu}{l_1m_2-l_2m_1}=\frac{1}{\sqrt{[\Sigma\,(m_1n_2-m_2n_1)^2]}} \quad ...(vi)$$

Also from (v) the coefficient of d

$$=\begin{vmatrix} l_1 & m_1 & n_1 \\ l_2 & m_2 & n_2 \\ \lambda & \mu & \nu \end{vmatrix}$$

$$= \lambda (m_1n_2 - m_2n_1) + \mu (n_1l_2 - n_2l_1) + \nu (l_1m_2 - l_2m_1)$$

$$= \frac{\Sigma (m_1n_2 - m_2n_1)}{\sqrt{[\Sigma (m_1m_2 - m_2n_1)^2]}}, \text{ putting the values of } \lambda, \mu, \nu \text{ from (vi)}$$

$$= \sqrt{[\Sigma (m_1m_2 - m_2n_1)^2]}.$$

$\therefore$ From (v), d the S.D. is given by

$$d = \begin{vmatrix} \alpha_1 - \alpha_2 & \beta_1 - \beta_2 & \gamma_1 - \gamma_2 \\ l_1 & m_1 & n_1 \\ l_2 & m_2 & n_2 \end{vmatrix} \div \sqrt{[\Sigma (m_1m_2 - m_2n_1)^2]}$$

Example 63:

If the lines

$$\frac{x-5}{4} = \frac{y-3}{6} = \frac{z-7}{5} \text{ and } \frac{x+1}{2} = \frac{y+4}{5} = \frac{z-9}{10} \text{ are in a plane then}$$

find the equation of plane containing them. If the lines are not in a plane then find the S.D. between them.

Solution:

Any point on the line $\frac{x-5}{4} = \frac{y-3}{6} = \frac{z-7}{5}$ is given by

$(5 + 4r_1, 3 + 6r_1, 7 + 5r_1)$

Similarly any point on the line $\frac{x+1}{2} = \frac{y+4}{5} = \frac{z-9}{10}$ is given as

$(-1 + 2r_2, -4 + 5 + 5r_2, 9 + 10r_2)$ ...(ii)

If the two given lines intersect, then for some values of r_1 and r_2 the two above points (i) and (ii) must coincide

i.e., $5 + 4r_1 = -1 + 2r_2,$

$3 + 6r_1 = -4 + 5r_2$

and $7 + 5r_1 = 9 + 10r_2$

Solving the first two of these equations we obtain $r_1 = -2$, $r_2 = -1$

But these values of r_1 and r_2 do not satisfy the above third equation viz. $7 + 5r_1 = 9 + 10r_2$. Hence the given lines do not lie on a plane *i.e.,* are not coplanar.

S.D. Between the Given Lines

Let l, m, n be the d.c.'s of the S.D. between the lines. Then we have

$$4l + 6m + 5n = 0 \text{ and } 2l + 5m + 10n = 0$$

Solving these equations we have $\dfrac{l}{60-25} = \dfrac{n}{10-40} = \dfrac{n}{20-12}$

$$\Rightarrow \frac{1}{35} = \frac{m}{-30} = \frac{n}{8} = \frac{\sqrt{(l^2 + m^2 + n^2)}}{\sqrt{\left[(35)^2 + (-30)^2 + (8)^2\right]}} = \frac{1}{\sqrt{(2189)}}$$

$\therefore$ The d.c.'s of S.D. are $\dfrac{35}{(\sqrt{2189})}, -\dfrac{30}{\sqrt{(2189)}}, \dfrac{8}{\sqrt{(2189)}}$ ...(iii)

Also A (5, 3, 7) is a point on the line $\dfrac{x-5}{4} = \dfrac{y-3}{6} = \dfrac{z-7}{5}$ and B (– 1, – 4, 9) is a point on the line $\dfrac{x+1}{2} = \dfrac{y+4}{5} = \dfrac{z-9}{10}$

$\therefore$ Required length of S.D. = projection of join of A and B on the line whose d.c.'s are given by (iii)

$$= \frac{35}{\sqrt{(2189)}}[5+1] - \frac{30}{\sqrt{(2189)}}[4+4] + \frac{8}{\sqrt{(2189)}}[7-9]$$

$$= \frac{210-210-16}{\sqrt{(2189)}} = \frac{16}{\sqrt{(2189)}}, \text{ numerically}$$ **Ans.**

Example 64:

Find the equations of the straight lines which bisects the angles between the lines $\dfrac{x}{l_1} = \dfrac{y}{m_1} = \dfrac{z}{n_1}; \dfrac{x}{l_2} = \dfrac{y}{m_2} = \dfrac{z}{n_2}$.

Solution:

We know that the d.c.'s of the lines bisecting the angles between the given lines are proportional to

$l_1 \pm l_2$, $m_1 \pm m_2$, $n_1 \pm n_2$.

Also both the given lines pass through (0, 0, 0) and hence their bisectors also pass through (0, 0, 0).

$\therefore$ The equations of the required bisectors are

$$\frac{x}{l_1 \pm l_2} = \frac{y}{m_1 \pm m_2} = \frac{z}{n_1 \pm n_2}.$$ **Ans.**

Example 65:

Prove that equation of the plane through the line

$$\frac{x-\alpha}{l}=\frac{y-\beta}{m}=\frac{z-\gamma}{n}$$

is $(x-\alpha)\frac{\lambda}{l}+(y+\beta)\frac{\mu}{m}+(z-\gamma)\frac{\nu}{n}=0$, where $\gamma+\mu+\nu=0$

Hence find the containing one given line and parallel to another given straight line.

Solution:

The given plane evidently passes through (α, β, γ), a point on the given line.

Also the given line is perpendicular to the normal to the plane if

$$l\frac{\lambda}{1}+m\frac{\mu}{m}+n\frac{\nu}{n}=0 \text{ i.e., } \lambda+\mu+\nu=0. \quad \text{...(i)}$$

Hence the planes $\Sigma(x-\alpha)\frac{\lambda}{l}=0$ passes through the line

$$\frac{x-\alpha}{l}=\frac{y-\beta}{m}=\frac{z-\gamma}{n}$$

Again if the plane $\Sigma(x-a)\frac{\lambda}{l}=0$ is parallel to another line

$$\frac{x}{l'}=\frac{y}{m'}=\frac{z}{n'} \text{ then } \frac{\lambda}{l}.l'+\frac{\mu}{m}.m'+\frac{\nu}{n}.n'=0$$

$$\Rightarrow \quad \lambda(l'/l)+\mu(m'/m)+\nu(n'/n)=0. \quad \text{...(ii)}$$

Solving (i) and (ii) we get

$$\frac{\lambda}{(n'/n)-(m'/m)}=\frac{\mu}{(l'/l)-(n'/n)}=\frac{\nu}{(m'/m)-(l'/l)}$$

$$\Rightarrow \quad \frac{\lambda\, mn}{mn'-m'n}=\frac{\mu\, nl}{nl'-n'l}=\frac{ml}{n'l-l'm}$$

$$\Rightarrow \quad \frac{(\lambda/l)}{mn'-m'n}=\frac{(\mu/l)}{nl-m'l}=\frac{(\nu/n)}{ln'-l'm}$$

Substituting these values of λ/l, μ/m, ν/n in the equation

$$\Sigma(x-\alpha)\frac{\lambda}{l}=0$$

we have the required equation as $\Sigma(x-\alpha)(mn'-m'n)=0$. **Ans.**

Example 66:

If the plane $ax+hy+gz=0$, $hx+by+fz=0$, $gx+fy+cz=0$ have a common line of intersection, then

$$\Delta \equiv \begin{vmatrix} a & h & g \\ h & b & f \\ g & f & c \end{vmatrix} = 0$$

and the direction ratios of the line satisfy the equations

$$\frac{l^2}{\frac{\partial \Delta}{\partial a}} = \frac{m^2}{\frac{\partial \Delta}{\partial b}} = \frac{n^2}{\frac{\partial \Delta}{\partial c}}.$$

Solution:

Equation of the given planes are $ax + hy + gz = 0$. ...(i)

$hx + by + fz = 0$ ...(ii)

and $gx + fy + cz = 0$ ...(iii)

If l, m, n be the d.c.'s of the common line of intersection, then from (i) and (ii) we have $al + hm + gn = 0$, $hl + bm + fn = 0$.

Solving these equation we have $\frac{l}{hf - bg} = \frac{m}{gh - af} = \frac{n}{ab - h^2}$ ...(iv)

Hence equations of the line of intersection of the planes (i) and (ii) are

$$\frac{x - 0}{hf - bg} = \frac{y - 0}{gh - af} = \frac{z - 0}{ab - h^2} \quad ...(v)$$

If this line lies on the plane (iii), then we have

$$g\,(hf - gb) + f\,(gx - af) + c\,(ab - h^2) = 0$$

$$\Rightarrow \quad abc + 2fgh - af^2 - bg^2 - ch^2 = 0 \quad ...(vi)$$

$$\Rightarrow \quad \Delta \equiv \begin{vmatrix} a & h & g \\ h & b & f \\ g & f & c \end{vmatrix} = 0,$$ **Hence proved.**

Also $\frac{\partial \Delta}{\partial a} = bc - f^2$, $\frac{\partial \Delta}{\partial b} = ca - g^2$, $\frac{\partial \Delta}{\partial c} = ab - h^2$. ...(viii)

Now $(hf - bg)^2 = h^2f^2 + b^2g^2 - 2hfgb$

$= h^2f^2 + b^2g^2 - b\,(af^2 + bg^2 + ch^2 - abc)$, from (vi)

$= h^2f^2 - abf^2 - bch^2 + ab^2c.$

$= f^2\,(h^2 - ab) - bc\,(h^2 - ab)$

$\Rightarrow \quad (hf - bg)^2 = (h^2 - ab)\,(f^2 - bc) = (ab - h^2)\,(bc - f^2)$ ...(viii)

Similarly $\quad (gh - af)^2 = (ab - h^2) = (ca - g^2)$ ...(ix)

$\therefore$ From (iv) with the help of (viii) and (ix) we have

$$\frac{l^2}{(ab-h^2)(bc-f^2)}=\frac{m}{(ab-h^2)(ca-g^2)}=\frac{n^2}{(ab-h^2)^2}$$

$$\Rightarrow \quad \frac{l^2}{(bc-f^2)}=\frac{m^2}{(ca-g^2)}=\frac{n^2}{(ab-h^2)}$$

$$\Rightarrow \quad \frac{l^2}{\left(\frac{\partial \Delta}{\partial a}\right)}=\frac{m^2}{\left(\frac{\partial \Delta}{\partial b}\right)}=\frac{n^2}{\left(\frac{\partial \Delta}{\partial c}\right)}, \text{ from (vii),} \qquad \textbf{Hence proved.}$$

Example 67:

Find the equations to the two planes through the origin which are parallel to the line $\frac{1}{2}(x-1)=-(y+3)=-\frac{1}{2}(z+1)$ *and distant 5/3 from it.*

Solution:

Equation of any plane through the origin is $ax + by + cz = 0$ ...(i)

The d. ratios of its normal are a, b, c

If this plane (i) is parallel to the given line, then the normal to (i) must be perpendicular to the given line whose d.c.'s are 2, – 1, – 2

$$\therefore \quad a.2 + b(-1) + c(-2) = 0 \qquad \text{...(ii)}$$

Also the plane (i) is at a distance 5/3 from the given line. Now any point on the given line is (1, – 3, – 1), so from above we have

$$\frac{a(1)+b(-3)+c(-1)}{\sqrt{(a^2+b^2+c^2)}}=\pm\frac{5}{3} \qquad \textbf{(Note)}$$

$$\Rightarrow \quad 3(a-3b-c)=\pm 5\sqrt{(a^2+b^2+c^2)}$$

$$\Rightarrow \quad 9(a-3b-c)^2=25(a^2+b^2+c^2)$$

$$\Rightarrow \quad 9\left[\frac{1}{2}(b+2c)-3b-c\right]^2=25\left[\frac{1}{4}(b+2c)^2+b^2+c^2\right],$$

$\therefore$ from (ii), $2a = b + 2c$

$$\Rightarrow \quad b^2-bc-2c^2=0 \text{ or } (b+c)(b-2c)=0$$

Either $b = -c$ or $b = 2c$

If $b = -c$, from (ii) we get $2a = c$ or $a=\frac{1}{2}c$

and the corresponding plane from (i) is $x - 2y + 2z = 0$

If $b = 2c$, from (ii) we get $2a = 2c + 2$ or $a = 2c$

and from (i), the corresponding plane is $2x + 2y + z = 0$

Ans. $x - 2y + 2z = 0,\ 2x + 2y + z = 0.$

Example 68:

Prove that the three lines drawn from origin with d.r.'s 2, 1, 5; 2, –1 1; 6, – 4, 1 are coplanar.

Solution:

Equations of these lines in the symmetric form can be written as

$$\frac{x}{2} = \frac{y}{1} = \frac{z}{5};\ \frac{x}{2} = \frac{y}{-1} = \frac{z}{1};\ \frac{x}{6} = \frac{y}{-4} = \frac{z}{1}$$

The equation of the plane passing through first two lines is

$$\begin{vmatrix} x & y & z \\ 2 & 1 & 5 \\ 2 & -1 & 1 \end{vmatrix} = 0$$

$\Rightarrow$ $x(1 + 5) + y(10 - 2) + z(-2 - 2) = 0$ or $3x + 4y - 2z = 0$ ...(i)

And the equation of the plane through last two lines is

$$\begin{vmatrix} x & y & z \\ 2 & -1 & 1 \\ 2 & -4 & 1 \end{vmatrix} = 0, \text{ as above}$$

$\Rightarrow$ $x(-1 + 4) + y(6 - 2) + z(-8 + 6) = 0$ or $3x + 4y - 2z = 0$, which is the same as given by the equation (i).

Hence the given lines are coplanar.

Example 69:

A square ABCD of diagonal 2a is folded along the diagonal AC so that the planes DAC, BAC are at right angles. Find the S.D. between DC and AB.

Solution:

Let O, the centre of the square, be taken as the origin and OA, OB and OD, be taken as x, y and z-axis respectively.

Then the co-ordinates of A, B, C and D are $(a, 0, 0)$, $(0, a, 0)$ $(-a, 0, 0)$ and $(0, 0, a)$ respectively.

Therefore equations of AB are given as $\frac{x-a}{a}=\frac{y-0}{-a}=\frac{z-0}{0}$ ...(i)

Equations of DC are given as $\frac{x-0}{a}=\frac{y-0}{0}=\frac{z-a}{a}$ ...(ii)

Now any point on the line DC is (0, 0, a).

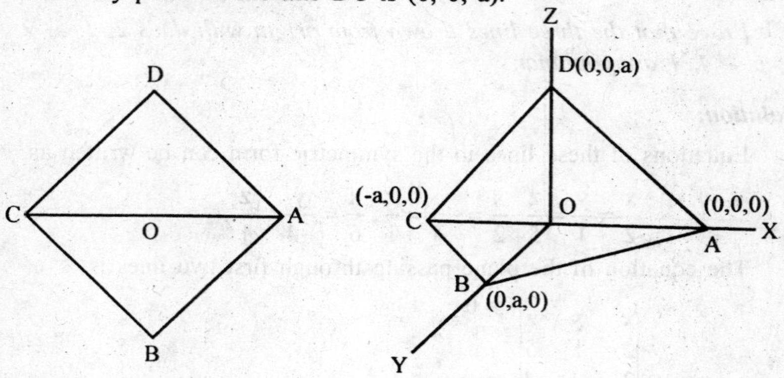

And the equation of the plane through the line AB and parallel to DC *i.e.,* through (i) and parallel to (ii) is given as

$$\begin{vmatrix} x-a & y-0 & z-0 \\ a & -a & 0 \\ a & 0 & a \end{vmatrix} = 0$$

$\Rightarrow$ (x – a) (– a.a) – y (a.a) + z (a.a) = 0 or x + y – z – a = 0 ...(iii)

$\therefore$ Required S.D. = length of perpendicular from (0, 0, a) to the plane (iii)

$$= \frac{0+0-a-a}{\sqrt{[1^2+1^2+(-1)^2]}} = \frac{2a}{\sqrt{3}} \text{ (numerically).}$$ **Ans.**

Example 70:

Show that the distance of the point (x_0, y_0, z_0) from $u \equiv ax + by + cz + d = 0$, $v \equiv a'x + b'y + c'z + d' = 0$ is

$$\left[\frac{(a'u_0 - av_0)^2 + (b'u_0 - bv_0)^2 + (c'u_0 - cv_0)^2}{(bc'-b'c)^2 + (ca'-c'a)^2 + (ab'-a'b)^2}\right]^{1/2}$$

Solution:

The equations of the given lines in the symmetric form can be written as

$$\frac{x-\dfrac{bd'-b'd}{ab'-a'b}}{bc'-b'c}=\frac{y-\dfrac{da'-d'a}{ab'-a'b}}{ca'-c'a}=\frac{z-0}{ab'-ab'} \quad ...(i)$$

∴ If p be the required perpendicular distance, the then we have

$$p^2=\frac{1}{A^2}\left[\begin{vmatrix} x_0-\dfrac{bd'-b'd}{ab'-a'b} & y_0-\dfrac{da'-d'a}{ab'-a'b} \\ bc'-b'c & ca'-c'a \end{vmatrix}^2\right.$$

$$\left.+\begin{vmatrix} y_0-\dfrac{da'-d'a}{ab'-a'b} & z_0-0 \\ ca'-c'a & ab'-a'b \end{vmatrix}^2+\begin{vmatrix} z_0-0 & x_0-\dfrac{bd'-b'd}{ab'-a'b} \\ ab'-a'b & bc'-b'c \end{vmatrix}^2\right] \quad ...(ii)$$

where $A^2 = (bc' - b'c)^2 + (ca' - c'a)^2 + (ab' - a'b)^2$ ...(iii)

Now the first determinant in (ii)

$$=\left(x_0-\frac{bd'-b'd}{ab'-a'b}\right)(ca'-c'a)-\left(y_0-\frac{da'-d'a}{ab'-a'b}\right)(bc'-b'c)$$

$$= x_0 (ca' - c'a) - y_0 (bc' - b'c)$$

$$-\frac{(da'-d'a)(bc'-b'c)-(bd'-b'd)(ca'-c'a)}{ab'-a'b}$$

$$= x_0 (ca' - c'a) - y_0 (bc' - b'c) - \frac{(dc'-d'a)(ab'-a'b)}{ab'-a'b}, \text{ on simplifying}$$

$$= x_0 (ca' - c'a) - y_0 (bc' - b'c) - (dc' - d'c)$$

$$= x_0 (a' x_0 + b' y_0 + c' z_0 + d') - c' (ax_0 + by_0 + cz_0 + d)$$

$$= cv_0 - c' u_0$$

∴ From (ii) we have

Similarly we can evaluate other determinates of (ii).

$$p^2 = \frac{1}{A^2} [(a' u_0 - av_0)^2 + (b' u_0 - bv_0)^2 + (c'u_0 - cv_0)^2],$$

there A^2 is given by (iii). **Hence proved.**

Example 71:

The direction cosines of OA, OB, OC are l_r, m_r, n_r, r = 1, 2 and 3 respectively and OA', OB', OC' bisect the angles BOC, COA, AOB. Show that the planes AOA', BOB', COC' pass through the line.

$$\frac{x}{l_1+l_2+l_3}=\frac{y}{m_1+m+m_3}=\frac{z}{n_1+n_2+n_3}.$$

Solution:

The d.c.'s of the line OA' which bisects the angle BOC are given as

$\frac{1}{2}(l_2 + l_3), \frac{1}{2}(m_2 + m_3), \frac{1}{2}(n_2 + n_3)$ **(Note)**

$\therefore$ The equations of OA' are $\frac{x}{l_2 + l_3} = \frac{y}{m_2 + m_3} = \frac{z}{n_2 + n_3}$ ...(i)

Also equations of OA are $\frac{x}{l_1} + \frac{y}{m_1} = \frac{z}{n_1}$. ...(ii)

$\therefore$ Equation of the plane AOA', which contains the line (i) and (ii) is

$$\begin{vmatrix} x & y & z \\ l_2 & m_1 & n_1 \\ l_2 + l_3 & m_2 + m_3 & n_2 + n_3 \end{vmatrix} = 0$$

$$\Rightarrow \begin{vmatrix} x & y & z \\ l_2 & m_1 & n_1 \\ l_1 + l_2 + l_3 & m_1 + m_2 + m_3 & n_3 + n_2 + n_3 \end{vmatrix} = 0,$$

adding second row to the third.

Similarly the equations to the planes BOB' and COC' are

$$\begin{vmatrix} x & y & z \\ l_2 & m_2 & n_2 \\ \Sigma l_1 & \Sigma m_1 & \Sigma n_1 \end{vmatrix} = 0 \text{ and } \begin{vmatrix} x & y & z \\ l_3 & m_3 & n_3 \\ \Sigma l_1 & \Sigma m_1 & \Sigma n_1 \end{vmatrix} = 0$$

where $\Sigma\, l_1 + l_2 + l_3$ etc.

Now the plane AOA' evidently passes through the line

$$\frac{x}{l_1 + l_2 + l_3} = \frac{y}{m_1 + m_2 + m_3} = \frac{z}{n_1 + n_2 + n_3} \quad \text{...(iv)}$$

as any point on (iv) viz. r $(l_1 + l_2 + l_3)$, r $(m_1 + m_2 + m_3)$, r $(n_1 + n_2 + n_3)$ satisfies (iii), which can be verified by putting these values of x, y, z in (iii) resulting in first and third rows being identical.

Similarly, we can prove that the planes BOB' and COC' also pass though line (iv). **Hence proved.**

Example 72:

Find the S.D. between z-axis and the line x + y + z + 3 = 0 = 3x + y + 2z + 2.

Solution:

Equations of z-axis in symmetric form are given by

$$\frac{x-0}{0}=\frac{y-0}{0}=\frac{z-0}{1} \quad \text{...(i)}$$

Also let l, m, n be the d.c.'s of the line are

$$x + y + z + 3 = 0 = 3x + y + 2z + 2 \quad \text{...(ii)}$$

Then as this line is perpendicular to the normals to both the planes x + y + z + 3 = 0 and 3x + y + 2z + 2 = 0, so we have

$l + m + n = 0$ and $3l + m + 2n = 0$

Solving these equation we have $\frac{l}{2-1}=\frac{m}{3-2}=\frac{n}{1-3}$ or $\frac{l}{1}=\frac{m}{2}=\frac{n}{-2}$ ...(iii)

Putting z = 0 in the given equations (iii) we get

$$x + y + z = 0 \qquad \text{and} \qquad 3x + y + 2 = 0$$

Solving these we get x + 1/2, y = – 7/2

∴ The line (ii) meets the plane z = 0 in (1/2, –7/2, 0) ...(iv)

∴ From (iii) and (iv), the equations of the line (ii) in this symmetric form are

$$\frac{x-(1/2)}{1}=\frac{y+(7/2)}{1}=\frac{z-0}{-2} \quad \text{...(v)}$$

Now we are to find S.D. between the lines (i) and (v)

Let l_1, m_1, n_1 be the d.c.'s of the line of S.D. to the lines (i) and (v), then S.D. being perpendicular to both these lines, we have

$l.0 + m.0- + n.1 = 0$ and $l.1 + m.1 + n.(-2) = 0$

Solving these we get $\frac{l}{1}=\frac{m}{-1}=\frac{n}{0}=\frac{\sqrt{(l^2+m^2+n^2)}}{\sqrt{[l^2+(-1)^2+0^2]}}=\frac{1}{\sqrt{2}}$

∴ d.c.'s of S.D. are $1/\sqrt{2}, -1/\sqrt{2}, 0$.

Also O (0, 0, 0) is a point on the line (i) and B (1/2, – 7/2, 0) is a point on the line (v).

∴ Required S.D. = Projection OB on a line whose d.c.s' are $1/\sqrt{2}, -1/\sqrt{2}, 0$

$$=\frac{1}{\sqrt{2}}\left(\frac{1}{2}-0\right)-\frac{1}{\sqrt{2}}\left(-\frac{7}{2}-0\right)+0\,(-0) \qquad \textbf{(Note)}$$

$$= \frac{1}{\sqrt{2}}\left(\frac{1}{2}+\frac{7}{2}\right) = 2\sqrt{2}\,.$$ **Ans.**

Example 73:

Show that the S.D. between the lines.

$$\frac{x-x_1}{\cos\alpha_1} = \frac{y-y_1}{\cos\beta_1} = \frac{z-z_2}{\cos\gamma_1}, \quad \frac{x-x_2}{\cos\alpha_2} = \frac{y-y_2}{\cos\beta_2} = \frac{z-z_2}{\cos\gamma_2}$$

meets the first line in a point whose distance from (x_1, y_1, z_1) *is* $[\Sigma(x_1 - x_2)(\cos\alpha_1 - \cos\theta\cos\alpha_2)]/\sin^2\theta$, *where* θ *is the angle between the lines.*

Solution:

Let the S.D. meet the first line at

P $[x_1 + r_1 \cos\alpha_1, y_1 + r_1 \cos\beta_1, z + r_1 \cos\gamma_1]$ and the second line at

Q $[x_2 + r_2 \cos\alpha_2, y_2 + r_2 \cos\beta_2, z_2 + r_2 \cos\gamma_2]$.

Then the distance of P from (x_1, y_1, z_1) is evidently r_1 which we are to evaluate.

The direction ratios of the line PQ are given as

$[x_1 + r_1 \cos\alpha_1 - x_2 - r_2 \cos\alpha_2, y_1 + r_1 \cos\beta_1 - y_2 - r_2 \cos\beta_2,$
$z_1 + r_1 \cos\gamma_1 - z_2 - r\text{–}2\text{–} \cos\gamma_2]$

Also the line PQ is perp. to both the given lines and hence we have

$\cos\alpha_1 \{x_1 + r_1 \cos\alpha_1 - x_2 - r_2 \cos\alpha_2)$
$+ \cos\beta_1 \{y_1 + r_1 \cos\beta_1 - y_2 - r_2 \cos\beta_2\}$
$+ \cos\gamma_1 \{z_1 + r_1 \cos\gamma_1 - z_2 - r_2 \cos\gamma_2) = 0$ ...(i)

and $\cos\alpha_2 \{x_1 + r_1 \cos\alpha_1 - x_2 - r_2 \cos\alpha_2\}$
$+ \cos\beta_2 \{y_1 + r_1 \cos\beta_1 - y_2 - r_2 \cos\beta_2\}$
$+ \cos\gamma_2 \{z_1 + r_1\text{–} \cos\gamma_1 - z_2 - r_2 \cos\gamma_2\} = 0$ ...(ii)

From (i) we get

$\{(x_1 - x_2)\cos\alpha_1 + (y_1 - y_2)\cos\beta_1 + (z_1 - z_2)\cos\gamma_1$
$+ r_1 \{\cos^2\alpha_1 \cos\beta_1 + \cos^2\gamma_1\}$
$- r_2 \{\cos\alpha_1 \cos\alpha_2 + \cos\beta_1 \cos\beta_2 + \cos\gamma_1 \cos\gamma_2\} = 0$ **(Note)**

$\Rightarrow \{\Sigma(x_1 - x_2)\cos\alpha_1\} + r_1 \{1\} - r_2 \{\cos\theta\} = 0,$

since $\Sigma\cos^2\alpha_1 = 1$ and $\Sigma\cos\alpha_1 \cos\alpha_2 = \cos\theta$

$\Rightarrow \quad r_1 - r_2 \cos\theta + \{\Sigma(x_1 - x_2)\cos\alpha_1\} = 0$ ...(iii)

Similarly from (ii) we can find that

$$-r_2 + r_1 \cos\theta + \{\Sigma (x_1 - x_2) \cos\alpha_2\} = 0 \qquad ...(iv)$$

Eliminating r_2 from (ii) and (iv) we get

$$r_1 (1 - \cos^2\theta) + \{\Sigma (x_1 - x_2) \cos\alpha_1\} (x_1 - x_2) \cos\alpha_2 \cos\theta = 0$$

$$\Rightarrow r_1 = \frac{\Sigma (x_1 - x_2)(\cos\alpha_1 - \cos\alpha_2 \cos\theta)}{\sin^2\theta}$$

Now r_1 being the distance of P from (x_1, y_1, z_1), neglecting is sign we have the required result.

Example 74:

If the three planes through P and the three given liens $y = 1, z = -1$; $z = 1, x = -1$; $x - 1, y = -1$ all pass through one line, P must lie on the surface $yz + zx + xy + 1 = 0$.

Solution:

Let P be the point (x_1, y_1, z_1).

The equations of any plane through the line $y = 1, z = -1$

i.e., $\quad y - 1 = 0, z + 1 = 0$ is $(y - 1) + \lambda (z + 1) = 0$ ...(i)

$\therefore$ This plane passes through P (x_1, y_1, z_1), is given by

$$\therefore (y_1 - 1) + 1 (z_1 + 1) = 0 \text{ or } \lambda = \frac{-(y_1 - 1)}{z_1 + 1}$$

$\therefore$ From (i), the equations of the plane through the line

$$y = 1, z = -1 \text{ is } (y - 1) - \frac{(y_1 - 1)}{(z_1 + 1)} (z + 1) = 0$$

$$\Rightarrow \quad (y - 1)(z_1 + 1) - (y_1 - 1)(z_1 + 1) = 0 \qquad ...(ii)$$

$$\Rightarrow \quad 0.x + (z_1 + 1) y - (y_1 - 1) z - (y_1 + z_1) = 0 \qquad ...(ii)$$

Similarly the equations of other two planes through P and the other two lines are

$$-(z_1 - 1) x + 0.y + (x_1 + 1) z - (x_1 + z_1) = 0 \qquad ...(iii)$$

and

$$(y_1 + 1) x - (x_1 - 1) y + 0.z - (x_1 + y_1) = 0 \qquad ...(iv)$$

If these three planes given by (ii), (iii) and (iv) pass through one line, then we must have $\Delta_4 = 0$.

i.e.,

$$\begin{vmatrix} 0 & z_1 + 1 & -(y_1 - 1) \\ -(z_1 - 1) & 0 & (x_1 + 1) \\ (y_1 + 1) & -(x_1 - 1) & 0 \end{vmatrix} = 0$$

$\Rightarrow (z_1 - 1)\,[-(x_1 - 1)(y_1 - 1)] + (y_1 + 1)\,[(z_1 + 1)(x_1 + 1)] = 0,$

expanding the determinant.

$\Rightarrow -(z_1 - 1)(x_1y_1 - x_1 - y_1 + 1) + (y_1 + 1)(x_1z_1 + x_1 + z_1 + 1) = 0$

$\therefore$ The locus of P (x_1, y_1, z_1) is $xy + yz + zx + 1 = 0$. **Hence proved.**

Example 75:

Show that the S.D. between the line $x + a = 2y = -12z$ and $x = y + 2a = 6z - 6a$ is 2a.

Solution:

The equation of given line

$$\frac{x+a}{2} = \frac{y}{1} = \frac{z}{-1/6}$$

and $$\frac{x}{1} = \frac{y+2a}{1} = \frac{z-a}{1/6}$$

$$\Rightarrow \quad \frac{x+a}{12} = \frac{y}{6} = \frac{z}{-1} \qquad \text{...(i)}$$

and $$\frac{x}{6} = \frac{y+2a}{6} = \frac{z-a}{1} \qquad \text{...(ii)}$$

Let l, m, n be the d.c.'s of the line S.D. to the given lines, then S.D. being perpendicular to both the lines (i) and (ii) we have

$12l + 6m - n = 0$ and $6l + 6m + n = 0$.

Solving these equations we have

$$\frac{l}{2} = \frac{m}{-3} = \frac{n}{6} = \frac{\sqrt{(l^2 + m^2 + n^2)}}{\sqrt{(2^2 + 3^2 + 6^2)}} = \frac{1}{7}$$

$\therefore$ d.c.'s of S.D. are 2/7, – 3/7, 6/7. ...(iii)

Also A (– a, 0, 0) is point on the line (i) and B (0, – 2a, a) is a point on the line (ii).

$\therefore$ Required S.D. = projection of AB on a line whose d.c.'s are given by (iii)

$$= \frac{2}{7}[0 + a] - \frac{3}{7}[-2a - 0] + \frac{6}{7}[a - 0] = 2a$$ **Hence proved.**

Example 76:

Evaluate 'a' so that the lines $ax - 4y + 7z + 16 = 0 = 4x + 3y - 2z + 3$ and $x - 3y + 4z + 6 = 0 = x + z + 1$ are coplanar.

Solution:

Given lines will be coplanar if

$$\begin{vmatrix} a & -4 & 7 & 16 \\ 4 & 3 & -2 & 3 \\ 1 & -3 & 4 & 6 \\ 1 & -1 & 1 & 1 \end{vmatrix} = 0$$

$$\Rightarrow \begin{vmatrix} a-4 & -4 & 3 & 12 \\ 7 & 3 & 1 & 6 \\ -2 & -3 & 1 & 3 \\ -2 & -3 & 1 & 3 \\ 0 & -1 & 0 & 0 \end{vmatrix} = 0,$$

$$\Rightarrow (-1) \begin{vmatrix} a-4 & 3 & 12 \\ 7 & 1 & 3 \\ -2 & 1 & 3 \end{vmatrix} = 0, \text{ expanding w.r.t. to } R_4$$

$$\Rightarrow \begin{vmatrix} a+2 & 0 & 3 \\ 9 & 0 & 3 \\ -2 & 1 & 3 \end{vmatrix} = 0, \text{ applying } R_1 - 3R_3 \text{ and } R_2 - R_3$$

$$\Rightarrow \begin{vmatrix} a+2 & 2 \\ 9 & 3 \end{vmatrix} = 0, \text{ expanding w.r.t. to } C_2$$

$\Rightarrow$ $3(a + 2) - (9 \times 3)$ or $3a = 27 - 6 = 21$ or $a = 7$. **Ans.**

Example 77:

If the axes are rectangular, find the S.D. between the lines $y = ax + b$, $z = \alpha x + \beta$ *and* $y = a'z + b'$, $z = \alpha' x + \beta'$.

Also deduce the condition for the lines to intersect. (Rohilkhand 90)

Solution:

The equations of the given lines can be written in the symmetric form

as $$\frac{x+(\beta/\alpha)}{(1/\alpha)} = \frac{y-b}{a} = \frac{z}{1} \text{ and } \frac{x+(\beta'/\alpha')}{(1/\alpha)} = \frac{y-b'}{a} = \frac{z}{1}$$

$\therefore$ If l, m, n be the d.c.'s of the required S.D., then we have

$l(1/\alpha) + m.a + n.1 = 0$ and $l(1/\alpha') + ma' + n.1 = 0$

Solving these we get

$$\frac{l}{(a-a')}=\frac{m}{(1/\alpha')-(1/\alpha)}=\frac{n}{(1/\alpha)\,a'-(1/\alpha').\alpha}$$

$$\Rightarrow \frac{l}{\alpha\alpha'\,(a-a')}=\frac{m}{(\alpha-\alpha')}=\frac{n}{(a'\alpha'-a\alpha)} \quad ...(i)$$

Also may A on the first given line is $\left(-\frac{\beta}{\alpha}, b, 0\right)$ and a point on the second given line is B $\left(-\frac{\beta'}{\alpha'}, b', 0\right)$

∴ The required S.D. = the projection of AB on the line whose d.c.'s are

l, m, n given by (i)

$$\Rightarrow \left[l\left\{\left(-\frac{b}{a}+\frac{b'}{a'}\right)\right\}+m\,\{b-b'\}+n\,\{0-0\}\right] \div \sqrt{(l^2+m^2+n^2)}$$

$$\Rightarrow \frac{\alpha\alpha'\,(a-a')\,(\alpha\beta'-\alpha'\beta)}{\alpha\alpha'\sqrt{(\Sigma l^2)}}+\frac{(\alpha-\alpha')\,(b-b')}{\sqrt{(\Sigma l^2)}}$$

$$\Rightarrow \frac{(a-a')\,(\alpha\beta'-\alpha'\beta)+(\alpha-\alpha')\,(b-b')}{\sqrt{[\alpha^2\,\alpha'^2\,(a-a')^2+(\alpha-\alpha')^2+(a'\alpha-a\alpha)^2]}} \quad \textbf{Ans.}$$

If the given lines intersect, then this S.D. = 0 and we have the required condition as $(a - a')\,(\alpha\beta' - \alpha'\beta) + (\alpha - \alpha')\,(b - b') = 0$. **Ans.**

Example 78:

Find the image of the line x – 3 – 6t, y 2t, z = 3 + 2t in the plane 3x + 4y – 5z + 26 = 0.

Solution:

The given line is x – 3 = – 6t, y = 2t, z – 3 = 2t.

$$\Rightarrow \frac{x-3}{-6}=\frac{y-0}{2}=\frac{z-3}{2}=t \quad \textbf{(Note)} \quad ...(i)$$

Any point on this is A (3 – 6t, 2t, 3 + 2t). ...(ii)

If A lies on the plane 3x + 4y – 5z + 26 = 0, then we have

3 (3 – 6t) + 4 (2t) – 5 (3 + 2t) + 26 = 0 or t = – 1

∴ From (ii), the point A is (9, – 2, 1), where A is the point of intersection of the given plane and the given line.

Also from (i) it is evident that any point on the given line is C (3, 0, 3). Let be the foot of the perpendicular from C on the given plane.

Now BC is a line perpendicular to the given plane *i.e.,* it is normal to the given plane and as such the direction ratios of BC are 3, 4, – 5 (the coefficients of x, y, z in the equation of the given plane).

∴ Equations of BC are given as $\frac{x-3}{3} = \frac{y-0}{4} = \frac{z-3}{-5}$.

Any point on this line is (3 + 3r, 4r, 3 – 5r). If this is B *i.e.,* if this point lies on the given plane then we have

$$3(3 + 3r) + 4(4r) - 5(3 - 5r) + 26 = 0 \qquad \text{or } r = -2/5$$

∴ The point B is $\left(3-\frac{6}{5}, -\frac{8}{5}, 3+2\right)$

i.e., $\left(\frac{9}{5}, \frac{-8}{5}, 5\right)$

∴ The direction ratios of the projection AB of the given line are given as

$$9-\frac{9}{5}, -2+\frac{8}{5}, 1-5 \quad i.e., \quad \frac{36}{5}, \frac{-2}{5}, -4$$

i.e., 36, – 2, – 20 *i.e.,* 18, – 1, – 10.

∴ The required equations of the projection AB are

$$\frac{x-9}{18} = \frac{y+2}{-1} = \frac{z-1}{-10}.$$ **Ans.**

Example 79:

A, B, C are (3, 2, 1), (– 2, 0, – 3), (0, 0, –2). Find the locus of P if the volume of tetrahedron PABC is 5.

Solution :

Let P be (x, y, z). Then if V be the volume of the tetrahedron PABC then we have

$$V = \text{"}-(1/6)\begin{vmatrix} x_1 & y_1 & z_1 & 1 \\ x_2 & y_2 & z_2 & 1 \\ x_3 & y_3 & z_3 & 1 \\ x_4 & y_4 & z_4 & 1 \end{vmatrix}\text{"} = -\frac{1}{6}\begin{vmatrix} x & y & z & 1 \\ 3 & 2 & 1 & 1 \\ -2 & 0 & -3 & 1 \\ 0 & 0 & -2 & 1 \end{vmatrix}$$ **(Note)**

$\Rightarrow \quad 5 = -(1/6)\begin{vmatrix} x & y & z+2 & 0 \\ 3 & 2 & 3 & 0 \\ -2 & 0 & -1 & 0 \\ 0 & 0 & -2 & 1 \end{vmatrix}$, subracting 4th row from rest remembering V is given as 5.

$\Rightarrow \quad -30 = \begin{vmatrix} x & y & z+2 \\ 3 & 2 & 3 \\ -2 & 0 & -1 \end{vmatrix}$, expanding with respect to 4th column

$\Rightarrow \quad -30 = \begin{vmatrix} x-2z-4 & y & z+2 \\ -3 & 2 & 3 \\ 0 & 0 & -1 \end{vmatrix}$,

$\Rightarrow \quad -30 = -\begin{vmatrix} x-2z-4 & y \\ -3 & 2 \end{vmatrix}$, expanding with respect to 3rd row

$\Rightarrow \quad -30 = -2(x - 2z - 4) - 3y$ or $2x + 3y - 4z = 38$. **Ans.**

Example 80:

Find the volume of the tetrahedron whose vertices are the points (2, –1, – 3), (4, 1, 3), (3, 2, – 1) and (1, 4, 2).

Solution :

We know that if (x_1, y_1, z_1), (x_2, y_2, z_2), (x_3, y_3, z_3) and (x_4, y_4, z_4) are the vertices of a tetrahedron, then its volume is given by

$$\begin{vmatrix} x_1 & y_1 & z_1 & 1 \\ x_2 & y_2 & z_2 & 1 \\ x_3 & y_3 & z_3 & 1 \\ x_4 & y_4 & z_4 & 1 \end{vmatrix}$$

$$\begin{vmatrix} 2 & -1 & -3 & 1 \\ 4 & 1 & 3 & 1 \\ 3 & 2 & -1 & 1 \\ 1 & 4 & 2 & 1 \end{vmatrix}$$

$$\begin{vmatrix} 6 & 0 & 0 & 2 \\ 4 & 1 & 3 & 1 \\ -1 & 1 & -4 & 0 \\ -3 & 3 & -1 & 0 \end{vmatrix},$$

$$= (1/6)\begin{vmatrix} 0 & 0 & 0 & 2 \\ 1 & 1 & 3 & 1 \\ -1 & 1 & -4 & 0 \\ -3 & 3 & -1 & 0 \end{vmatrix}$$, subtracting 3 times 4th column from lst

$$= (1/6)(-2)\begin{vmatrix} 1 & 1 & 3 \\ -1 & 1 & -4 \\ -3 & 3 & -1 \end{vmatrix}$$, expanding with respect to lst row

$$\begin{vmatrix} 1 & 1 & 3 \\ 0 & 2 & -1 \\ 0 & 6 & 8 \end{vmatrix}$$

$$\begin{vmatrix} 2 & -1 \\ 6 & 8 \end{vmatrix},$$

$$= -\frac{1}{3}[16 + 6] = (22/3), \text{ numerically.}$$ **Ans.**

Example 81:

Show that the volume of the tetrahedron formed by the planes $my + nz = 0$, $nz + lx = 0$, $lx + my = 0$ and $lx + my + nz = p$ is

$$\frac{2}{3}p^3/(lmn).$$

Hence deduce the area of the triangle formed on the plane

$$lx + my + nz = p.$$

Solution :

The planes are given as

$$my + nz = 0, \quad \text{...(i)}$$

$$nz + lx = 0, \quad \text{...(ii)}$$

$$lx + my = 0, \quad \text{....(iii)}$$

and $$lx + my + nz = p \quad \text{...(iv)}$$

The point of intersection of the planes (i), (ii) and (iii) is obviously (0, 0, 0) *i.e.*, the origin.

Solving (ii), (iii) and (iv), we can easily and that

$$x = -p/l,\ y = p/m$$

and $$z = p/n$$

i.e., these planes intersect in the point $(-p/l, p/m, p/n)$,

Similarly the other two vertices of tetrahedron are given as (p/l, – p/m, p/n) and (p/l, p/m, – p/n),

∴ The volume of the tetrahedron

$$= \text{"}(1/6)\begin{vmatrix} x_1 & y_1 & z_1 \\ x_2 & y_2 & z_2 \\ x_3 & y_3 & z_3 \end{vmatrix}\text{"}$$

$$= (1/6)\begin{vmatrix} -p/l & p/m & p/n \\ p/l & -p/m & p/n \\ p/l & p/m & -p/n \end{vmatrix} = \frac{p^3}{6\,lmn}\begin{vmatrix} -1 & 1 & 1 \\ 1 & -1 & 1 \\ 1 & 1 & -1 \end{vmatrix}$$

$$= \frac{p^3}{6\,lmn}\begin{vmatrix} -1 & 0 & 0 \\ 1 & 0 & 2 \\ 1 & 2 & 0 \end{vmatrix}$$, adding 1st column to 2nd and 3rd

$$= \frac{p^3}{6\,lmn}(4) = \frac{2}{3}.\frac{p^3}{lmn}.$$ **Hence proved.**

Let Δ be the area of the triangle formed by the plane (iv) and d be the length of perpendicular from the opposite vertex (0, 0, 0) to the plane (iv).

i.e., d = length of perp. from (0, 0, 0) to $lx + my + nz = p$

$$= \frac{p}{\sqrt{(l^2 + m^2 + n^2)}}$$

Also volume of tetrahedron $= \frac{1}{3}.\Delta.d$

i.e., $$\frac{2p^3}{3\,lmn} = \frac{1}{3}\Delta.\frac{p}{\sqrt{(l^2 + m^2 + n^2)}}$$

$$\Rightarrow \quad \Delta = \frac{2p^2\sqrt{(l^2 + m^2 + n^2)}}{lmn}$$ **Ans.**

Example 82:

Find the locus of the centroid of the tetrahedron of constant volume $64k^3$, formed by the three co-ordinate planes and a variable plane.

Solution :

Let the equation of the variable plane be $lx + my + nz = p$ ...(i)

Also the equations of the co-ordinate planes are x = 0, y = 0, z = 0.

Solving these four equations taking three at a time, we get co-ordinates of the vertices of the tetrahedron as follows

(0, 0, 0) (0, 0, p/n); (p/l, 0, 0) and (0, p/m, 0).

Let the centroid of this tetrahedron be (x_1, y_1, z_1), then we have

$$x_1 = \frac{1}{4}\left(0+0+\frac{p}{l}+0\right) = \frac{p}{4l}$$

Similarly, $y_1 = \frac{p}{4m}$ and $z_1 = \frac{p}{4n}$

$$\Rightarrow \quad \frac{p}{l} = 4x_1, \; \frac{P}{m} = 4y_1 \text{ and } \frac{P}{n} = 4z_1. \quad \text{...(ii)}$$

Also the volume of tetrahedron = $64k^3$. (given)

i.e., $(1/6)\begin{vmatrix} p/l & 0 & 0 \\ 0 & p/m & 0 \\ 0 & 0 & p/n \end{vmatrix} = 64k^3,$

$$\Rightarrow \quad \frac{p^3}{6lmn} = 64k^3 \text{ or } \frac{1}{6}\left(\frac{p}{l}\right)\left(\frac{p}{m}\right)\left(\frac{p}{n}\right) = 64k^3 \quad \text{...(iii)}$$

$\therefore$ From (ii) $(1/6)(4x_1)(4y_1)(4z_1) = 64k^3$ or $x_1y_1z_1 = 6k^3$,

$\therefore$ Required locus of the centroid (x_1, y_1, z_1) is $xyz = 6k^3$. **Ans.**

Example 83:

In the above example find the locus of the foot of the perpendicular from the origin to the variable plane.

Solution :

Let (x_2, y_2, z_2) be the foot of the perpendicular from the origin to the plane (i) of last example.

Then the direction cosines of this perpendicular *i.e.,* the line joining (0, 0, 0) and (x_2, y_2, z_2) are proportional to x_2, y_2, z_2.

But the direction cosines of the perpendicular which is also the normal to the plane (i) are proportional to l, m, n.

Hence we have $\frac{x_2}{l} = \frac{y_2}{m} = \frac{z_2}{n}$. ...(iv)

Also (x_2, y_2, z_2) lies on (i), so we have $lx_2 + my_2 + nz_2 = p$. ...(v)

From (iv), we have $\frac{x_2^2}{lx_2} = \frac{y_2^2}{my_2} = \frac{z_2^2}{nz_2}$ **(Note)**

(multiplying num, and denom. of each fraction by the same quantity)

We have

$$\Rightarrow \quad \frac{x_2^2}{lx_2} = \frac{y_2^2}{my_2} = \frac{z_2^2}{nz_2} = \frac{x_2^2+y_2^2+z_2^2}{lx_2+my_2+nz_2} = \frac{x_2^2+y_2^2+z_2^2}{p}, \text{ from (v)}$$

$$\Rightarrow \quad \frac{x_2}{l} = \frac{x_2^2+y_2^2+z_2^2}{p} \text{ etc.} \qquad \textbf{(Note)}$$

$$\Rightarrow \quad \frac{p}{l} = \frac{x_2^2+y_2^2+z_2^2}{x_2}, \frac{p}{m} = \frac{x_2^2+y_2^2+z_2^2}{y_2}, \frac{p}{n} = \frac{x_2^2+y_2^2+z_2^2}{z_2}$$

Substituting these value in (iii) of last example, we have

$$\frac{1}{6}.\frac{\left(x_2^2+y_2^2+z_2^2\right)^3}{x_2y_2z_2} = 64k^3$$

$$\Rightarrow \quad \left(x_2^2+y_2^2+z_2^2\right)^3 = 384k^3\, x_2y_2z_2$$

$\therefore$ The required locus of (x_2, y_2, z_2) is

$(x^2 + y^2 + z^2)^3 = 384k^3\, xyz.$ **Ans.**

Example 84:

If the volume of the tetrahedron whose vertices are (a, 1, 2), (3, 0, 1) (4, 3, 6), (2, 3, 2) is 6, find the value of 'a'.

Solution :

If V be the volume of the given tetrahedron, then

$$V = (1/6)\begin{vmatrix} a & 1 & 2 & 1 \\ 3 & 0 & 1 & 1 \\ 4 & 3 & 6 & 1 \\ 2 & 2 & 2 & 1 \end{vmatrix}$$

$$\Rightarrow \quad (1/6)\begin{vmatrix} a-2 & -2 & 0 & 0 \\ 1 & -3 & -1 & 0 \\ 2 & 0 & 4 & 0 \\ 2 & 3 & 2 & 1 \end{vmatrix}, \text{ subtracting } R_4 \text{ from } R_1, R_2, R_3$$

$$\Rightarrow \quad = (1/6)\begin{vmatrix} a-2 & -2 & 0 \\ 1 & -3 & -1 \\ 2 & 0 & 4 \end{vmatrix}, \text{ expanding w.r. to } C_4$$

$$\Rightarrow \quad = (1/6)\begin{vmatrix} a-2 & -2 & 0 \\ 1 & -3 & -1 \\ 2 & -12 & 0 \end{vmatrix}, \text{ replacing } R_4 \text{ by } R_4 + 4R_3$$

$\Rightarrow \quad \begin{vmatrix} 9-2 & -2 \\ 6 & -12 \end{vmatrix},$

$\Rightarrow \quad = (1/6)\ [12\ (a - 2) - 6.2] = (1/6)\ [12a - 36] = 2a - 6$

$\Rightarrow \quad V = 2a - 6.$

But $\quad V = 6$ (given)

Hence

$\therefore\ 6 = 2a - 6$ or $2a = 12$ or $a = 6.$ **Ans.**

Example 85:

A, B and C are three fixed points and a variable point P moves so that the volume of the tetrahedron PABC is constant. Find the locus of P.

Solution :

Let A, B and C be the points (a, 0, 0), (0, b, 0) and (0, 0, c) respectively and P be (x_1, y_1, z_1).

Now the volume of tetrahedron PABC = constant (given)

Hence we have $\begin{vmatrix} x_1 & y_1 & z_1 & 1 \\ a & 0 & 0 & 1 \\ 0 & b & 0 & 1 \\ 0 & 0 & c & 1 \end{vmatrix}$

$\Rightarrow \quad \begin{vmatrix} x_1 & y_1 & z_1 - c & 0 \\ a & 0 & -c & 0 \\ 0 & b & -c & 0 \\ 0 & 0 & c & 1 \end{vmatrix} =$ constant, substracting 4th row from the rest.

$\Rightarrow \quad \begin{vmatrix} x_1 & y_1 & z_1 - c \\ a & 0 & -c \\ 0 & b & -c \end{vmatrix} =$ constant

$\Rightarrow \quad x_1\ (bc) - y_1\ (-\ ac) + (z_1 - c)\ (ab) =$ constant

$\Rightarrow \quad bcx_1 + cay_1 + abz_1 =$ constant

$\Rightarrow \quad \frac{x_1}{a} + \frac{y_1}{b} + \frac{z_1}{c} =$ constant

$\therefore$ The locus of P (x_1, y_1, z_1) is $\frac{x}{a} + \frac{y}{b} + \frac{z}{c} =$ constant, which is a plane parallel to the plane ABC whose equation is given as

$$\frac{x}{a}+\frac{y}{b}+\frac{z}{c}=1.$$

Example 86:

A point moves to that three mutually perpendicular lines PA, PB, PC may be drawn cutting the axes OX, OY, OZ at A, B, C and the volume of the tetrahedron OABC is constant and equal to (1/6) k^3. Prove that P lies on the surface $(x^2 + y^2 + z^2)^3 = 8k^3 xyz$.

Solution :

Let the coordinates of P (x_1, y_1, z) and those of A, B, and C on the coordinate axes be (a, 0, 0), (0, b, 0) and (0, 0, c) respectively.

$\therefore$ The direction ratios of the lines PA, PB and PC are

$x_1 - a, y_1, z_1$ $x_1, y_1 - b, z_1$ and $x_1, y_1, z_1 - c$ respectively given as

Since these lines PA, PB and PC are mutually perpendicular so from the formula "$a_1a_2 + b_1b_2 + c_1c_2$" = 0 we have

$$(x_1 - a).x_1 + y_1 . (y_1 - b) + z_1 . z_1 = 0$$

$$x_1 . x_1 + (y_1 - b) . y_1 + z_1 . (z_1 - c) = 0$$

and $$x_1 . (x_1 - a) + y_1 . y_1 + (z_1 - c) . z_1 = 0$$

$\Rightarrow$ $$x_1^2 + y_1^2 + z_1^2 = ax_1 + by_1; \quad ...(i)$$

$\Rightarrow$ $$x_1^2 + y_1^2 + z_1^2 = by_1 + cz_1 \quad ...(ii)$$

and $$x_1^2 + y_1^2 + z_1^2 = cz_1 + ax_1. \quad ...(iii)$$

Also volume of tetrahedron OABC = (1/6) k^3 (given)

$\Rightarrow$ $$(1/6)\begin{vmatrix} a & 0 & 0 \\ 0 & b & 0 \\ 0 & 0 & z \end{vmatrix} = (1/6)\,k^3 \text{ or } abc = k^3 \quad ...(iv)$$

Now to get the locus of P (x_1, y_1, z_1) we are to eliminate a, b, c from (i), (ii) and (iii).

Adding (i), (ii) and (iii) we get $3\left(x_1^2 + y_1^2 + z_1^2\right) = 2\,(ax_1 + by_1 + cz_1)$...(v)

Subtracting twice (i) from (v) we get $x_1^2 + y_1^2 + z_1^2 = 2cz_1$. ...(vi)

Similarly from (ii) and (v) we get $x_1^2 + y_1^2 + z_1^2 = 2ax_1$ ...(vii) and from (iii) and (v) we get $x_1^2 + y_1^2 + z_1^2 = 2by_1$ ...(viii)

Multiplying (vi), (vii) and (viii) we get

$$\left(x_1^2 + y_1^2 + z_1^2\right)^3 = 8abc\ x_1\ y_1\ z_1$$

$$\Rightarrow \quad \left(x_1^2 + y_1^2 + z_1^2\right)^3 = 8k^3\ x_1\ y_1\ z_1, \text{ from (iv).}$$

∴ The required locus of P (x_1, y_1, z_1) is

$$(x^2 + y^2 + z^2)^3 = 8k^3\ xyz.$$ **Hence proved.**

Example 87:

If O, A, B, C, D are any five points and $p_1, p_2, p_3\ p_4$ are the projections of OA, OB, OC, OD on any given line, prove that p_1 vol. OBCD $-p_2$ vol. OCDA + p_3 vol. ODAB – p_4 vol. OABC = 0.

Solution :

Let O be taken as origin and the given line be taken as x-axis or a line parallel to x-axis. Let the co-ordinates of A, B, C and D be (x_1, y_1, z_1), (x_2, y_2, z_2), (x_3, y_3, z_3) and (x_4, y_4, z_4) respectively.

∴ Projection of OA on the given line = x_1 *i.e.*, $p_1 = x_1$.

Similarly, $p_2 = x_2$, $p_3 = x_3$ and $p_4 = x_4$.

∴ p_1 vol. OBCD – p_2 vol. OCDA + p_3 vol. ODAB – p_4 vol. OABC

$$\Rightarrow x_1 \begin{vmatrix} x_2 & y_2 & z_2 \\ x_3 & y_3 & z_3 \\ x_4 & y_4 & z_4 \end{vmatrix} - x_2 \begin{vmatrix} x_3 & y_3 & z_3 \\ x_4 & y_4 & z_4 \\ x_1 & y_1 & z_1 \end{vmatrix} + x_3 \begin{vmatrix} x_4 & y_4 & z_4 \\ x_1 & y_1 & z_1 \\ x_2 & y_2 & z_2 \end{vmatrix} - x_4 \begin{vmatrix} x_1 & y_1 & z_1 \\ x_2 & y_2 & z_2 \\ x_3 & y_3 & z_3 \end{vmatrix}$$

$$\Rightarrow x \begin{vmatrix} x_2 & y_2 & z_2 \\ x_3 & y_3 & z_3 \\ x_4 & y_4 & z_4 \end{vmatrix} - x_2 \begin{vmatrix} x_1 & y_1 & z_1 \\ x_3 & y_3 & z_3 \\ x_4 & y_4 & z_4 \end{vmatrix} + x_3 \begin{vmatrix} x_1 & y_1 & z_1 \\ x_2 & y_2 & z_2 \\ x_4 & y_4 & z_4 \end{vmatrix} - x_4 \begin{vmatrix} x_1 & y_1 & z_1 \\ x_2 & y_2 & z_2 \\ x_3 & y_3 & z_3 \end{vmatrix}$$

passing a row over the rest of the rows in second and third det.

(Note)

$$\Rightarrow \begin{vmatrix} x_1 & x_1 & y_1 & z_1 \\ x_2 & x_2 & y_2 & z_2 \\ x_3 & x_3 & y_3 & z_3 \\ x_4 & x_4 & y_4 & z_4 \end{vmatrix}$$

= 0, as first two columns are identical. **Hence proved.**

Example 88:

A, B, C, D are coplanar and A', B', C', D' are their projections on any plane, prove that

$$vol.\ AB'\ C'D' = -\ vol.\ A'BCD.$$

Solution :

Let the coordinates of A, B, C and D be (x_1, y_1, z_1); (x_2, y_2, z_2), (x_3, y_3, z_3) and (x_4, y_4, z_4) respectively.

Then the coordinates of their projection A', B', C', D' on yz-plane (say) are $(0, y_1, z_1)$, $(0, y_2, z_2)$, $(0, y_3, z_3)$ and (y_4, z_4).

$$\therefore \text{vol. } AB'\ C'\ D' = (1/6)\begin{vmatrix} x_1 & y_1 & z_1 & 1 \\ 0 & y_2 & z_2 & 1 \\ 0 & y_3 & z_3 & 1 \\ 0 & y_4 & z_4 & 1 \end{vmatrix}$$

and $$\text{vol. } A'\ BCD = (1/6)\begin{vmatrix} 0 & y_1 & z_1 & 1 \\ x_2 & y_2 & z_2 & 1 \\ x_3 & y_3 & z_3 & 1 \\ x_4 & y_4 & z_4 & 1 \end{vmatrix}$$

$$\therefore \text{vol. } A\ B'\ C'\ D' + \text{vol. } A'\ BCD$$

$$= (1/6)\left[\begin{vmatrix} x_1 & y_1 & z_1 & 1 \\ 0 & y_2 & z_2 & 1 \\ 0 & y_3 & z_3 & 1 \\ 0 & y_4 & z_4 & 1 \end{vmatrix} + \begin{vmatrix} 0 & y_1 & z_1 & 1 \\ x_2 & y_2 & z_2 & 1 \\ x_3 & y_3 & z_3 & 1 \\ x_4 & y_4 & z_4 & 1 \end{vmatrix}\right]$$

$$= (1/6)\begin{vmatrix} x_1+0 & y_1 & z_1 & 1 \\ 0+x_2 & y_2 & z_2 & 1 \\ 0+x_3 & y_3 & z_3 & 1 \\ 0+x_4 & y_4 & z_4 & 1 \end{vmatrix} \quad \text{...(i)}$$

$$= (1/6)\begin{vmatrix} x_1 & y_1 & z_1 & 1 \\ x_2 & y_2 & z_2 & 1 \\ x_3 & y_3 & z_3 & 1 \\ x_4 & y_4 & z_4 & 1 \end{vmatrix}$$

Again if A, B, C, D are coplanar, then

$$\begin{vmatrix} x_1 & y_1 & z_1 & 1 \\ x_2 & y_2 & z_2 & 1 \\ x_3 & y_3 & z_3 & 1 \\ x_4 & y_4 & z_4 & 1 \end{vmatrix} = 0$$

∴ From (i), we have vol. AB' C' D' + vol. A' BCD = 0

vol. AB' C' D' = – vol. A' BCD. **Hence proved.**

Example 89:

Find the volume of the tetrahedron formed by planes whose equations are $y + z = 0$, $z + x = 0$, $x + y = 0$ and $x + y + z = 1$.

Solution :

The planes are $y + z = 0$ ...(i), $z + x = 0$ ...(ii)

$x + y = 0$ ...(iii) and $x + y + z = 1$ ...(iv)

The point of intersection of the planes (i), (ii) and (iii) is obviously the origin *i.e.,* (0, 0, 0).

Solving (ii), (iii) and (iv) we get y = 1, z = 1 and therefore from (iv) we have x = – 1.

i.e., these planes intersecting in (– 1, 1, 1).

Similarly the other two vertices of the tetrahedron are (1, –1, 1) and (1, 1, –1).

∴ Required volume

$$\Rightarrow \quad (1/6)\begin{vmatrix} x_1 & y_1 & z_1 \\ x_2 & y_2 & z_2 \\ x_3 & y_3 & z_3 \end{vmatrix} = (1/6)\begin{vmatrix} -1 & 1 & 1 \\ 1 & -1 & 1 \\ 1 & -1 & 1 \end{vmatrix}$$

$$\Rightarrow \quad (1/6)\begin{vmatrix} -1 & 0 & 0 \\ 1 & 0 & 2 \\ 1 & 2 & 0 \end{vmatrix}, \text{ adding first column to the rest.}$$

$$\Rightarrow \quad (1/6)\begin{vmatrix} 0 & 2 \\ 2 & 0 \end{vmatrix} = -(1/6)(-4) = 4/6 = 2/3.$$ **Ans.**

Example 90:

Find the volume of the tetrahedron formed by the coordinate planes and the plane $lx + my + nz = p$.

Solution :

The planes are $x = 0$ ...(i) $y = 0$ ...(ii)

$z = 0$ ...(iii) and $lx + my + nz = p$ ...(iv)

The point of intersection of the planes (i), (ii) and (iii) is obviously (0, 0, 0) *i.e.,* the origin.

Solving (ii), (iii) and (iv) we find $x = p/l, y = 0, z = 0$, i.e these planes intersect in the point $(p/l, 0, 0)$.

Similarly the other two vertices of the tetrahedron are $(0, p/m, 0)$ and $(0, 0, p/n)$.

∴ Required volume

$$= (1/6)\begin{vmatrix} x_1 & x_2 & z_1 \\ x_2 & x_2 & z_2 \\ x_3 & x_3 & z_3 \end{vmatrix} = (1/6)\begin{vmatrix} p/l & 0 & 0 \\ 0 & p/m & 0 \\ 0 & 0 & p/m \end{vmatrix} = p^2/(6\,lmn). \quad \textbf{Ans.}$$

Example 91:

If the lengths of two opposite edges of a tetrahedron are a, b, their shortest distance is equal to d and the angle between them is θ, then prove that its volume is (1/6) abd sin θ.

Solution :

Let the vertex A be taken as origin *i.e.,* (0, 0, 0) and let AB = a. Let the direction cosines of AB be taken as l_1, m_1, n_1. Then the co-ordinates of B are (l_1a, m_1a, z_1a), and the equations of the line AB are given as

$$\frac{x}{l_1} = \frac{y}{m_1} = \frac{z}{n_1} \quad \text{...(i)}$$

The edge opposite to AB is CD. Let the vertex C be (α, β, γ) and the direction cosines of the line CD be l_2, m_2, n_2. Given CD = b.

∴ The equations of the line CD are given by

$$\frac{x-\alpha}{l_2} = \frac{y-\beta}{m_2} = \frac{z-\gamma}{n_2} \quad \text{...(ii)}$$

= b for D.

∴ The co-ordinates of D are $(\alpha + l_2b, \beta + m_2b, \gamma + n_2b)$.

Also if θ be the angle between the lines AB and CD whose d.c.'s are l_1, m_1, n_1 and l_2, m_2, n_2 respectively, then

$$\sin\theta = \sqrt{[\Sigma\,(m_1n_2 - m_2n_1)^2]} \quad ...(iii)$$

Now d = the shortest distance between lines (i) and (ii) then we have

$$= \begin{vmatrix} \alpha-0 & \beta-0 & \gamma-0 \\ l_1 & m_1 & n_1 \\ l_2 & m_2 & n_2 \end{vmatrix} \div \sqrt{[\Sigma\,(m_1n_2 - m_2n_1)]}$$

$$= \begin{vmatrix} \alpha & \beta & \gamma \\ l_1 & m_1 & n_1 \\ l_2 & m_2 & n_2 \end{vmatrix} \div \sin\theta, \text{ from (iii)}$$

$$\Rightarrow d\sin\theta = \begin{vmatrix} \alpha & \beta & \gamma \\ l_1 & m_1 & n_1 \\ l_2 & m_2 & n_2 \end{vmatrix} \quad ...(iv)$$

Now the volume of the tetrahedron ABC.

$$= \text{"}(1/6) \begin{vmatrix} x_1 & y_1 & z_1 \\ x_2 & y_2 & z_2 \\ x_3 & y_3 & z_3 \end{vmatrix}\text{"}$$, since one of the vertices viz. A is taken as origin.

$$= (1/6) \begin{vmatrix} \alpha & \beta & \gamma \\ al_1 & am_1 & an_1 \\ \alpha+l_2b & \beta+m_2b & \gamma+n_2b \end{vmatrix}$$, substracting the co-ordinates of B, C, D

(Note)

$$= (1/6)\,a \begin{vmatrix} \alpha & \beta & \gamma \\ l_1 & m_1 & n_1 \\ l_2b & m_2b & n_2b \end{vmatrix}$$, substracting 1st row from 3rd.

$$= (1/6)\,ab \begin{vmatrix} \alpha & \beta & \gamma \\ l_1 & m_1 & n_1 \\ l_2 & m_2 & n_2 \end{vmatrix} = (1/6)\,ab.d\sin\theta, \text{ from (iv)}$$ **Hence proved.**

Example 92:

AB is the S.D. between two given lines and A', B' are variable points on them, such that the volume of the tetrahedron ABA' B' is constant. Prove that the locus of the mid-point of A'B' is a hyperbola whose asymptotes are parallel to lines.

Solution :

Let the given lines be

$$\frac{x}{1}=\frac{y}{m}=\frac{z-c}{0}=r \qquad ...(i)$$

and $$\frac{x}{1}=\frac{y}{-m}=\frac{z+c}{0}=r' \qquad ...(ii)$$

Let AB = 2c, so that A is (0, 0, c) and B (0, 0, – c)

Also any point A' on (i) is (r, mr, c) and B' on (ii) is (r', – mr', – c)

∴ Volume of the tetrahedron ABA 'B' is given by

$$= (1/6)\begin{vmatrix} 0 & 0 & c & 1 \\ 0 & 0 & -c & 1 \\ r & mr & c & 1 \\ r' & -mr' & -c & 1 \end{vmatrix} = (1/6)\begin{vmatrix} 0 & 0 & c & 1 \\ 0 & 0 & -2c & 0 \\ r & mr & 0 & 0 \\ r' & -mr' & -2c & 0 \end{vmatrix},$$

subtracting 1st row from the rest.

$$= -(1/6)\begin{vmatrix} 0 & 0 & -2c \\ r & mr & 0 \\ r' & -mr' & -2c \end{vmatrix},$$ expanding with respect last column.

= (1/6) . 2c. (– 2mrr') = – (2/3) mcrr' = λ (say), a constant (given)

∴ rr' = – (3λ)/(2mc). ...(iii)

Now let P (x_1, y_1, z_1) be the mid-point of A'B', then we have

$$x_1 = \frac{1}{2}(r + r);\ y_1 = \frac{1}{2}(mr - mr);\ z_1 = \frac{1}{2}(c - c)$$

$$\Rightarrow \qquad r + r' = 2x_1,\ r - r' = 2y_1/m;\ z_1 = 0 \qquad ...(iv)$$

Now the required locus of P is obtained by eliminating r and r' from (ii) and (iv) and the procedure for the same is as follows:

We know 4rr' = $(r + r')^2 - (r - r')^2$

$$\therefore\ 4\left[\frac{-3\lambda}{2mc}\right] = (2x_1)^2 - \left(\frac{2y_1}{m}\right)^2,$$ from (iii) and (iv).

Also $$z_1 = 0$$

∴ The locus of P (x_1, y_1, z_1) is $\left(\frac{y}{m}\right)^2 - x^2 = \frac{3\lambda}{2mc},\ z = 0$

$$\Rightarrow \qquad y^2 - m^2x^2 = (3m\lambda)/(2c),\ z = 0.$$

These equations represent to hyperbola on the xy-plane and whose asymptotes are parallel to the lines $y = \pm mx$, $z = 0$ **(Note)**

i.e., parallel to the given lines (i) and (ii) whose equations can be written as $y = \pm mx$, $z = \pm c$.

Example 93:

A line of constant length has its extremities on two fixed straight lines. Prove that the locus of its mid-point is an ellipse whose axes are equally inclined to the lines.

Solution :

If $2l$ be the constant length of the variable line AB then we can have

$(2l)^2 = (AB)^2 = (r - r')^2 + (mr + mr')^2 + (c + c)^2$

$\Rightarrow \quad 4l = (r - r')^2 + m^2 (r + r')^2 + 4c^2 = (2y_1/m)^2 + m^2 (2x_1)^2 + 4c^2$

$\Rightarrow \quad l^2 = \left(y_1^2 / m^2\right) + m^2 x_1^2 + c^2$ Also $z_1 = 0$

$\therefore$ The locus of P (x_1, y_1, z_1) is $m^2x^2 + (y^2/m^2) = l^2 - c^2$, $z = 0$ which represents an ellipse on the xy-plane whose axes lie along the x and y-axes which by definition, are the bisectors of the angles between the given lines.

Example 94:

Prove that the locus of a variable line which intersects the three given lines $y = mx$, $z = c$; $y = -mx$, $z = -c$; $y = z$, $mx = -c$ is the surface $y^2 - m^2x^2 = z^2 - c^2$.

Solution :

The equation of the plane through the line $y = mx$, $z = c$ is

$$(y - mx) + \lambda_1 (z - c) = 0 \qquad ...(i)$$

The equation of the plane through the line $y = -mx$, $z = -c$ is

$$(y + mx) + \lambda_2 (z + c) = 0 \qquad ...(ii)$$

Now any line intersecting the first two given lines is given by plane (i) and (ii). The above two planes intersect in a line and as it meets the third line $y = z$, $mx = -c$, so putting $mx = -c$ and $z = y$ in (i) and (ii) we get

$(y + c) + \lambda_1 (y - c) = 0$ and $(y - c) + \lambda_2 (y + c) = 0$

$$\Rightarrow \quad \left(\frac{y+c}{y-c}\right) = -\lambda_1 \text{ and } \left(\frac{y+c}{y-c}\right) = -\frac{1}{\lambda_2} \qquad \textbf{(Note)}$$

$$\therefore -\lambda_1 = -1/\lambda_2 \text{ or } \lambda_1\lambda_2 = 1 \qquad ...(iii)$$

Eliminating λ_1 and λ_2 between (i), (ii) and (iii) we get

$$\left(\frac{y-mx}{z-c}\right)\left(\frac{y+mx}{z+c}\right)=1 \qquad \textbf{(Note)}$$

$$\Rightarrow \qquad y^2 - m^2x^2 = z^2 - c^2. \qquad \textbf{Hence proved.}$$

Example 95:

Prove that the locus of a line which meets the lines $y = \pm mx$, $z = \pm c$ and the circle $x^2 + y^2 = a^2$, $z = 0$ is

$$c^2m^2 (cy - mxz)^2 + c^2 (yz - cmx)^2 = a^2.m^2 (z^2 - c^2)^2.$$

Solution :

Given lines are $y - mx = 0$, $z - c = 0$; ...(i)

$y + mx = 0$, $z + c = 0$ ...(ii)

and the circle is$x^2 + y^2 = a^2$, $z = 0$ ...(iii)

Any line intersecting (i) and (ii) is

$$(y - mx) + k (z - c) = 0, (y + mx) + k_2 (z + c) = 0 \qquad ...(iv)$$

If it meets the circle (iii), then we are to eliminate k_1, k_2 from (iii) and (iv).

Putting $z = 0$ in (iv) we get $(y - mx) - k_1c = 0$, $(y + mx) + k_2c = 0$

$$\Rightarrow \qquad mx - y + k_1c = 0, mx + y + k_2c = 0$$

Adding and subtracting these, we get

$$x = -\frac{(k_1+k_2)\,c}{2m}, \; y = \frac{c\,(k_1-k_2)}{2}$$

Substituting these values of x and y in (iii), we get

$$\frac{(k_1+k_2)\,c^2}{4m^2} + \frac{(k_1-k_2)^2\;c^2}{4} = a^2$$

$$\Rightarrow \qquad [(k_1 + k_2)^2 + m^2 (k_1 - k_2)^2]\, c^2 = 4a^2m^2. \qquad ...(v)$$

$$\Rightarrow \left[\left\{\left(\frac{mx-y}{z-c}\right)+\left(-\frac{mx+y}{z+c}\right)\right\}^2 + m^2 \left\{\left(\frac{mx-y}{z-c}\right)-\left(-\frac{mx+y}{z+c}\right)\right\}^2\right] c^2$$

$$= 4a^2m^2 \text{ from (iv)}$$

Now simplify and get the result.

Example 96:

Find the locus of the mid-points of the line whose extremities are on two given lines and which are parallel to a given plane.

Solution :

Let the equations of the two lines be taken as

$$\frac{x}{l} = \frac{y}{m} = \frac{z-c}{0} = r \qquad ...(i)$$

and $$\frac{x}{l} = \frac{y}{-m} = \frac{z+c}{0} = r' \qquad ...(ii)$$

The coordinates of any two points on these are given as

A (r, mr, c) and B (r', – mr', – c) ...(iii)

∴ The direction ratios of the variable line AB are

r – r', mr + mr', c + c *i.e.,* r – r', m (r + r'), 2c ...(iv)

Let this line AB be parallel to a given plane

Ax + By + Cz + D = 0 ...(v)

∴ From (iv) and (v) we get

A (r – r') + Bm (r + r') + C.2c = 0 ...(vi)

Let P (x_1, y_1, z_1) be the mid-point of AB, then from (iii) we get

$$x_1 = \frac{1}{2}(r + r');\ y_1 = \frac{1}{2}(mr - mr')'\ z = \frac{1}{2}(c - c) = 0 \qquad ...(vii)$$

Eliminating r and r' from (vi) with the help of (vii) we get

$A\,[2y_1/m] + Bm\,[2x_1] + 2c\,C = 0;\ z_1 = 0$ **(Note)**

∴ The required locus of P (x_1, y_1, z_1) is

$Bm^2x + Ay + mcC = 0,\ z = 0$

which evidently, represents, a line in the plane z = 0 (*i.e.,* xy-plane).

Example 97:

If QQ' is equal to the distance of P from the origin O, then prove that P lies on the surface

$$(1 - m^2)^2 (x^2 + y^2) = (1 + m^2)^2 (4c^2 - z^2).$$

Solution :

We know that

Q and Q' are (r, mr, c) and (r', – mr', – c),

where $r = \dfrac{(x_1 + my_1)}{(1+m^2)}$, $r' = \dfrac{x_1 - my_1}{(1+m^2)}$ and P is (x_1, y_1, z_1).

Now $(QQ')^2 = (r - r')^2 + (mr + mr')^2 + (c + c)^2$

$$= r^2 (1 + m^2) + r'^2 (1 + m^2) - 2 r r' (1 - m^2) + 4c^2$$

$$= \frac{(x_1 + my_1)^2}{(1+m^2)} + \frac{(x_1 - my_1)^2}{(1+m^2)} - \frac{2(x_1^2 - y_1^2 m^2)(1-m^2)}{(1+m^2)^2} + 4c^2$$

$$\Rightarrow (QQ')^2 (1 + m^2)^2 = (1 + m^2) [(x_1 + my_1)^2 + (x_1 - my_1)^2]$$

$$- 2 (1 - m^2) (x_1^2 - m^2 y_1^2) + 4c^2 (1 + m^2)^2$$

$$= 4m^2 x_1^2 + 4m^2 y_1^2 + 4c^2 (1 + m^2)^2 \quad ...(iv)$$

Also $OP^2 = x_1^2 + y_1^2 + z_1^2 = (QQ')^2$ (given)

$\therefore$ From (iv) above we have

$$(x_1^2 + y_1^2 + z_1^2) (1 + m^2)^2 = 4m^2 x_1^2 + 4m^2 y_1^2 + 4c^2 (1 + m^2)^2$$

$$\Rightarrow (x_1^2 + y_1^2) [(1 + m^2)^2 - 4m^2] = (1 + m^2)^2 (4c^2 - z_1^2)$$

$$\Rightarrow (x_1^2 + y_1^2) (1 - m^2)^2 = (1 + m^2)^2 (4c^2 - z_1^2)$$

$\therefore$ The required locus of P is

$$(x^2 + y^2) (1 - m^2)^2 = (1 + m^2)^2 (4c^2 - z^2).$$

Example 98:

Find the locus of a point which moves so that the ratio of its distance from two given lines is constant.

Solution :

Let the variable point be P (x_1, y_1, z_1) and the equation of the lines be

$$\frac{x}{1} = \frac{y}{m} = \frac{z-c}{0} \text{ and } \frac{x}{1} = \frac{y}{-m} = \frac{z+c}{0}.$$

Let p_1 and p_2 be the lengths of the perpendiculars from P to these lines, then we have

$$p_1^2 = \frac{1}{1+m^2} \left[\begin{vmatrix} x_1 & y_1 \\ 1 & m \end{vmatrix}^2 + \begin{vmatrix} y_1 & z_1 - c \\ m & 0 \end{vmatrix}^2 + \begin{vmatrix} z_1 - c & x_1 \\ 0 & 1 \end{vmatrix}^2 \right]$$

$$\Rightarrow p_1^2 = [1/(1 + m^2)^2] [(mx_1 - y_1)^2] + m^2 (z_1 - c)^2 + (z_1 - c)^2] \quad ...(i)$$

$$\text{Similarly } p_2^2 = \frac{1}{1+m^2} [(- mx_1 - y_1)^2 + m^2 (z_1 - c)^2 + (z_1 + c)^2], \quad ...(ii)$$

substituting $- m$ for m and $- c$ for c in (i).

Also p_1/p_2 = constant = λ, say

Then $p_1^2 = \lambda^2 p_2^2$

$$\Rightarrow \quad (mx_1 - y_1)^2 + m^2(z_1 - c)^2 + (z_1 - c)^2 = \lambda[(-mx_1 - y_1)^2 + m^2(z_1 + c)^2 + (z_1 + c)^2]$$

$$\Rightarrow \quad m^2x_1^2(1-\lambda^2) + y_1^2(1-\lambda^2) - 2m(1+\lambda^2)x_1y_1 + (1-\lambda^2)(1+m^2)z_1^2 - 2c(m^2+l)(1+\lambda^2)z_1 + (1-\lambda^2)c^2(1+m^2) = 0$$

$\therefore$ The required locus of P (x_1, y_1, z_1) is

$$m^2x^2 + y^2 + (1+m^2)z^2 + c^2(1+m^2) - 2m\frac{(1+\lambda^2)}{(1-\lambda^2)}xy - \frac{2c(m^2+1)(1+\lambda^2)z}{(1-\lambda^2)} = 0$$

Example 99:

A and B are variable points on two given non-intersecting lines CD and EF. Find the locus of a point P such that PA, PB are perpendicular to one another and perpendicular to CD and EF respectively.

Solution :

If we take the equations of CD and EF as

$$\frac{x}{1} = \frac{y}{m} = \frac{z-c}{0} \text{ and } \frac{x}{1} = \frac{y}{-m} = \frac{z+c}{0}$$

Then the coordinates of A and B are (r, mr c) and (r', – mr', – c) respectively. Let P be (x_1, y_1, z_1).

Then the direction ratios of PA and PB are given as

$x_1 - r$, $y_1 - mr$, $z_1 - c$ and $x_1 - r'$, $y_1 + mr'$, $z_1 + c$ respectively.

If PA is perpendicular to PB, then we have

$$(x_1 - r)(x_1 - r') + (y_1 + mr)(y_1 + mr')(z_1 - c)(z_1 + c) = 0 \quad ...(i)$$

If PA is perpendicular to CD and PB is perpendicular to EF

then we have $(x_1 - r) \cdot 1 + (y_1 - mr) \cdot m + (z_1 - c)\, 0 = 0$ ...(ii)

and $(x_1 - r') \cdot 1 + (y_1 + mr') \cdot (-m) + (z_1 + c) \cdot 0 = 0$ ...(iii)

The required locus of P can be obtained by eliminating r and r' from (i), (ii) and (iii).

From (ii) we get $r = (x_1 + my_1)/(1 + m^2)$

From (iii) we get $r' = (x_1 - my_1)/(1 + m^2)$

$$\therefore\ x_1 - r = x_1 - \frac{(x_1 + my_1)}{(1+m^2)} = \frac{m\,(mx_1 - y_1)}{(1+m^2)};$$

$$x_1 - r' = x_1 - \frac{(x_1 + my_1)}{(1+m^2)} = \frac{m\,(mx_1 - y_1)}{(1+m^2)}$$

$$y_1 - mr = y_1 - \frac{m\,(x_1 + my_1)}{1+m^2} = \frac{y_1 - mx_1}{1+m^2};$$

$$\text{and } y + mr' = y_1 + \frac{m\,(x_1 + my_1)}{1+m^2} = \frac{y_1 + mx_1}{1+m^2}$$

Substituting these values in (i), we get

$$\frac{m^2\left(m^2 x_1^2 - y_1^2\right)}{\left(1+m^2\right)^2} + \frac{y_1^2 - m^2 x_1^2}{\left(1+m^2\right)^2} + \left(z_1^2 - c^2\right) = 0$$

$$\Rightarrow \quad (1 - m^2)\left(y_1^2 - m^2 x_1^2\right) + \left(z_1^2 - c^2\right)(1 + m^2)^2 = 0$$

$\therefore$ The locus of P (x_1, y_1, z_1) is

$$(1 - m^2)(y^2 - m^2x^2) + (z^2 - c^2)(1 + m^2)^2 = 0.$$ **Ans.**

Example 100:

A, B are variable points on two given non-intersecting lines and AB is of constant length 2k. Find the surface generated by AB.

Solution :

Let the given lines be

$$\frac{x}{1} = \frac{y}{m} = \frac{z-c}{0} = r;$$

$$\frac{x}{1} = \frac{y}{-m} = \frac{z+c}{0} = r'$$

Then the coordinates of A and B are (r, mr, c) and $(r', -mr', -c)$ respectively. Also AB = 2k (given) as

$$\therefore (2k)^2 = (AB)^2 = (r - r')^2 + (mr + mr')^2 + (c + c)^2$$

$$\Rightarrow \quad (r - r')^2 + m^2 (r + r')^2 + 4c^2 = 4k^2 \qquad \text{...(i)}$$

Also the equations of the line AB are

$$\frac{x-r}{r-r'} = \frac{x-mr}{m\,(r+r')} = \frac{z-c}{2c} \qquad \text{...(iii) (Note)}$$

Now the required surface generated by AB is obtained by eliminating r and r' from (i) and (ii).

From (ii) we have

$$\frac{z-c}{2c}=\frac{(y-mr)+m(x-r)}{m(r+r')+m(r-r')}=\frac{(y-mr)-m(x-r)}{m(r+r')-m(r-r')} \quad \textbf{(Note)}$$

$$\Rightarrow \quad \frac{z-c}{2c}=\frac{y+mx-2mr}{2mr}=\frac{y-mx}{2mr'},$$

hence we have $\frac{z-c}{2c}=\frac{y+mx}{2mr}-1, \frac{z-c}{2c}=\frac{y-mx}{2mr'}$

$$\Rightarrow \quad \frac{z+c}{c}=\frac{y+mx}{mr}, \frac{z-c}{c}=\frac{y-mc}{mr'} \quad \textbf{(Note)}$$

$$\Rightarrow \quad r=\frac{c(y+mx)}{m(z+c)}; r'=\frac{c(y-mx)}{m(z-c)}$$

Substituting these values of r and r' in (i) and simplifying we get the required surface generated by AB as

$$c^2(mzx-cy)^2+c^2m^2(yz-mcx)^2=m^2(z^2-c^2)^2(k^2-c^2). \quad \textbf{Ans.}$$

Example 101:

Find the surface generated by a straight line which intersects the lines $x+y=0=z$, $x-y-z=0=x+y-2a$ and the parabola $y=0=x^2-2az$.

Solution :

Given lines are $x+y=0=z$, ...(i)

$$x-y-z=0=x+y-2a \quad ...(ii)$$

and the parabola is $y=0, x^2=2ab$ ...(iii)

Also line intersecting (i) and (ii) is given as

$$(x+y)+k_1z=0, (x-y-z)+k_2(x+y-2a)=0 \quad ...(iv)$$

If it meets the parabola (iii), we have to eliminate x, y, z from (iii) and (iv).

Putting y = 0 from (iii), in (iv), we get

$$x+k_1z=0, (x-z)+k_2(x-2a)=0$$

$$\Rightarrow \quad x+k_1z=0$$

and $\quad (1+k_2)x-z-2ak_2=0$

Solving these simultaneously, we get

$$\frac{x}{-2ak_1k_2} = \frac{z}{2ak_2} = \frac{1}{-1-k_1(1+k_2)}$$

$$\Rightarrow \quad x = \frac{2ak_1k_2}{1+k_1+k_1k_2} \quad z = \frac{-2ak_2}{1+k_1+k_1k_2}$$

Substituting these values of x and z in (iii) we get

$$\left(\frac{2ak_1k_2}{1+k_1+k_1k_2}\right)^2 = 2a\left(\frac{-2ak_2}{1+k_1+k_1k_2}\right)$$

$$\Rightarrow \quad k_1^2k_2 + (1 + k_1 + k_2k_2) = 0, \text{ on simplifying}$$

$$\Rightarrow \quad k_1^2k_2 + k_1k_2 + k_1 + 1 = 0 \qquad ...(v)$$

To find the required locus we are to eliminate k_1 and k_2 from (iv) and (v)

Substituting $k_1 = -\left(\frac{x+y}{z}\right)$, $k_2 = \frac{y+z-x}{x+y-2a}$ in (v) we get the required locus $\left(\frac{x+y}{z}\right)^2\left(\frac{x+z-x}{x+y-2a}\right) - \left(\frac{x+y}{z}\right)\left(\frac{x+z-x}{x+y-2a}\right) - \left(\frac{x+y}{z}\right) + 1 = 0$

$$\Rightarrow \quad (x + y)^2 (y + z - x) - z (x + y) (y + z - x)$$

$$- z (x + y) (x + y - 2a) + z^2 (x + y - 2a) = 0$$

$$\Rightarrow \quad (x + y) (y + z - x) [(x + y) - z] - z (x + y - 2a) [(x + y) - z] = 0$$

$$\Rightarrow \quad (x + y - z) [xy + xz - x^2 + y^2 + yz - xy - zx - 2y + 2az] = 0$$

$$\Rightarrow \quad (x + y - z) (2ax - x^2 + y^2) = 0.$$

Ans.

Example 102:

Find the surface generated by a straight line which meets the two lines $y = mx$, $z = c$ and $y = -mx$, $z = -c$ at the same angle.

Solution :

The given lines are $y - mx = 0$, $z - c = 0$ ...(i)

$$y + mx = 0, \; z + c = 0 \qquad ...(ii)$$

Any line intersecting (i) and (ii) is given by

$$(y - mx) - k_1(z - c) = 0, \; (y + mx) - k_2(z + c) = 0 \qquad ...(iii)$$

Now we are to find the d-cosines of this line. For this omitting the constant terms we have $mx - y + k_1z = 0$, $mx + y - k_2z = 0$.

Solving these simultaneously, we have

$$\frac{x}{k_2 - k_1} = \frac{y}{mk_1 + mk_2} = \frac{z}{m+m}$$

$\therefore$ d-ratios of the line (iii) are $k_2 - k_1$, $m(k_1 + k_2)$, 2m. ...(iv)

Again the equations of the line (i) and (ii) can be written in the symmetric form as

$$\frac{x}{1} = \frac{y}{m} = \frac{z-c}{0}$$

and $$\frac{x}{1} = \frac{y}{-m} = \frac{z+c}{0}$$

Since these lines make equal angles α (say) with the line (iii) whose d-ratios are given by (iv), so we have

$$\frac{1.(k_2 - k_1) + m\,m\,(k_1 + k_2) + 0.(2m)}{\sqrt{(1+m^2+0)}\sqrt{[(k_2-k_1)^2 + m^2(k_1+k_2)^2 + (2m)^2]}}$$

$$\frac{1.(k_2 - k_1) + m(-m)(k_1 + k_2) + 0.(2m)}{\sqrt{(1+m^2+0)}.\sqrt{[(k_2-k_1)^2 + m^2(k_1+k_2)^2 + (2m)^2]}}$$

$\Rightarrow$ $2m^2(k_1 + k_2) = 0$, on simplifying

$\Rightarrow$ $k_1 + k_2 = 0$

$\Rightarrow$ $\left(\frac{y-mx}{z-c}\right) + \left(\frac{y+mx}{z+c}\right) = 0$, substituting values of k_1, k_2, from (iii)

$\Rightarrow$ $(y - mx)(z + c)(y + mx)(z - c) = 0$

$\Rightarrow$ $y[(z + c) + (z - c)] - mx[(z + c) - (z - c)] = 0$

$\Rightarrow$ $2yz - 2cmx = 0$ or $yz = cmx$, which is the required surface.

Example 103:

Find the locus of a straight line that intersects two given lines and makes a right angle with one of them.

Solution :

Let the given lines be $\frac{x}{1} = \frac{y}{m} = \frac{z-c}{0}$ and $\frac{x}{1} = \frac{y}{-m} = \frac{z+c}{0}$

$\Rightarrow$ $y = mx$, $z = c$ and $y = -mx$, $z = -c$.

Any line that intersects these lines are given by the planes

$(y - mx) + k_1 (z - c) = 0$ and $(y + mx) + k_2 (z + c) = 0$ ...(i)

If λ, μ, ν are the direction ratios of this line, then

$- m \lambda + \mu + k_1 \nu = 0$ and $m\lambda + \mu + k_2 \nu = 0$

Solving these, we get $\dfrac{\lambda}{k_2 - k_1} = \dfrac{\mu}{m(k_1 + k_2)} = \dfrac{\nu}{-m-m}$

$\therefore$ The direction ratios of the line intersecting the given lines are

$(k_2 - k_1), m (k_1 + k_2), - 2m.$

If this line is perpendicular to the first given line then

$(k_2 - k_1) . 1 + m (k_1 + k_2) m + (- 2m) . 0 = 0$

$\Rightarrow$ $(k_2 - k_1) + m^2 (k_1 + k_2) = 0.$ ...(ii)

Also from (ii), $k_1 = \dfrac{mx - y}{z - c}$ and $k_2 = \dfrac{mx + y}{z + c}$

$\therefore$ From (ii), the required locus is

$$\left[-\left\{\frac{mx+y}{z+c}\right\}-\left\{\frac{mx-y}{z-c}\right\}\right]+m^2\left[\frac{mx-y}{z-c}-\frac{mx+y}{z+c}\right]=0$$

$\Rightarrow (mx + y) (z - c) + (mx - y) (z + c) = m^2 [(mx - y) (z + c) - (mx + y) (z - c)]$

$\Rightarrow$ $(mxz - cy) = m^2 (- yz + mcx).$ **Ans.**

Example 104:

Find the surface generated by a line which intersects the lines $y = a = z$ and $x + 3z = a = y + z$ and is parallel to the plane $x + y = 0$.

Solution :

Any line that intersects the given line is given by the planes

$(y - a) + k_1 (z - a) = 0$ and $(x + 3z - a) + k_2 (y + z - a) = 0$ ...(i)

If λ, μ, ν are direction ratios of this line, then we have

$\mu + k_1 \nu = 0$ and $\lambda + k_2 \mu + (3 + k_2) \nu = 0$

Solving these we get $\dfrac{\lambda}{(3+k_2)-k_1k_2} = \dfrac{\mu}{k_1} = \dfrac{\nu}{-1}$

$\therefore$ The direction ratios of the line intersecting the given lines are

$3 + k_2 - k_2k_1, k_1, - 1$

If this line is parallel to the plane $x + y = 0$, then this is perpendicular to the normal to this plane

We have $(3 + k_2 - k_1k_2).1 + k_1.1 = 0$ **(Note)**

$\Rightarrow \quad 3 + k_1 + k_2 - k_1k_2 = 0$ ...(ii)

Also from (i), we get $k_1 =$

$\therefore$ From (ii), the required locus is

$$3+\frac{a-y}{z-a}+\frac{a-x-3z}{y+z-a}-\left\{\frac{a-y}{z-a}\right\}\left\{\frac{a-x-3z}{y+z-a}\right\}=0$$

$\Rightarrow 3(z - a)(y + z - a) + (a - y)(y + z - a) + (a - x - 3z)(z - a)$
$- (a - y)(a - x - 3z) = 0$

$\Rightarrow \quad (y + z - a)[3z - 3a + a - y] = (a - x - 3z)[a - y - z + a]$

$\Rightarrow \quad y^2 + yz + xy + xz - 2az - 2ax = 0$, on simplifying

$\Rightarrow \quad x^2 + yz + xy + xz = 2a(x + z)$. **Ans.**

Example 105:

Show that the locus of lines which meet the lines at the same angle is $xy \cos \alpha - az \sin \alpha)(zx \sin \alpha - ay \cos \alpha) = 0$.

Solution :

Let a line meet the given lines in the points

$(-a, r \sin \alpha, -r \cos \alpha)$ and $(a, r' \sin \alpha, r' \cos \alpha)$ **(Note)**

Then the direction ratios of this line are

$a + a, r' \sin \alpha - r \sin \alpha, r' \cos \alpha + r \cos \alpha$

$\therefore$ The equations of this line are

$$\frac{x+a}{2a}=\frac{y-r \sin \alpha}{(r'-r) \sin \alpha}=\frac{z+r \cos \alpha}{(r+r') \cos \alpha} \quad ...(i)$$

This line meets the given line at the same angle and so on equating the cosine of the angles which this line subtends with the given line, we have

$2a.0 + (r' - r) \sin \alpha.\sin \alpha + (r' + r) \cos \alpha(-\cos \alpha)$

$= \pm [2a.0 + (r' - r) \sin \alpha \sin \alpha + (r' + r) \cos \alpha \cos \alpha]$ **(Note)**

$\Rightarrow (r' + r) \cos^2 \alpha = 0$, if + sign is taken

and $\quad (r' - r) \sin^2 \alpha = 0$, if – sign is taken

i.e., $\quad r' = \pm r$

If $r' = r$, then (i) becomes $\dfrac{x+\alpha}{2\alpha}=\dfrac{y-r \sin \alpha}{0}=\dfrac{z+r \cos \alpha}{2r \cos \alpha}$

whence $y = r \sin \alpha$ and $(z + r \cos \alpha) 2\alpha = (2r \cos \alpha)(x + a)$

$\Rightarrow$ $y = r \sin\alpha$ and $ax = xr \cos\alpha$

Eliminating r we get, $az \sin\alpha = xy \cos\alpha$

$\Rightarrow$ $xy\cos\alpha - az\sin\alpha = 0$ ...(ii)

If $r' = -r$, then from (i) eliminating r as above we can get

$zx\sin\alpha - ay\cos\alpha = 0$ ...(iii)

Hence from (ii) and (iii) we get the required result.

Example 106:

Prove that the straight lines which intersects the three lines $y - z = 1$, $x = 0$; $z - x = 1$ $y = 0$ and $x - y = 1$, $z = 0$ lies on the surface whose equation is $x^2 + y^2 + z^2 - 2yz - 2zx - 2xy = 1$.

Solution :

The given lines are $y - z - 1 = 0$, $x = 0$, ...(i)

$z - x - 1 = 0$, $y = 0$ ...(ii) and $x - y - 1 = 0$, $z = 0$ ...(iii)

Equations of any line intersecting the lines (i) and (ii) are

$(y - z - 1) - \lambda_1 x = 0, (z - x - 1) - \lambda_2 y = 0.$ ...(iv)

If the line (iv) intersects the line (iii), then we are to eliminate x, y, z from (iii) and (iv).

From (iii) we get $z = 0$ and $y = x - 1$.

Substracting these values in (iv) we get

$(x - 1 - 0 - 1) - \lambda_1 x = 0, (0 - x - 1) - \lambda_2 (x - 1) = 0$

$\Rightarrow$ $(1 - \lambda_1) x = 2, (1 + \lambda_2) x = -(1 - \lambda_2)$

$\Rightarrow$ $x = 2/(1 - \lambda_1), x = -(1 - \lambda_2)/(1 + \lambda_2).$

Equating two values of x, we get x, y, z eliminant of (iii) and (iv) as

$$\frac{2}{1-\lambda_1} = -\left(\frac{1-\lambda_2}{1+\lambda_2}\right) \text{ or } 2(1 + \lambda_2) = -(1 - \lambda_1)(1 - \lambda_2)$$

$\Rightarrow$ $2 + 2\lambda_2 = -(1 - \lambda_2 - \lambda_1 + \lambda_1\lambda_2)$

$\Rightarrow$ $\lambda_1 \lambda_2 - \lambda_1 + \lambda_2 + 3 = 0$ ...(v)

The required locus is obtained by eliminating λ_1, λ_2 from (iv) and (v)

From (iv) we have $\lambda_1 = \dfrac{y-z-1}{x}, \lambda_2 = \dfrac{z-x-1}{y}$

Substituting these in (v) we get the required locus as

$$\left(\frac{y-z-1}{x}\right)\left(\frac{z-x-1}{y}\right)-\left\{\frac{y-z-1}{x}\right\}+\left\{\frac{z-x-1}{y}\right\}+3=0$$

$\Rightarrow$ $(y-z-1)(z-x-1)-y(y-z-1)+x(z-x-1)+3xy=0$

$\Rightarrow$ $x^2+y^2+z^2-2yz-2zx-2xy=1.$ **Hence proved.**

Example 107:

Find the locus of the variable line which cuts three lines $y = b$, $z = -c$; $z = c$, $x = -a$ and $x = a$, $y = -b$.

Solution :

The equation of the plane through the line $y = b$, $z = -c$ is

$$(y-b)+\lambda_1(z+c)=0 \quad ...(i)$$

And the equation of the plane through the line $z = c$, $x = -a$ is

$$(z-c)+\lambda_2(x+a)=0 \quad ...(ii)$$

Now any line intersecting the first two given lines is given by planes (i) and (ii). The above two planes intersect in a line and as it meets the third line $x = a$, $y = -b$ so putting $x = a$ and $y = -b$ in (i) and (ii) we get

$$(-b-b)+\lambda_1(z+c)=0 \text{ and } (z-c)+\lambda_2(a+a)=0$$

$\Rightarrow$ $z+c=2b/\lambda_1$ and $z-c=-2a\lambda_2$

Subtracting we get $2c=2[(b/\lambda_1)+a\lambda_2]$

$\Rightarrow$ $c=(b+a\lambda_1\lambda_2)/\lambda_1$ or $c\lambda_1=a\lambda_1\lambda_2+b$ ...(iii)

Eliminating λ_1 and λ_2 between (i), (ii) and (iii) we get

$$c\left[-\left(\frac{y-b}{z+c}\right)\right]=a\left[\left(\frac{y-b}{z+c}\right)\left(\frac{z-c}{x+a}\right)\right]+b$$

$\Rightarrow$ $-c(y-b)(x+a)=a(y-b)(z-c)+b(x+a)(z+c)$

$\Rightarrow$ $ayz+byz+cxy+abc=0$, on simplifying

This is the required locus. **Ans.**

Example 108:

Find the surface generated by the lines which intersect the lines $y = mx$, $z = c$; $y = -mx$, $z = -c$ and x-axis.

Solution :

As the line intersecting the lines $y = mx$, $z = c$ and $y = -mx$, $z = -c$ is given by the planes

$$(y - mx) + \lambda_1 (z - c) = 0 \quad ...(i)$$

and $$(y + mx) + \lambda_2 (z + c) = 0 \quad ...(ii)$$

It meets x-axis *i.e.*, $y = 0 = z$, where

$$(0 - mx) + \lambda_1 (0 - c) = 0 \text{ and } (0 + mx) + \lambda_2 (0 + c) = 0$$

$$\therefore \lambda_2 = \lambda_1 \quad ...(iii)$$

Eliminating λ_1, λ_2 between (i), (ii) and (iii) we get

$$-\left\{\frac{y - mx}{z - c}\right\} = -\left\{\frac{y + mx}{z + c}\right\}$$

$\Rightarrow$ $(y - mx)(z + c) = (y + mx)(z - c)$ or $cy = mzx$. **Ans.**

Example 109:

(a) Given a point P and a straight line AB, show how to find the equations of the perpendicular PN drawn from P on AB as the line of intersection of two suitable planes.

(b) If P is (1, 2, 3) and AB is $\frac{1}{3}(x - 2) = \frac{1}{4}(y - 3) = \frac{1}{5}(z - 4)$, *find the line PN in this manner and obtain the coordinates of N.*

Solution :

(a) Let P be (x_1, y_1, z_1) and the equations of the straight line AB be given as

$$\frac{x - \alpha}{l} = \frac{y - \beta}{m} = \frac{z - \gamma}{n}$$

Now PN can be looked upon as the line of intersection of the plane through the line AB and the point P and the plane perpendicular to AB and the point P.

Now the equation of any plane through the line AB is given by

$$A(x - \alpha) + B(y - \beta) + C(z - \gamma) = 0 \quad ...(i)$$

where $$Al + Bm + Cn = 0 \quad ...(ii)$$

If (i) passes through P (x_1, y_1, z_1), then we have

$$A(x_1 - \alpha) + B(y_1 - \beta) + C(z_1 - \gamma) = 0 \quad ...(iii)$$

Solving (ii) and (iii) simultaneously we get

$$\frac{A}{m(z_1 - \gamma) - n(y_1 - \beta)} = \frac{B}{n(x_1 - \alpha) - l(z_1 - \gamma)} = \frac{C}{l(y_1 - \beta) - m(x_1 - \alpha)}$$

Substituting these proportionate values of A, B, C in (i) we get the equation of the plane through the line AB and the point P as

$$\Sigma\ [\{m\ (z_1 - \gamma) - n\ (y_1 - \beta)\}\ (x - \alpha)] = 0 \qquad \text{....(iv)}$$

Also equation of any plane through P (x_1, y_1, z_1) is

$$a\ (x - x_1) + b\ (y - y_1) + c\ (z - z_1) = 0 \qquad \text{...(v)}$$

If it is perpendicular to the line AB whose d.c.'s are proportional to l, m, n then we must have $a/l = b/m = c/n$.

∴ The equation of the plane through the point P and perpendicular to the line AB is $l\ (x - x_1) + m\ (y - y_1) + n\ (z - z_1) = 0$ ...(vi)

Hence the equations of PN, the line of intersection of the planes through AB and P and prep. to AB and P, are given by (iv) and (vi).

(b) Here P is (1, 2, 3) and the line AB is given by

$$\frac{1}{3}\ (x - 2) = \frac{1}{4}\ (y - 3) = \frac{1}{5}\ (z - 4) \qquad \text{...}(\alpha)$$

And plane through the line AB is given by

$$A\ (x - 2) + B\ (y - 3) + C\ (z - 4) = 0, \qquad \text{...(I)}$$

where $A.3 + B.4 + C.5 = 0$ ...(II)

Also if (I) passes through P (1, 2, 3), then

$$A\ (1 - 2) + B\ (2 - 3) + C\ (3 - 4) = 0 \text{ or } A + B + C = 0 \qquad \text{...(III)}$$

Solving (II) and (III) we get

$$\frac{A}{4-5} = \frac{B}{5-3} = \frac{C}{3-4} \text{ or } \frac{A}{1} = \frac{B}{-2} = \frac{C}{1} = k \text{ (say)}$$

$\therefore\ A = k,\ B = -2k,\ C = k.$

Substituting these values of A, B, C in (I) we get the equation of the plane through the line AB and the point P as

$$(x - 2) - 2\ (y - 3) - (z + 4) = 0 \text{ or } x - 2y + z = 0 \qquad \text{...(IV)}$$

Also equation of any plane through P (1, 2, 3) is

$$a\ (x - 1) + b\ (y - 2) + c\ (z - 3) = 0 \qquad \text{...(V)}$$

If it is perpendicular to the line AB, then

$$a/3 = b/4 = c/5 = \lambda \text{ (say)} \Rightarrow a = 3\lambda,\ b = 4\lambda,\ c = 5\lambda.$$

∴ From (V) the equation of the plane through P (1, 2, 3) and perpendicular to the line AB is $3(x - 1) + 4\ (y - 2) + 5\ (z - 3) = 0$

$$\Rightarrow \quad 3x + 4y + 5z = 26 \qquad \text{...(VI)}$$

Hence, the equations of the line PN are

$$x - 2y + z = 0,\ 3x + 4y + 5z = 26$$

Also the foot N of the perpendicular PN is the point of intersection of the plane (VI) and the line AB given by (α).

Any point on the line AB given by (α) is

$$(2 + 3r, 3 + 4r, 4 + 5r) \quad \text{...(VII)}$$

If it lies on the plane given by (VI), we get

$$3(2 + 3r) + 4(3 + 4r) + 5(4 + 5r) = 26$$

$$\Rightarrow \quad 50r = 26 - 38 = -12 \text{ or } r = -(6/25)$$

$\therefore$ The coordinates of N from (VII) are

$$\left(2-\frac{18}{25}, 3-\frac{24}{25}, 4-\frac{30}{25}\right) \text{ or } \left(\frac{32}{25}, \frac{51}{25}, \frac{70}{25}\right).$$ **Ans.**

Example 110:

Find the foot and hence the length of the perpendicular from (5, 7, 3) to the line $\frac{1}{3}(x - 15) = \frac{1}{8}(y - 29) = -\frac{1}{5}(z - 5)$. *Find also the equation of the plane in which the perpendicular and the given straight line lie.*

Solution :

The given point is P (5, 7, 3) and the line is

$$\frac{x-15}{3} = \frac{y-29}{8} = \frac{z-5}{-5} \quad \text{...(i)}$$

Any point on it is N (15 + 3r, 29 + 8r, 5 – 5r) is given by ...(ii)

$\therefore$ The d.r.'s of line PN are (15 + 3r) – 5, (29 + 8r) – 7, (5 – 5r)– 3

i.e 10 + 3r, 22 + 8r, 2 – 5r ...(iii)

Let N be the foot of the perpendicular from P to (i), then PN is perpendicular to (i) and so we have

$$3(10 + 3r) + 8(22 + 8r) - 5(2 - 5r) = 0 \quad \text{or } r = -2$$

$\therefore$ From (ii) the foot N of the perpendicular PN is given by

$$(15 - 6, 29 - 16, 5 + 10) \text{ or } (9, 13, 15)$$

$\therefore$ PN = distance between P(5, 7, 3) and N (9, 13, 15)

$$= \sqrt{\left[(9-5)^2 + (13-7)^2 + (15-3)^2\right]} = \sqrt{(16+36+144)} = 14.$$

Again the equation of the plane containing the given line (i) is given by

$$A(x - 15) + B(y - 29) + C(z - 5) = 0 \quad \text{...(iv)}$$

where $\quad A.3 + B.8 + C.(-5) = 0$ ...(v)

If the plane (iv) passes through P (5, 7, 3) then we have

$A(5-15) + B(7-29) + C(3-5 = 0$ or $5A + 11B + C = 0.$ ...(vi)

Solving (v) and (vi) we get

$$\frac{A}{8+55} = \frac{B}{-25-3} = \frac{C}{33-40} \text{ or } \frac{A}{9} = \frac{B}{-4} = \frac{C}{-1}$$

∴ From (iv) the required equation of the plane through the line (i) and the perpendicular PN [or the point P as N is a point on (i)] is $9(x-15) - 4(y-29) - (z-5) = 0$ or $9x - 4y - z = 14.$ **Ans.**

Example 111:

Find the distance of the point P (4, 3, 5) from the axis of y.

Solution :

The equations of y-axis are $\frac{x}{0} = \frac{y}{1} = \frac{z}{0}$,

∴ Any point N on y-axis is (0, r, 0) ...(i)

∴ The direction cosines of the line PN are $0-4,\ r-3,\ 0-5$

i.e., $\quad -4,\ r-3,\ -5.$ ...(ii)

Let N be the foot of the perpendicular from P to y-axis, then PN is perpendicular to the y-axis whose direction cosines are 0, 1, 0 and so from (ii) we have $0.(-4) + 1.(r-3) + 0.(-5) = 0$ or $r = 3$

∴ From (i) the coordinates of N are (0, 3, 0)

∴ Required distance = PN = $\sqrt{\left[(4-0)^2 + (3-3)^2 (5-0)^2\right]}$

$= \sqrt{(16+0+25)} = \sqrt{(41)}$. **Ans.**

Example 112:

Find the perpendicular distance of P (1, 2, 3) from the line

$$\frac{1}{3}(x-6) = \frac{1}{2}(y-7) = -\frac{1}{2}(z-7).$$

Solution :

The equations of the given line are

$$\frac{x-6}{3} = \frac{y-7}{2} = \frac{z-7}{-2}.$$

∴ Any point N on this line is $(3r+6,\ 2r+7,\ -2r+7)$ ...(i)

$\therefore$ d.r.'s of the line PN are

$3r + 7 - 1, 2r + 7 - 2, -2r + 7 - 3$ or $3r + 5, 2r + 5, -2r+4$ ...(ii)

Let N be the foot of the perpendicular from P to the given line, then PN is perpendicular to the given line whose d.r.'s are 3, 2, – 2 and from (ii) we have $3(3r + 5) + 2(2r + 5) - 2(-2r + 4) = 0$ or $r = -1$

$\therefore$ From (i) the coordinates of N are (3, 5, 9)

$\therefore$ Required distance PN $= \sqrt{\left[(3-1)^2 + (5-2)^2 + (9-3)^2\right]}$

$= \sqrt{[4+9+36]} = 7$. **Ans.**

Example 113:

Find the distance of the point (1, 2, 3) from the line joining the points (–1, 2, 5) and (2, 3, 4).

Solution :

Equation of the line joining the points (–1, 2, 5) and (2, 3, 4) is given by

$$\frac{x+1}{2+1} = \frac{y-2}{3-2} = \frac{z-5}{4-5} \text{ or } \frac{x+1}{3} = \frac{y-2}{1} = \frac{z-5}{-1} \quad ...(i)$$

Also the given point is P (1, 2, 3). ...(ii)

Any point on the line (i) is N $(-1 + 3r, 2 + r, 5 - r)$

Let N be the foot of the perpendicular from P to (i), then the d.r.'s of PN are $(-1 + 3r) - 1, (2 + r) - 2, (5 - r) - 3$ or $3r - 2, r, 2 - r$ ...(iii)

Also if N be the foot of the perpendicular from P to (i), then PN is perpendicular to (i) and so we have

$(3r - 2).3 + (r).1 + (2 - r).(-1) = 0$ or $r = 8/11$

$\therefore$ From (ii), N is $\left(-1+\frac{24}{11}, 2+\frac{8}{11}, 5-\frac{8}{11}\right)$ *i.e.,* $\left(\frac{13}{11}, \frac{30}{11}, \frac{47}{11}\right)$

$\therefore$ Required distance = PN

$$= \sqrt{\left[\left(\frac{13}{11}-1\right)^2 + \left(\frac{30}{11}-2\right)^2 + \left(\frac{47}{11}-3\right)^2\right]}$$

$$= \frac{1}{11}\sqrt{[4+64+196]} = \frac{1}{11}\sqrt{(264)} = \frac{2\sqrt{6}}{\sqrt{(11)}}.$$ **Ans.**

Example 114:

Find the length of the perpendicular from a given point (1, – 2, 3) on a line through (2, – 3, 5) which makes equal angles with the axes.

Solution :

The d.c.'s of the line which makes equal angles with the axes can be taken as l, l, l, where $l^2 + l^2 + l^2 = 1$.

∴ The equations of the given line can be written as

$$\frac{x-2}{l} = \frac{y-(-3)}{l} = \frac{z-5}{l} \text{ or } \frac{x-2}{1} = \frac{y+3}{1} = \frac{z-5}{1} \quad ...(i)$$

Now proceed as in Ex. 2 (c) above. **Ans.** $\sqrt{(14/3)}$.

Example 115:

Find the equations of the perpendicular from the origin to the line ax + by + cz + d = 0 = a'x + b'y + c'z + d'.

Solution :

The perpendicular from the origin to the given line is the line of intersection of the following two planes :

(i) the plane through origin and containing the given line and (ii) the plane through origin and perpendicular tot he given line. **(Note)**

Now the equation of the plane through the given line is

$$(ax + by + cz + d) + \lambda (a'x + b'y + c'z + d') = 0. \quad ...(i)$$

If this plane passes through the origin (0, 0, 0), then from (i) we have

$$d + \lambda d' = 0 \text{ or } \lambda = - d/d'$$

∴ From (i) the equations of the plane through the origin and the given lines is $d' (ax + by + cz) - d (a'x + b'y + c'z) = 0$

$$\Rightarrow (d'a - da') x + (d'b - db') y + (d'c - dc') z = 0 \quad ...(ii)$$

Again the equation of any plane through the origin is

$$Ax + By + Cz = 0 \quad ...(iii)$$

Also if l, m, n be the d.c.'s of the given line, then we have

$$al + bm + cn = 0,$$

$$a'l + b'm + c'n = 0.$$

Solving these we get $\frac{l}{bc'-b'c} = \frac{m}{ca'-c'a} = \frac{n}{ab'-a'b}$ ...(iv)

Now if the plane (iii) is perpendicular to the given line then the normal to the plane (iii) is parallel to the given line.

∴ We have $A/l = B/m = C/n$

$$\Rightarrow \quad \frac{A}{bc'-b'c} = \frac{B}{ca'-c'a} = \frac{C}{ab'-a'b'} \text{ from (iv),}$$

∴ From (iii) the equation of the plane through the origin and perpendicular to the given line is

$$(bc' - b'c)\,x + (ca' - c'a)\,y + (ab' - a'b)\,z = 0. \quad \text{...(v)}$$

Hence planes (ii) and (v) together give the required equations of the perpendicular from the origin to the given line.

Example 116:

Find the equations of the perpendicular from the point (3, –1, 11) to line $\frac{1}{2}x = \frac{1}{3}(y-2) = \frac{1}{4}(z-3)$. *Find also the coordinates of the foot of the perpendicular.*

Solution :

The given point is P (3, – 1, 11) and the equations of the line are

$$\frac{x}{2} = \frac{y-2}{3} = \frac{z-3}{4} \quad \text{...(i)}$$

Any point on this line is N (2r, 2 + 3r, 3 + 4r) ...(ii)

Let N be the foot of the perpendicular from P (3, – 1, 11) to (i).

∴ The d.r.'s of the line PN are

$2r - 3,\ 2 + 3r + 1,\ 3 + 4r - 11$ or $2r - 3,\ 3r + 3,\ 4r - 8,$ ...(iii)

Also if N be the foot of the perpendicular from P to (i), then PN is perpendicular to (i) and we have

$$(2r - 3).2 + (3r + 3).3 + (4r - 8).4 = 0$$

which gives $\quad 29r - 29 = 0$ or $r = 1$.

∴ From (ii), the coordinates of N, the foot of perpendicular from P to (ii) is (2, 5,7).

And from (iii), the d.r.'s of the perpendicular PN are

– 1, 6, – 4 or 1, – 6, 4.

∴ The required equations of the perpendicular PN are given by

$$\frac{x-3}{1} = \frac{y+1}{-6} = \frac{z-11}{4}.$$

Ans.

Example 117:

Find the equations to the perpendicular from the origin to the line x + 2y + 3z + 4 = 0, 2x + 3y + 4z + 5 = 0. Find also the coordinates of the foot of the perpendicular.

Solution :

The equation of any plane through the given line is given as

$$(x + 2y + 3z + 4) + \lambda (2x + 3y + 4z + 5) = 0$$

If it passes through the origin (0, 0, 0) then $4 + 5\lambda = 0$ or $\lambda = -4/5$.

∴ From (i), the equation of the plane through the origin and the given line is $5 (x + 2y + 3z - 4) - 4 (2x + 3y + 4z + 5) = 0$

$$\Rightarrow \quad 3x + 2y + z = 0. \quad ...(ii)$$

Also if l, m, n be the d.c.'s of the given line, then we have

$$l + 2m + 3n = 0 \quad \text{and } 2l + 3m + 4n = 0$$

Solving, we get $\frac{l}{8-9} = \frac{m}{6-4} = \frac{n}{3-4}$ or $\frac{l}{1} = \frac{m}{-2} = \frac{n}{1}$ ...(iii)

Also any plane through the origin is $Ax + By + Cz = 0$

If this plane is perpendicular to the given line, then the normal to this plane must be parallel to the given line.

∴ We have $A/l = B/m = C/n$ or $A/1 = B/-2 = C/1$, from (iii)

Hence the equation of the plane through the origin and perpendicular to the given line is $1.x - 2.y + 1.z = 0$ or $x - 2y + z = 0$ ...(iv)

The line of intersection of the planes (ii) and (iv) is the perpendicular from the origin to the given line, hence the required equation of the perpendicular are $3x + 2y + z = 0$, $x - 2y + z = 0$. ...(v)

Ans.

∴ The d.r.'s l_1, m_1, n_1 of this perpendicular are given by

$$3l_1 + 2m_1 + n_1 = 0 \text{ and } l_1 - 2m_1 + n_1 = 0$$

Solving these we have $\frac{l_1}{2+2} = \frac{m_1}{1-3} = \frac{n_1}{-6-2}$ or $\frac{l_2}{2} = \frac{m_2}{-1} = \frac{n_1}{-4}$

∴ The equations (v) of the perpendicular from the origin to the given line can be written in the symmetric form as

$$\frac{x-0}{2} = \frac{y-0}{-1} = \frac{z-0}{-4} = r \text{ (say)} \quad ...(vi)$$

Example 118:

Find the locus of points equidistant from the lines

$y - mx = 0 = z - c; \; y + mx = 0 = z + c.$

Solution :

The given lines are written as

$$\frac{x}{1/m} = \frac{y}{1} = \frac{z-c}{0} \quad \text{...(i);}$$

$$\frac{x}{1/m} = \frac{y}{-1} = \frac{z+c}{0} \quad \text{...(ii)}$$

Let P (x_1, y_1, z_1) be a point equidistant from these lines. Let PN_1 and PN_2 be the perpendiculars from P to these lines (i) and (ii) respectively.

Any point N_1 on the line (i) is [(r/m), r, c] ...(iii)

$\therefore$ The d.r.'s of the line PN_1 are $x_1 - (r/m)$, $y_1 - r$, $z_1 - c$.

If N_1 is the foot of the perpendicular from P on the line (i), then PN_1 is perpendicular to the line (i) and so we have

$$(1/m)\,[x_1 - (r/m)] + 1.\,(y_1 - r) + 0.\,(z_1 - c) = 0$$

$$\Rightarrow (x_1/m) + y_1 = r\,[1 + 1/m^2)] \text{ or } mx_1 + m^2 y_1 = r\,(m^2 + 1)$$

$$\Rightarrow \quad r = (mx_1 + m^2y_1)/(m^2 + 1)$$

$$\therefore \text{ From (iii), } N_1 \text{ is } \left[\frac{x_1 + my_1}{m^2+1}, \frac{mx_1 + m^2y_1}{m^2+1}, c\right]$$

$$\therefore \; PN_1^2 = \left(x_1 - \frac{x_1 + my_1}{m^2+1}\right)^2 + \left(y_1 - \frac{mx_1 + m^2y_1}{m^2+1}\right)^2 + (z_1 - c)^2$$

$$= \frac{m^2\,(y_1 - mx_1)^2}{(m^2+1)^2} + \frac{(y_1 - mx_1)^2}{(m^2+1)^2} + (z_1 - c)^2$$

$$\Rightarrow PN_1^2 = \frac{(y_1 - mx_1)^2}{(m^2+1)} + (z_1 - c)^2, \text{ on simplifying} \quad \text{...(iv)}$$

Example 119:

Find the distance of the point (3, 8, 2) from the line

$$\frac{1}{2}(x-1) = \frac{1}{4}(y-3) = \frac{1}{3}(z-2)$$

measured parallel to the plane $3x + 2y - 2z + 15 = 0$.

Solution :

Let the given point (3, 8, 2) be P.

Any point N on the given line is (1 + 2r, 3 + 4r, 2 + 3r) ...(i)

∴ The direction ratios of the line PN are given as

(1 + 2r) – 3, (3 + 4r) – 8, (2 + 3r) – 2 *i.e.*, 2r – 2, 4r –5, 3r. ...(ii)

If PQ is parallel to the plane 3x + 2y – 2z + 15 = 0, then PQ is perpendicular to the normal to this plane, consequently we have

$$3(2r-2)+2(4r-5)-2(3r)=0 \text{ or } r=2.$$

∴ From (i) the coordinates of N are given by

[1 + 2 (2), 3 + 4 (2), 2 + 3 (2)] *i.e.*, (5, 11, 8)

∴ Required distance = distance between P (3, 8, 2) and N (5, 11, 8)

$$= \sqrt{[(5-3)^2+(11-8)^2+(8-2)^2]} = 7.$$ **Ans.**

Similarly we can prove that $PN_2^2 = \dfrac{(y_1+mx_1)^2}{(m^2+1)} + (z_1+c)^2$ ...(v)

∴ The point P (x_1, y_1, z_1) is equidistant from the given lines,

so $PN_1 = PN_2$ *i.e.*, $PN_1^2 = PN_2^2$

$$\Rightarrow \frac{(y_1-mx_1)^2}{m^2+1}+(z_1-c)^2 = \frac{(y_1+mx_1)^2}{m^2+1}+(z_1+c)^2, \text{ from (iv), (v)}$$

$$\Rightarrow [(z_1-c)^2-(z_1+c)^2](m^2+1) = (y_1+mx_1)^2-(y_1-mx_1)^2$$

$$\Rightarrow [(2z_1)(-2c)](m^2+1) = (2y_1)(2mx_1)$$

or $mx_1y_1 + cz_1(m^2+1) = 0$

∴ The locus of P is $mxy + c(m^2+1)z = 0$. **Ans.**

Example 120:

If L is the line $\frac{1}{2}(x-1) = -y = (z+2)$, *find the direction cosines of the projection of L on the plane 2x + y – 3z = 4 and the equation of the plane through L parallel to the line 2x + 5y + 3z = 4.* x – y – 5z = 6.

Solution :

Any point on the line L is A (1 + 2r, – r, 2 + r). ...(i)

If A lies on the plane 2x + y – 3z = 4, then we have ...(ii)

2 (1 + 2r) + (–r) – 3 (– 2 + r) = 4, which does not give any value of r as the coefficient of r is zero.

This shows that the line L is parallel to the plane (ii).

In such case we have

Any plane through the line L is A (x – 1) + By + C (z + 2) = 0,...(iii)

Where $A.(2) + B(-1) + C(1) = 0.$ ...(iv)

If the plane (iii) is perpendicular to the given plane (ii), then we have

$2.A + 1.B - 3.C = 0$ ...(v)

Solving (iv) and (v), we get $\frac{A}{3-1} = \frac{B}{2+6} = \frac{C}{2+2}$ or $\frac{A}{1} = \frac{B}{4} = \frac{C}{2}$.

∴ From (iii) the equation of the plane through L and perpendicular to the given plane (ii) is $1.(x - 1) + 4(y) + 2(z + 2) = 3$

⇒ $x + 4y + 2z + 3 = 0$ ...(vi)

The plane (ii) and (vi) together give the projection of L on the plane (ii).

If l, m, n be the d.c.'s of the projection, then from (iii) and (iv) we have

$2l + m - 3n = 0$ and $l + 4m + 2n = 0$

Solving these equation we have

$$\frac{l}{14} = \frac{m}{-7} = \frac{n}{7} \text{ or } \frac{l}{2} = \frac{m}{-1} = \frac{n}{1} = \frac{\sqrt{(l^2 + m^2 + n^2)}}{\sqrt{(2^2 + 1^2 + 1^2)}} = \frac{1}{\sqrt{6}}$$

⇒ $l = 2/\sqrt{6}$, $m = -1/\sqrt{6}$, $n = 1/\sqrt{6}$. **Ans.**

Let l_1, m_1, n_1 be the d.c.'s of the line given by

$2x + 5y + 3z = 4,\ x - y - 5z = 6.$ ...(vii)

Then $2l_1 + 5m_1 + 3n_1 = 0$, $l_1 - ml - 5n_1 = 0$

Solving, $\frac{l_1}{-25+3} = \frac{m_1}{3+10} = \frac{n_1}{-2-5}$ or $\frac{l_1}{22} = \frac{m_1}{-13} = \frac{n_1}{7}$ ...(viii)

Now the equation of any plane through L is (iii) where (iv) holds.

If this plane is parallel to the line (vii) whose d.c.'s are given by (viii), then the normal to the plane (iii), must be perpendicular to the line (vii) and so we have $Al_1 + Bm_1 + Cn_1 = 0$ or $22A - 13B - 7C = 0.$ ...(ix)

Solving (iv) and (ix) we get $\frac{A}{6} = \frac{B}{8} = \frac{C}{-4}$ or $\frac{A}{3} = \frac{B}{4} = \frac{C}{-2}$

∴ From (iii), the equation of the plane through L parallel to (vii) is $3(x - 1) + 4(y) - 2(z + 2) = 0$ or $3x + 4y - 2z = 7.$ **Ans.**

Example 121:

Find the projection (or 'image') of the line

$$(x - 1)/3 = (y - 2)/4 = (x - 3)/5$$

on (or 'in' if image is used in place of 'projection' above) the plane

$$x - y + z + 2 = 0.$$

Solution :

The given line is $\frac{x-1}{3} = \frac{y-2}{4} = \frac{z-3}{5}$ ...(i)

Any point on this line A is (1 + 3r, 2 + 4r, 3 + 5r) ...(ii)

If A lies on the given plane x – y + z + 2 = 0 then we have

$$(1 + 3r) - (2 - 4r) + (3 + 5r) + 2 = 0, \text{ or } r = -1$$

∴ From (ii) the point A is (– 2, – 2, – 2), where A is the point of intersection of the given plane and the given line.

Now from (ii) it is evident that a point on the given line is C (1, 2, 3) for r = 0). Let B be the foot of the perpendicular from C to the given plane.

Now BC is a line perpendicular to given plane, the direction ratios of BC are 1, –1, 1 (the coefficients of x, y and z in the equation of the given plane).

∴ The equations of BC a line passing through C (1, 2, 3) and having 1, – 1, 1 as direction ratios are $\frac{x-1}{1} = \frac{y-2}{-1} = \frac{z-3}{1}$.

Any point on this lie BC is (1 + r, 2 – r, 3 + r). If the point is B *i.e.*, if this point lies on the given plane, then we have

$$(1 + r) - (2 - r) + (3 - r) + 2 = 0 \text{ or } 3r + 4 = 0 \text{ or } r = -(4/3)$$

∴ The point B is $\left(1-\frac{4}{3}, 2+\frac{4}{3}, 3-\frac{4}{3}\right)$

i.e., $\left(-\frac{1}{3}, \frac{10}{3}, \frac{5}{3}\right)$

∴ The direction ratios of the projection AB of the given line are

$$-\frac{1}{3}+2, \frac{10}{3}+2, \frac{5}{3}+2$$

i.e., – 5, 16, 11

∴ The required equations of the projection AB are

$$\frac{x+2}{-5} = \frac{y+2}{6} = \frac{z+2}{11}.$$ **Ans.**

Example 122:

Find the equations of the plane through the line $\frac{x}{l}=\frac{y}{m}=\frac{z}{n}$ *and perpendicular to the plane containing the lines.*

$$\frac{x}{m}=\frac{y}{n}=\frac{z}{l} \text{ and } \frac{x}{n}=\frac{y}{l}=\frac{z}{m}$$

(Agra 91 : Kanpur 95; Purvanchal 93; Rohilkhand 91)

Solution :

The equation of the given plane contains the line

$\frac{x}{m}=\frac{y}{n}=\frac{z}{l}$ and $\frac{x}{n}=\frac{y}{l}=\frac{z}{m}$, both of which pass through the origin is

$$\begin{vmatrix} x & y & z \\ m & n & l \\ n & l & m \end{vmatrix} = 1$$

$$\Rightarrow \quad (mn - l^2)\,x + (ln - m^2)\,y + (m - n^2)\,z = 0. \qquad \text{...(i)}$$

Also the equation of the plane through the line $x/l = y/m = z/n$ is

$$ax + by + cz = 0 \qquad \text{...(ii)}$$

where $\quad al + bm + cn = 0 \qquad \text{...(iii)}$

If the planes (i) and (ii) are perpendicular, then we have

$$a\,(mn - l^2) + b\,(nl - m^2) + c\,(ml - n^2) = 0 \qquad \text{...(iv)}$$

Solving (iii) and (iv), we have

$$\frac{a}{m\,(ml-n^2)-n\,(nl-m^2)} = \frac{}{n\,(nm-l^2-l(ml-n^2)}$$

$$= \frac{c}{l\,(nl-m^2)-m\,(nm-l^2)}$$

$$\Rightarrow \frac{a}{(m^2-n^2)l+mn\,(n-m)} = \frac{b}{(n^2-l^2)\,m+nl\,(n-l)} = \frac{c}{(l^2-m^2)\,n+lm\,(l-m)}$$

$$\Rightarrow \frac{a}{(m-n)\,(lm+mn+nl)} = \frac{b}{(n-l)\,(nm+lm+nl)} = \frac{c}{(l-m)\,(ln+nm+lm)}$$

$$\Rightarrow \quad \frac{a}{m-n}=\frac{b}{n-l}=\frac{c}{l-m}$$

$\therefore$ from (ii), equation of the required plane is

$$(m - n)\,x + (n - l)\,y + (l - m)\,z = 0. \qquad \textbf{Ans.}$$

Example 123:

Prove that the three lines drawn from the origin O with d.c.'s l_1, m_1, n_1; l_2, m_2, n_2 *and* l_3, m_3, n_3 *i.e.,* $x/l_1 = y/m_1 = z/n_1$, $x/l_2 = y/m_2 = z/n_2$ *and* $x/l_3 = y/m_3 = z/n_3$ *are coplanar, if*

$$\begin{vmatrix} l_1 & m_1 & n_1 \\ l_2 & m_2 & n_2 \\ l_3 & m_3 & n_3 \end{vmatrix}$$

Solution :

Since the three lines whose d.c.'s are given pass through the origin, so if they are coplanar then they must be perpendicular to the same line through the origin. Let the d.c.'s of this line through the origin be l, m, n.

Then as this line is perpendicular to the given lines, so we have

$$ll_1 + mm_1 + nn_1 = 0$$

$$ll_2 + mm_2 + nn_2 = 0$$

and $$ll_3 + mm_3 + nn_3 = 0$$

Eliminating l, m, n between these we have the required condition as

$$\begin{vmatrix} l_1 & m_1 & n_1 \\ l_2 & m_2 & n_2 \\ l_3 & m_3 & n_3 \end{vmatrix}$$

Hence proved.

Example 124:

Show that the lines

$$\frac{x+1}{-3} = \frac{y-3}{2} = \frac{z+2}{1}$$

and $$\frac{x}{1} = \frac{y-7}{-3} = \frac{z+7}{2}$$

intersect. Find the coordinates of the point of intersection and the equation of the plane containing them.

Solution :

Any point on the line

$$\frac{x+1}{-3} = \frac{y-3}{2} = \frac{z+2}{1}$$ is given by

$$(-1 - 3r, 3 + 2r, -2 + r) \qquad ...(i)$$

Similarly any point on the line $\frac{x}{1} = \frac{y-7}{-3} = \frac{z+7}{2}$ is given as

$$(r', 7 - 3r', -7 + 2r') \qquad ...(ii)$$

If the two given lines intersect then for some value of r and r' the two above points (i) and (ii) must coincide *i.e.,*

$$-1 - 3r = r'; \ 3 + 2r = 7 - 3r'; \ -2 + r = -7 + 2r'$$

Solving the first two of these equations we get r = – 1, r' = 2.

These values of r and r' satisfy the third equation also, hence the given lines intersect.

Substituting these values r and r' in (i) or (ii) we get the required coordinates of the point of intersection as (2, 1, – 3). **Ans.**

Also the equation of the plane containing the given lines is

$$\begin{vmatrix} x+1 & y-3 & z+2 \\ -3 & 2 & 1 \\ 1 & -3 & 2 \end{vmatrix}$$

$$\Rightarrow \quad (x + 1)(4 + 3) - (y - 3)(-6 - 1) + (z + 2)(9 - 2) = 0$$

$$\Rightarrow \quad x + y + z = 0.$$ **Ans.**

Example 125:

Prove that the lines

$$\frac{x-a+d}{\alpha-\delta} = \frac{y-a}{\alpha} = \frac{z-a-d}{\alpha+\delta}$$

and $$\frac{x-b+c}{\beta-\gamma} = \frac{y-b}{\beta} = \frac{z-b-c}{\beta+\gamma}$$

are coplanar and find the equation to the plane in which they lie.

Solution :

Given lines are coplanar, if

$$\begin{vmatrix} 2(a-b) & a-b & (a+d)-(b+c) \\ 2\alpha & \alpha & \alpha+\delta \\ 2\beta & \beta & \beta+\gamma \end{vmatrix}$$

The first column being twice the second column, the determinant on the left vanishes, hence the given lines are coplanar.

Also the equation of the plane in which the two given lines lie is

$$\begin{vmatrix} x-a+d & y-a & z-a-d \\ \alpha-\delta & \alpha & \alpha+\delta \\ \beta-\gamma & \beta & \beta+\gamma \end{vmatrix}$$

$$\Rightarrow \begin{vmatrix} x+z-2a & y-a & z-a-d \\ 2\alpha & \alpha & \alpha+\delta \\ 2\beta & \beta & \beta+\gamma \end{vmatrix}$$

$$\Rightarrow \begin{vmatrix} (x+z-2a)-2(y-a) & y-a & z-a-d \\ 2\alpha-2(\alpha) & \alpha & \alpha+\delta \\ 2\beta-2(\beta) & \beta & \beta+\gamma \end{vmatrix}$$

$$\Rightarrow \begin{vmatrix} x+z-2y & y-a & z-a-d \\ 0 & \alpha & \alpha+\delta \\ 0 & \beta & \beta+\gamma \end{vmatrix}$$

$\Rightarrow$ $(x + z - 2y)\,[\alpha\,(\beta + \gamma) - \beta\,(\alpha + \delta)] = 0$ or $x + z - 2y = 0$. **Ans.**

Example 126:

Prove that lines

$$(x - 1) = \frac{1}{2}\,(y - 1) = \frac{1}{3}\,(z - 1);\ \frac{1}{2}\,(x - 4) = \frac{1}{3}\,(y - 6) = \frac{1}{3}\,(z - 7)$$

are coplanar and find the coordinates of the point of intersection.

Solution :

Any point on the line $\dfrac{x-1}{1} = \dfrac{y-1}{2} = \dfrac{z-1}{3}$ is given as

$$(1 + r,\ 1 + 2r,\ 1 + 3r) \qquad \text{...(i)}$$

Similarly any point on the line $\dfrac{x-4}{2} = \dfrac{y-6}{3} = \dfrac{z-7}{3}$ is given as

$$(4 + 2r',\ 6 + 3r',\ 7 + 3r') \qquad \text{...(ii)}$$

If the two given lines are coplanar, then they intersect and so for some values of r and r' the points (i) and (ii) must coincide

i.e., $1 + r = 4 + 2r',\ 1 + 2r = 6 + 3r',\ 1 + 3r = 7 + 3r$

Solving second and third of these, we get $r = 1$, $r' = -1$ which satisfy first equation also. Hence the lines intersect *i.e.*, given lines are coplanar.

Putting $r = 1$ in (i), the required point is

$(1 + 1,\ 1 + 2,\ 1 + 3)$ *i.e.*, $(2, 3, 4)$. **Ans.**

Example 127:

Prove that the two lines

$$x - 3 = -\frac{1}{3}\,(y + 4) = \frac{1}{3}\,(z - 5)$$

and $x - 4 = \frac{1}{3}(y - 5) = -\frac{1}{4}(z + 6)$ *intersect, and find the coordinates of the point of intersection.*

Solution :

Any point on the first line is (3 + r, – 4 – 3r, 5 + 3r) and any point on the second line is (4 + r', 5 + 3r' – 6 – 4r').

If these two lines intersect then for some values of r and r', these points must concide *i.e.,* 3 + r = 4 + r', –4 – 3r = 5 + 3r', 5 + 3r = – 6 – 4r'

All these three equations are satisfied by r = –1 and r' = –2.

Hence the given lines intersect and putting r = – 1 or r' = – 2 in the above coordinates, we have the required point of intersection as

(3 – 1, – 4 + 3, 5 – 3) or (2, – 1, 2).

Example 128:

Prove that the lines

$$\frac{x-1}{2} = \frac{y-2}{3} = \frac{z-3}{4} \text{ and } \frac{x-2}{3} = \frac{y-3}{4} = \frac{z-4}{5} \text{ are coplanar.}$$

Also find their point of intersection.

Solution :

Any point on the line $\frac{x-1}{2} = \frac{y-2}{3} = \frac{z-3}{4}$ is given by

(1 + 2r, 2 + 3r, 3 + 4r) ...(i)

Similarly any point on the line $\frac{x-2}{3} = \frac{y-3}{3} = \frac{z-4}{5}$ is given as

(2 + 3r', 3 + 4r', 4 + 5r') ...(ii)

If the two given lines are coplanar, then they intersect and so for some values of r and r', the points (i) and (ii) must coincide.

i.e., 1 + 2r = 2 + 3r'; 2 + 4r = 3 + 4r'; 3 + 4r = 4 + 5r'

Solving first and third of these we get r = –1, r' = –1.

Also these values of r and r' satisfy the second equation hence the lines intersect *i.e.,* the given lines are coplanar.

Putting r = –1 in (i), required point of intersection is (– 1, – 1, – 1).

Ans.

Example 129:

Prove that the lines

$$\frac{x-a}{a'}=\frac{y-b}{b'}=\frac{z-c}{c'}$$

and $$\frac{x-a'}{a}=\frac{y-b'}{b}=\frac{z-c'}{c}$$

intersect and find the coordinates of the point of intersection and the equation of the plane in which they lie.

Solution :

Any point on the first line is (a + a' r, b + b' r, c + c' r) and any point on the second line is (a' + ar', b' + br', c' + cr').

If these two lines intersect then for some values of r' and r these points must coincide *i.e.,* a + a'r = a' + ar'; b + b'r = b' + br'; c + c'r + cr' = c' + cr'

Evidently all these equations are satisfied by r = 1 = r'.

Hence the given lines intersect and putting r = 1 or r' = 1 in the above coordinates we have the required point of intersection as

(a + a', b + b', c + c') **Ans.**

Also the equation of the plane in which the given lines lie is

$$\begin{vmatrix} x-a & y-b & z-c \\ a & b & c \\ a' & b' & c \end{vmatrix} \qquad \begin{vmatrix} x & y & z \\ a & b & c \\ a' & b' & c' \end{vmatrix}$$

adding the second row to the first.

Example 130

Show that the lines

$$\frac{x}{\alpha}=\frac{y}{\beta}=\frac{z}{\gamma},\ \frac{x}{a\alpha}=\frac{y}{b\beta}=\frac{z}{c\gamma},\ \frac{x}{\alpha/a}=\frac{y}{\beta/b}=\frac{z}{\gamma/c}$$

are coplanar if a = b

or *b = c or c = a.*

Solution :

The given lines are coplanar if Δ = 0

$$\begin{vmatrix} \alpha & \beta & \gamma \\ a\alpha & b\beta & c\gamma \\ \alpha/a & \beta/b & \gamma/c \end{vmatrix} \qquad \begin{vmatrix} 1 & 1 & 1 \\ a & b & c \\ 1/a & 1/b & 1/c \end{vmatrix}$$

$$\Rightarrow \begin{vmatrix} a & b & c \\ a^2 & b^2 & c^2 \\ 1 & 1 & 1 \end{vmatrix}$$ multiplying 1st, 2nd and 3rd columns by a, b and c respectively.

$$\Rightarrow \begin{vmatrix} a & b-a & c-a \\ a^2 & b^2-a^2 & c^2-a^2 \\ 1 & 0 & 0 \end{vmatrix}$$ substracting first column from sec ond and third columns.

$$\Rightarrow \begin{vmatrix} b-a & c-a \\ b^2-a^2 & c^2-a^2 \end{vmatrix} \quad \text{or } (b-a)(c-a)\begin{vmatrix} 1 & 1 \\ b+a & c+a \end{vmatrix}$$

$\Rightarrow (b-a)(c-a)(c-b) = 0$ or $a = b$; $c = a$; $c = b$. **Ans.**

Example 131:

Show that the lines $x - 4 = -\dfrac{1}{2}(y + 1) = z$ *and* $4x - y + 5z - 7 = 0 = 2x - 5y - z - 3$ *are coplanar. Find the equation of the plane containing them.*

Solution :

Any plane through the second line is given as

$$(4x - y + 5z - 7) + \lambda (2x - 5y - z - 3) = 0$$

or $(4 + 2\lambda) x - (1 + 5\lambda) y + (5 - \lambda) z - (7 + 3\lambda) = 0$ ...(i)

If it is parallel to the line $\dfrac{x-4}{1} = \dfrac{y+1}{-2} = \dfrac{z}{1}$, then we have $1.(4 + 2\lambda) - 2.\{-(1 + 5\lambda)\} + 1.(5 - \lambda) = 0$ or $11\lambda + 11 = 0$ or $\lambda = -1$

Hence from (i) the equation of the plane through the second line and parallel to the first is $(4 - 2) x - (1 - 5) y + (5 + 1) z - (7 - 3) = 0$

$\Rightarrow$ $2x + 4y + 6z - 4 = 0$ or $x + 2y + 3z - 2 = 0$ ...(ii)

Also from the equation of the first line it is evident that $(4, -1, 0)$ is a point on this line. And from (ii) we find that the point $(4, -1, 0)$ lies on the plane given by (ii). Hence the given line are coplanar and the equation of the plane containing them is given by (ii).

Example 132:

Prove that the line $(1/2)(x - 9) = -(y + 4) = (z - 5)$ *and* $6x + 4y - 5z = 4$, $x - 5y + 2z = 12$ *are coplanar. Find the also their point of intersection.*

Solution :

Any plane through the second line is given as

$$(6x + 4y - 5z - 4) + \lambda\,(x - 5y + 2z - 12) = 0$$

or $(6 + \lambda)\,x + (4 - 5\lambda)\,y + (2\lambda - 5)\,z - (4 + 12\lambda) = 0$ ...(i)

If it is parallel to the first line viz. $\frac{x-9}{2} = \frac{y+4}{-1} = \frac{z-5}{1}$, then we have

$$2.\,(6 + \lambda) - 1.(4 - 5\lambda) + 1.(2\lambda - 5) = 0 \text{ or } \lambda = -1/3$$

Hence from (i) the equation of the plane through the second line and parallel to the first is given as

$$\left(6-\frac{1}{3}\right)x+\left(4+\frac{5}{3}\right)y+\left(-\frac{2}{3}-5\right)z-(4-4)=0$$

$\Rightarrow$ $17x + 17y - 17z = 0$ or $x + y - z = 0$ ...(ii)

Also from the equation of the first line it is evident that $(9, -4, 5)$ is a point on this line. And from (ii) we find that the point $(9, -4, 5)$ lies on the plane given by (ii). Hence the given lines are coplaner *i.e.,* they intersect.

Now any poin on the first line is $(9 + 2r, -4 - r, 5 + r)$ ...(iii)

As the two given lines intersect, therefore for some value of r, the point given by (iii) lies on the second line and so satisfies the equations of the two planes which constitute the second line and so we have

$$6\,(9 + 2r) + 4\,(-4 - r) - 5\,(5 + r) = 4$$

and $$(9 + 2r) - 5\,(-4 - r) + 2\,(5 + r) = 12.$$

Both of these give $r = -3$ and so from (iii) the required point of intersection is $(9 - 6, -4 + 3, 5 - 3)$ *i.e.,* $(3, -1, 2)$. **Ans.**

Example 133:

A, A'; B, B'; C, C' are points on the coordinates axes. Prove that the lines of intersection of the planes A'BC, AB'C; B'CA, BC'A and C'AB, CA'B' are coplanar.

Solution :

Let A and A' be (a, 0, 0) and (a', 0, 0); B and B' be (0, b, 0) and (0, b', 0); C and C' be (0, 0, c) and (0, 0, c').

Then the equations (intercept form) of the planes A'BC and AB'C' are given as

$$\frac{x}{a'}+\frac{y}{b}+\frac{z}{c}=1$$

and $$\frac{x}{a}+\frac{y}{b'}+\frac{z}{c'}=1$$

∴ The equation of the plane through the line of intersection of these two planes is as $\left(\frac{x}{a'}+\frac{y}{b}+\frac{z}{c}-1\right)+\lambda\left(\frac{x}{a}+\frac{y}{b'}+\frac{z}{c'}-1\right)=0$, for the some value of λ.

If we take λ = 1, the line of intersection lies in the plane is

$$\left(\frac{x}{a'}+\frac{y}{b}+\frac{z}{c}-1\right)+\left(\frac{x}{a}+\frac{y}{b'}+\frac{z}{c'}-1\right)=0$$

$$\Rightarrow\left(\frac{1}{a}+\frac{1}{a'}\right)x+\left(\frac{1}{b}+\frac{1}{b'}\right)y+\left(\frac{1}{c}+\frac{1}{c'}\right)z=2$$

The symmetry in this equation indicates that the lines of intersection of the other two pairs of planes also lie in this plane and hence the lines of intersection of the given pairs of planes are coplanar.

Example 134:

Prove that the straight lines

$\frac{x}{\alpha}=\frac{y}{\beta}=\frac{z}{\gamma}, \frac{x}{l}=\frac{y}{m}=\frac{z}{n}$ *and* $\frac{x}{a\alpha}=\frac{y}{b\beta}=\frac{z}{c\gamma}$ *will lie in one plane if*

$$\frac{l}{\alpha}(b-c)+\frac{m}{\beta}(c-a)+\frac{n}{\gamma}(a-b)=0$$

Solution :

Here we observe that all the three given lines pass through the origin O and hence if they are coplanar they must be perpendicular to some line through O. Let d.c.'s of this line through O be l_1, m_1, n_1.

Then as this line is perpendicular to the given lines. Thus we have

$$l_1\alpha + m_1\beta + n_1\gamma = 0 \quad \text{...(i)}$$

$$l_1 l + m_1 m + n_1 n = 0 \quad \text{...(ii)}$$

$$l_1 a\alpha + m_1 b\beta + n_1 c\gamma = 0 \quad \text{...(iii)}$$

Eliminating l_1, m_1 and n_1 from (i), (ii) and (iii), then we have the required condition as follow

$$\begin{vmatrix} \alpha & \beta & \gamma \\ l & m & n \\ a\alpha & b\beta & c\gamma \end{vmatrix} \qquad \begin{vmatrix} 1 & 1 & 1 \\ l/\alpha & m/\beta & n/\gamma \\ a & b & c \end{vmatrix}$$

taking α, β and γ common from first, second and third columns.

or $-(l/\alpha)(c-b)+(m/\beta)(c-a)-(n/\gamma)(b-a)=0$, expanding the det. with respect to second row.

$$\Rightarrow \quad (l/\alpha)(b - c) + (m/\beta)(c - a) + (n/\gamma)(a - b) = 0.$$

Hence the result.

Example 135:

Show that the line of intersection of the planes

7x – 4y + 7z + 16 = 0, 4x + 3y + 3y – 2z + 3 = 0 is coplanar with the line intersection of x – 3y + 4z + 6 = 0, x – y + z + 1 = 0.

Solution :

The given lines will be coplanar if

$$\begin{vmatrix} 7 & -4 & 7 & 16 \\ 4 & 3 & -2 & 3 \\ 1 & -3 & 4 & 6 \\ 1 & -1 & 1 & 1 \end{vmatrix}$$

$$\Rightarrow \begin{vmatrix} 3 & -4 & 3 & 12 \\ 7 & 3 & 1 & 6 \\ -2 & -3 & 1 & 3 \\ 0 & -1 & 0 & 1 \end{vmatrix}$$

$c_1 = c_1 + c_2$

$c_3 = c_2 + c_3, \; c_4 = c_4 + c_3$

$$\Rightarrow \begin{vmatrix} 3 & 3 & 12 \\ 7 & 1 & 6 \\ -2 & 1 & 3 \end{vmatrix}$$

$$\Rightarrow \begin{vmatrix} 9 & 3 & 3 \\ 9 & 0 & 3 \\ -2 & 1 & 3 \end{vmatrix}$$

The determinant of the left vanishes as two rows are identical and hence given lines are coplanar.

Example 136:

Show that the lines x + y + z – 3 = 0 = 2x + 3y + 4z – 5 and 4x – y + 5z – 7 = 0 = 2x – 5y – z – 3 are coplanar and find the plane in which they lie.

Solution :

Let *l*, m, n be the d.c.'s of the line x + y + z = 3, 2x + 3y + 4z = 5

Then we must have $l + m + n = 0$, $2l + 3m + 4n = 0$

Solving above equation we have

$$\Rightarrow \quad \frac{l}{4-3} = \frac{m}{2-4} = \frac{n}{3-2} \text{ or } \frac{l}{1} = \frac{m}{-2} = \frac{n}{1}$$

which gives the direction ratios of the first line as

Let $(x_1, y_1, 0)$ be any point on this line, then we have

$x_1 + y_1 = 3$ and $2x_1 + 3y_1 = 5$ **(Note)**

Solving these we get $x_1 = 4$, $y_1 = -1$

∴ Any point on the first line is (4, –1, 0), so the equation of the first line in the symmetric form can be written as follows

$$\frac{x-4}{1} = \frac{y+1}{-2} = \frac{z-0}{1} \qquad ...(i)$$

Similarly the equations of the other given line in the symmetric form can be written as

$$\frac{x_1-(11/7)}{13} = \frac{y_1-0}{7} = \frac{z_1-(1/7)}{-9} \qquad ...(ii)$$

Any point on the line (i) is $(4 + r, -2r - 1, r)$ ...(iii)

and on the line (ii) is $\left(13r'+\frac{11}{7}, 7r', -9r'+\frac{1}{7}\right)$ ...(iv)

If these two lines meet in a point, then for some values of r and r', the points given by (iii) and (iv) must be identical. Then we have

i.e., $4 + r = 13r' + (11/7)$, $-2r - 2 = 7r'$, $r = -9r' + (1/7)$

Solving last two of these we get $r = -10/11$, $r' = 9/77$ which satisfy first viz. $4 + r = 13r' + (11/7)$ also.

Hence the two given intersect *i.e.,* are coplanar.

[Putting r' = 9/77 in (iv), the point of intersection of the given lines can be found to be (34/11, 9/11, 10/11).

The equation of the plane in which these lines lie is

$$\begin{vmatrix} x-4 & y+1 & z-0 \\ 1 & -2 & 1 \\ 13 & 7 & -9 \end{vmatrix}$$

$\Rightarrow (x - 4) [18 - 7] - (y + 1) [-9 - 13] + z [7 + 26] = 0$

$\Rightarrow (x - 4) + (y + 1) (2) + z (3) = 0$ or $x + 2y + 3z = 2$. **Ans.**

Example 137:

Show that the line x + 2y – z = 3, 3x – y + 2z = 1 is coplanar with the line 2x – 2y + 3z = 2, x – y + z + 1 = 0 and find the plane in which these two lines lie.

Solution :

Do your self. **Ans.** 7x – 7y + 8z + 3 = 0

Example 138:

Prove that the lines x = ay + b = cz + d and x = αy + β = γz + δ are coplanar if (γ – c) (aβ – bα) – (α – a) (cδ – dγ) = 0 (Kanpur91)

Solution :

The equations of the given lines can be written in the symmetric form as

$$\frac{x-0}{1}=\frac{y+(b/a)}{(1/a)}=\frac{z+(d/c)}{(1/c)};\ \frac{x-0}{1}=\frac{y+(\beta/\alpha)}{(1/\alpha)}=\frac{z+(\delta/\gamma)}{(1/\gamma)}$$

These lines will be coplanar if

$$\begin{vmatrix} 0-0 & \frac{b}{a}-\frac{\beta}{\alpha} & \frac{d}{c}-\frac{\delta}{\gamma} \\ 1 & \frac{1}{a} & \frac{1}{c} \\ 1 & \frac{1}{\alpha} & \frac{1}{\gamma} \end{vmatrix}$$

$$\Rightarrow \begin{vmatrix} 0 & (b\alpha-a\beta)/a\alpha & (d\gamma-c\delta)/c\gamma \\ 0 & \frac{1}{a}-\frac{1}{\alpha} & \frac{1}{c}-\frac{1}{\gamma} \\ 1 & \frac{1}{\alpha} & \frac{1}{\beta} \end{vmatrix}$$

$$\Rightarrow \begin{vmatrix} (b\alpha-a\beta)/a\alpha & (d\gamma-c\delta)/c\gamma \\ (\alpha-a)/a\alpha & (\gamma-c)/c\gamma \end{vmatrix}$$

$$\Rightarrow \begin{vmatrix} b\alpha-a\beta & d\gamma-c\delta \\ \alpha-a & \gamma-c \end{vmatrix}$$

$\Rightarrow$ (bα – aβ) (γ – c) – (α – a) (dγ – cδ) = 0

$\Rightarrow$ (γ – c) (aβ – bα) – (α – a) (cδ – dγ) = 0. Hence proved.

Example 139:

Prove that the lines $3x - 5 = 4y - 9 = 3z$ and $x - 1 = 2y - 4 = 3z$ meet in a point and the equation of the plane in which they lie is $3x - 8y + 13 = 0$.

Solution :

The equations of the line $3x - 5 = 4y - 9 = 3z$ can be rewritten as

$$\frac{x-(5/3)}{\left(\frac{1}{3}\right)} = \frac{y-(9/4)}{\left(\frac{1}{4}\right)} = \frac{z}{\left(\frac{1}{3}\right)} \text{ or } \frac{x-(5/3)}{4} = \frac{y-(9/4)}{3} = \frac{z}{4} \quad \text{....(i)}$$

Similarly the equation of the second line can be rewritten as

$$\frac{x-1}{1} = \frac{y-2}{\left(\frac{1}{2}\right)} = \frac{z}{\left(\frac{1}{3}\right)}$$

$$\Rightarrow \quad \frac{x-1}{6} = \frac{y-2}{3} = \frac{z}{2} \quad \text{...(ii)}$$

Any point on the line (i) is given as $[(5/3) + 4r, (9/4) + 3r, 4r]$ ...(iii)

and any point on the line (ii) is $(1 + 6r, 2 + 3r, 2r')$ ...(iv)

If the two lines meet in a point, then for some values of r and r' the points given by (iii) and (iv) must be identical.

i.e., $(5/2) + 4r = 1 + 6r'$: $(9/4) + 3r = 2 + 3r'$; $4r = 2r'$.

Solving first and third of these we get $r = 1/12$, $r' = 1/6$ which satisfy the second viz. $(9/4) + 3r = 2 + 3r'$ also. Hence the two lines intersect *i.e.*, are coplanar.

[Putting $r = (1/12)$ in (iii) we can get the point of intersection as $\left(2, \frac{5}{2}, \frac{1}{3}\right)$. Also the equation of the plane through the line given by (i) and (ii) is

$$\begin{vmatrix} x-1 & y-2 & z \\ 6 & 3 & 2 \\ 4 & 3 & 4 \end{vmatrix} = 0$$

$$\Rightarrow \quad 6(x - 1) - 16(y - 2) + 6(z) = 0$$

$$\Rightarrow \quad 3x - 8y + 3z + 13 = 0.$$

Hence proved.

Example 140:

Find the equations of the line through the point (– 4, 3, 1) parallel to the plane x + 2y – z = 5 so as to intersect the line $-\frac{1}{3}(x+1)=\frac{1}{2}(y-3)$ $= -(z-2)$. *Find also the point of intersection.*

Solution :

Let P be the point (– 4, 3, 1).

Let the line through P (– 4, 3, 1) parallel to the given plane meet the given line $\frac{x+1}{-3}=\frac{y-3}{2}=\frac{z-2}{-1}=r$ (say) in Q.

Then the coordinates of Q may be taken as (–1 – 3r, 3 + 2r, 2 – r)....(i)

∴ Direction ratios of PQ are [– 4 – (– 1 – 3r), 3 – (3 + 2r), 1–(2– r)]

⇒ (3r – 3 – 2r, r – 1) ...(ii)

But PQ is parallel to the plane x + 2y – z = 5. ...(iii)

the direction ratios of whose normal are 1, 2, –1.

Since PQ is parallel to the plane (iii), so PQ is perp. to the normal to the plane (iii) and so we have (3r – 3).1 + (– 2r).2 + (r – 1) (– 1) = 0

⇒ – 2r – 2 = 0 or r = – 1

Hence from (i) the point of intersection Q is (2, 1, 3). **Ans.**

Also from (ii) the direction ratios of PQ are – 6.2.–2 or 3, – 1, 1.

∴ The equation of the line PQ are

$$\frac{x-(-4)}{3}=\frac{y-3}{-1}=\frac{z-1}{1} \text{ or } \frac{x+4}{3}=\frac{y-3}{-1}=\frac{z-1}{1}$$ **Ans.**

Example 141:

Find the equations of the straight line drawn through the origin which will intersect both the lines

$\frac{1}{2}(x-1)=\frac{1}{3}(y-2)=\frac{1}{4}(z-3)$, $\frac{1}{4}(x+2)=\frac{1}{3}(y-3)=\frac{1}{2}$ $(z-4)$.

Solution :

The equation of the plane containing the first line is given as

A (x – 1) + B (y – 2) + C (z – 3) = 0, ...(i)

where we have 2A + 3B + 4C = 0 ...(ii)

If this plane passes through the origin, then we know

$$A(-1) + B(-2) + C(-3) = 0 \quad ...(iii)$$

Solving (ii) and (iii) we obtain $\frac{A}{-1} = \frac{B}{2} = \frac{C}{-1}$

$\therefore$ From (i) the plane through the first line and the origin is given as

$$-(x-1) + 2(y-2) - (z-3) = 0 \quad \text{or } x - 2y + z = 0 \quad ...(iv)$$

Also the equation of the plane through the given second line is given by

$$a(x+2) + b(y-3) + c.(z-4) = 0 \quad ...(v)$$

where we have $a.4 + b.3 + c.2 = 0$ ...(vi)

If this plane given by (v) passes through the origin, then we have

$$a(2) + b(-3) + c(-4) = 0 \quad ...(vii)$$

Solving (vi) and (vii) we obtain $\frac{a}{-6} = \frac{b}{20} = \frac{c}{-18}$

$\therefore$ From (v), equation of the plane through the second line and origin is

$$-6(x+2) + 20(y-3) - 18(z-4) = 0 \text{ or } 3x - 10y + 9z = 0 \quad ...(viii)$$

$\therefore$ The required line is given by (iv) and (viii).

Example 142:

Find the equations of the line intersecting the lines $x - a = y = z - a$, $x + a = y = \frac{1}{2}(z + a)$ *and parallel to the line*

$$\frac{1}{2}(x - a) = y - a = \frac{1}{3}(z - 2a).$$

Solution :

Any point on the line $x - a = y = z - a = r$ (say)

is given as $P(a + r, r, a + r)$ ...(i)

and any point on the line $x + a = y = \frac{1}{2}(z + a) = r'$ (say)

is given as $P'(-a + r', r' - a + 2r')$ (say) ...(ii)

$\therefore$ The direction ratios of PP' are given by

$$\{(a + r) - (-a + r'), r - r', (a + r) - (-a - 2r')\}$$

$$\Rightarrow \quad \{r - r' + 2a, r - r', r - 2r' + 2a\} \quad ...(iii)$$

But the line PP' is parallel to the line $\frac{x-a}{2}=\frac{y-a}{1}=\frac{z-2a}{3}$

i.e., the direction ratios of PP' are 2, 1, 3.

$\therefore$ From (iii) we get $\frac{r-r'+2a}{2}=\frac{r-r'}{1}=\frac{r-2r'+2a}{3}$

which gives $r - r' + 2a = 2(r - r');\ 3(r - r') = r - 2r' + 2a$

$\Rightarrow$ $r - r' - 2a = 0;\ 2r - r' - 2a = 0$

Solving these we get $r = 0$ and $r' = -2a$.

Substituting the values of r and r' in (i) and (ii) we find that P is (a, 0, a) and P' is (– 3a, – 2a, – 5a).

$\therefore$ The required equation of PP' is $\frac{x-a}{2}=\frac{y-0}{1}=\frac{z-a}{3}$ **Ans.**

which gives $7(-r + 4r' + 7) = -5(r + 2r' - 4)$

and $2(r + 2r' - 4) = 7(-3r - 3r' - 8)$

$\Rightarrow$ $2r - 38r' - 29 = 0$ and $23r + 25r' + 48 = 0$

Solve these for *r* and *r'*. Substituting the values so obtained in (iii) we get the coordinates of the required points of intersection P and P' and then find the length of PP'.

Example 143:

A line with direction ratios 7, – 5, 2 is drawn to intersect the lines $\frac{x-7}{-1}=\frac{y+2}{1}=\frac{z-5}{3}, \frac{x-3}{2}=\frac{y-5}{4}=\frac{z+3}{-3}$.

Find the coordinates of the points of intersection and the length intercepted on it.

Solution :

Given line are $\frac{x-7}{-1}=\frac{y+2}{1}=\frac{z-5}{3}=r$ (say) ...(i)

and $\frac{x-3}{2}=\frac{y-5}{4}=\frac{z-3}{-3}=r'$ (say) ...(ii)

Any point on (i) given as P (7 – r, – 2 + r, 5 + 3r) and any point on (ii) is given as

P' (3 + 2r', 5 + 4r', – 3 – 3r'). ...(iii)

$\therefore$ The direction ratios of the line PP' are given by

$[(3 + 2r') - (7 - r), (5 + 4r') - (-2 + r), (-3 - 3r') - (5 + 3r)]$

or $(r + 2r' - 4, -r + 4r' + 7, -3r - 3r' - 8)$

But the direction ratios of PP' are given to be 7, – 5, 2 so we have

$$\frac{r+2r'-4}{7} = \frac{-r+4r'+7}{-5} = \frac{-3r-3r'-8}{2}.$$

Example 144:

Find the equation to the line drawn parallel to $\frac{1}{4}x = y = z$ *so as to meet the lines* $5x - 6 + 4y + 3 = z$ *and* $2x - 4 = 3y + 5 = z$.

Solution :

The equations of the lines intersecting the given lines are given as

$$\{(5x - 6) - (4y + 3)\} + k\,\{(4y + 3) - z\} = 0$$

and $$\{(2x - 4) - (3y + 5)\} + k'\,\{(3y + 5) - z\} = 0$$

$\Rightarrow 5x + 4(k - 1)y - kz + (3k - 9) = 0$

and $$2x + 3(k' - 1)y - k'z + (5k' - 9) = 0 \qquad ...(i)$$

If the line given by (i) is parallel to the line $x/4 = y/1 = z/1$ ...(ii)

then the line (ii) is perpendicular to the normals to each of the planes given by (i) and so we have

$4.5 + 1.4(k - 1) + 1(-k)\ 0$ or $k - 16/3$

and $4.2 + 1.3(k' - 1) + 1(-k') = 0$ or $k' = -5/2.$

Substituting these values of k and k' in (i) we have the required equations as $15x - 76y + 16z - 75 = 0$, $4x - 21y + 5z - 43 = 0$. **Ans.**

Example 145:

Find the equations to the line that intersects the lines $2x + y - 1 = 0 = x - 2y + 3z$; $3x - y + z + 2 = 0 = 6x + 5y - 2z - 3$ *and is parallel to the line* $x = \frac{1}{2}y = \frac{1}{3}z$.

Solution :

The equations of the lines intersecting the given lines are as

$$(2x + y - 1) + \lambda(x - 2y + 3z) = 0$$

and $$(3x - y + z + 2) + \mu(6x + 5y - 2z - 3) = 0$$

$\Rightarrow$ $$(2 + \lambda)x + (1 - 2\lambda)y + 3\lambda z + (3\lambda - 1) = 0$$

and $$(3 + 6\mu)x + (5\mu - 1)y + (1 - 2\mu)z + (2 - 3\mu) = 0 \quad ...(i)$$

If the line given by (i) is parallel to the line $x/1 = y/2 = z/3$ ...(ii)

the line (ii) is perpendicular to the normals to each of the planes given by (i) and so we have

$$1\,(2 + \lambda) + 2\,(1 - 2\lambda) + 3\,(3\lambda) = 0 \quad \text{or } \lambda = -2/3$$

and $$1\,(3 + 6\mu) + 2\,(5\mu - 1) + 3\,(1 - 2\mu) = 0 \text{ or } \mu = -2/5$$

Substituting these values of λ and μ in (i), the required equations are

$$4x + 7y - 6z - 9 = 0,\ 3x - 15y + 9z + 16 = 0.$$ **Ans.**

Example 146:

Find the equations to the line which intersect the lines $2x + y - 4 = 0 = y + 2z$ and $x + 3z = 4$, $2x + 5z = 8$ and passes through the point $(2, -1, 1)$.

Solution :

The equations of the required line are given by

$$(2x + y - 4) + \lambda\,(y + 2z) = 0 \quad ...(i)$$

and $$(x + 3z - 4) + \mu\,(2x + 5z - 8) = 0 \quad ...(ii)$$

If the above lines pass through $(2, -1, 1)$, then we have

$$(4 - 1 - 4) + \lambda\,(-1 + 2) = 0 \text{ and } (2 + 3 - 4) + \mu\,(4 + 5 - 8) = 0$$

i.e. $-1 + \lambda = 0$ and $1 + \mu = 0$ *i.e.,* $\lambda = 1$. $\mu = -1$

$\therefore$ From (i) and (ii), the required equations are given as

$$(2x + y - 4) + (y + 2z) = 0,\ (x + 3z - 4) - (2x + 5z - 8) = 0$$

$$\Rightarrow \quad 2x + 2y + 2z = 4,\ -x - 2z + 4 = 0$$

$$\Rightarrow \quad x + y + z = 2,\ x + 2z = 4.$$ **Ans.**

Example 147:

Find the equations of the straight line drawn from the origin to intersect the lines $3x + 2y + 4z - 5 = 0 = 2x - 3y + 4z + 1$ and $2x - 4y + z + 6 = 0 = 3x - 4y + z - 3$.

Solution :

The equation of the required line are given by

$$(3x + 2y + 4z - 5) + \lambda\,(2x - 3y + 4z + 1) = 0 \quad ...(i)$$

and $$(2x - 4y + z + 6) + \mu\,(3x - 4y + z - 3) = 0 \quad ...(ii)$$

If the above line passes through the origin $(0, 0, 0)$, then we have

$$-5 + \lambda = 0 \text{ and } 6 - 3\mu = 0 \text{ i.e., } \lambda = 5,\ \mu = 2$$

∴ From (i) and (ii) the required equations are given by

$$(3x + 2y + 4z - 5) + 5\ (2x - 3y + 4z + 1) = 0$$

and $$(2x - 4y + z + 6) + 2\ (3x - 4y + z - 3) = 0$$

⇒ $$13x - 13y + 24z = 0 \quad \text{and } 8x - 12y + 3z = 0.$$ **Ans.**

Example 148:

Show that the equation of the straight line through the origin cutting each of the lines.

$$\frac{x-x_1}{l_1} = \frac{y-y_1}{m_1} = \frac{z-z_1}{n_1}$$

and $$\frac{x-x_2}{l_2} = \frac{y-y_2}{m_2} = \frac{z-z_2}{n_2}$$

are $$\begin{vmatrix} x & y & z \\ x_1 & y_1 & z_1 \\ l_1 & m_1 & n_1 \end{vmatrix} \quad \begin{vmatrix} x & y & z \\ x_2 & y_2 & z_2 \\ l_2 & m_2 & n_2 \end{vmatrix}.$$

Solution :

Equation of the plane through the first line is given by

$$A\ (x - x_1) + B\ (y - y_1) + C\ (z - z_1) = 0, \quad \text{...(i)}$$

where we have $Al_1 + Bm_1 + Cn_1 = 0$ ...(ii)

If this plane passes through the origin then from (i) we get

$$A\ (-x_1) + B\ (-y_1) + C\ (-z_1) = 0 \text{ or } Ax_1 + By_1 + Cz_1 = 0 \quad \text{...(iii)}$$

Eliminating A, B, C from (i), (ii) and (iii) we obtain

$$\begin{vmatrix} x-x_1 & y-y_1 & z-z_1 \\ l_1 & m_1 & n_1 \\ x_1 & y_1 & z_1 \end{vmatrix}$$

⇒ $$\begin{vmatrix} x & y & z \\ x_1 & y_1 & z_1 \\ l_1 & m_1 & n_1 \end{vmatrix}$$

$$R_1 = R_1 + R_3 \quad \text{...(iv)}$$

$$R_2 \Leftrightarrow R_3$$

Similarly the equation of the plane through the origin and second line is

$$\begin{vmatrix} x & y & z \\ x_2 & y_2 & z_2 \\ l_1 & m_2 & n_2 \end{vmatrix} \quad \text{...(v)}$$

The planes (iv) and (v) taken together give the required line.

Hence proved.

Example 149:

Show that the planes $2x + 4y + 4z = 7$; $5x + y - z = 9$, $x - y - z = 6$ form a triangular prism.

Solution :

Here the rectangular array is given by

$$\left\|\begin{matrix} 2 & 4 & 2 & -7 \\ 5 & 1 & -1 & -9 \\ 1 & 1 & -1 & -6 \end{matrix}\right\|$$

$$\therefore \Delta_4 = \begin{vmatrix} 2 & 4 & 2 \\ 5 & 1 & -1 \\ 1 & -1 & -1 \end{vmatrix} = \begin{vmatrix} 2 & 6 & 4 \\ 5 & 6 & 4 \\ 1 & 0 & 0 \end{vmatrix} \text{ adding 1st column to 2nd and 3rd.}$$

$$\begin{vmatrix} 2 & 1 & 1 \\ 5 & 1 & 1 \\ 1 & 0 & 0 \end{vmatrix} \text{ two columns being identical}$$

$$\begin{vmatrix} 4 & 2 & -7 \\ 1 & -1 & -9 \\ -1 & -1 & -6 \end{vmatrix} = \begin{vmatrix} 6 & 0 & -25 \\ 1 & -1 & -9 \\ -2 & 0 & 3 \end{vmatrix}, \text{ adding twice 2nd row to 1st and subtracting 2nd row from 3rd.}$$

$$\Rightarrow \quad \Delta_1 = -\begin{vmatrix} 6 & -25 \\ -2 & 3 \end{vmatrix}, \text{ expanding w.r. to 2nd column}$$

$$= -[18 - 50] = 32 \neq 0$$

Thus we have $\Delta_4 = 0$, $\Delta_1 \neq 0$ and so the given planes form a triangular prism.

Example 150:

Prove that the planes $ny - mz = \lambda$, $lz - nx = \mu$, $mx - ny = \nu$ have a common line if and only if $l\lambda + m\mu + n\nu = 0$.

Solution :

Given planes are $\quad 0.x + ny - mz - \lambda = 0,$

$$-nx + 0.y + lz - \mu = 0$$

and $\quad mx - ny + 0 . z - \nu = 0$

$\therefore$ The rectangular array is $\left\|\begin{matrix} 0 & n & -m & -\lambda \\ -n & 0 & l & -\mu \\ m & -n & 0 & -\nu \end{matrix}\right\|$

$$\begin{vmatrix} 0 & n & -m \\ -n & 0 & l \\ m & -n & 0 \end{vmatrix} \quad \begin{vmatrix} n & -m \\ -n & 0 \end{vmatrix} \quad \begin{vmatrix} n & -m \\ 0 & l \end{vmatrix}.$$

$$= n(-mn) + m(nl) = mn(l - n) \qquad \text{...(i)}$$

$$\begin{vmatrix} n & -m & -\lambda \\ 0 & l & -\mu \\ -n & 0 & -\nu \end{vmatrix} \quad \begin{vmatrix} n & -m & -\lambda \\ 0 & l & -\mu \\ 0 & -m & -\nu-\lambda \end{vmatrix}, \text{ adding } R_1 \text{ to } R_3$$

$$= n\begin{vmatrix} l & -\mu \\ -m & -\nu-\lambda \end{vmatrix} = n(-l\nu - l\lambda - m\mu). \qquad \text{...(ii)}$$

Now if the given plane intersects in a line, then Δ_4 must be zero

i.e., $mn(l - n) = 0$ and $\Delta_1 = 0$ (or $\Delta_2 = 0$ or $\Delta_3 = 0$)

i.e., $l - n = 0$ and $l\nu + l\lambda + m\mu = 0$ $\quad\because m \neq 0, n \neq 0$

i.e., $l = n$ and $l\nu + l\lambda + m\mu = 0$

i.e., $n\nu + l\lambda + m\mu = 0$ *i.e.*, $l\lambda + m\mu + n\nu = 0$. **Hence proved.**

Example 151:

The plane $(x/a) + (y/b) + (z/c) = 1$ meets the axes in A, B and C. Prove that the planes through the axes and the internal bisector of the angles of the triangle ABC pass through the line

$$\frac{x}{a\sqrt{(b^2+c^2)}} = \frac{y}{b\sqrt{(c^2+a^2)}} = \frac{z}{c\sqrt{(a^2+b^2)}}.$$

Solution:

The coordinates of the point A, B and C are (a, 0, 0), (0, b, 0) and (0, 0, c) respectively.

The d.c.'s of the sides AB and AC of the Δ ABC are given by

$$\frac{a}{\sqrt{(a^2+b^2)}}, \frac{-b}{\sqrt{(a^2+b^2)}}, 0 \text{ and } \frac{a}{\sqrt{(a^2+c^2)}}, 0\, \frac{-c}{\sqrt{(a^2+c^2)}} \text{ respectively.}$$

Also we know that the d.r.'s of internal bisector of the two lines whose d.c.'s are l_1, m_1, n_1 and l_1, m_2, n_2 are given as

$$\frac{1}{2}(l_1 + l_2), \frac{1}{2}(m_1 + m_2), \frac{1}{2}(n_1 + n_2)$$

∴ The d.c.'s of the internal bisector of AB and AC are given by

$$\frac{1}{2}\left[\frac{a}{\sqrt{(a^2+b^2)}}-\frac{a}{\sqrt{(a^2+c^2)}}\right], \frac{1}{2}\left[-\frac{b}{\sqrt{(a^2+b^2)}}-0\right], \frac{1}{2}\left[0+\frac{c}{\sqrt{(a^2+c^2)}}\right]$$

$$\Rightarrow \quad \frac{1}{2}a\left[\frac{1}{\sqrt{(a^2+b^2)}}-\frac{1}{\sqrt{(a^2+c^2)}}\right], -\frac{1}{2}\frac{b}{\sqrt{(a^2+b^2)}}, \frac{1}{2}\frac{c}{\sqrt{(a^2+c^2)}}$$

$\Rightarrow$ l, m, n (say) ...(i)

Any plane through x-axis *i.e.*, y = 0, z = 0 is $y + \lambda z = 0$ ...(ii)

If the internal bisector of AB and AC whose d.c.'s l, m, n are given by (i) above lies on the plane (ii) then we have $l.0 + m.1 + n.\lambda = 0$ or $\lambda = -m/n$.

∴ From (ii), the equation of the plane through x-axis and the internal bisector of AB and AC is $y - (m/n)\, z = 0$ or $y/m = z/n$.

$$\Rightarrow \quad \frac{2y\sqrt{(a^2+b^2)}}{b}=\frac{2z\sqrt{(a^2+c^2)}}{c}, \text{ putting the values of m and n.}$$

$$\Rightarrow \quad \frac{y}{b\sqrt{(c^2+a^2)}}=\frac{z}{c\sqrt{(a^2+b^2)}} \qquad \text{...(iii)}$$

Similarly the equations of the other planes are given by

$$\frac{z}{c\sqrt{(a^2+b^2)}}=\frac{x}{a\sqrt{(b^2+c^2)}} \qquad \text{...(iv)}$$

and

$$\frac{x}{a\sqrt{(b^2+c^2)}}=\frac{y}{b\sqrt{(c^2+a^2)}} \qquad \text{...(v)}$$

Evidently the line of intersection of the planes (iii), (iv) and (v) is

$$\frac{x}{a\sqrt{(b^2+c^2)}}=\frac{y}{b\sqrt{(c^2+a^2)}}=\frac{z}{c\sqrt{(a^2+b^2)}}$$ **Hence proved.**

Example 152:

The plane x/a + y/b + z/c = 1 meets the axes OX, OY and OZ in A, B, C respectively. Prove that the planes through the axes perpendicular to the sides of the triangle ABC pass through ax = by = cz.

Find also the coordinates of the orthocentre of the triangle ABC.

Solution :

The coordinates of A, B and C are (a, 0, 0), (0, b, 0) and (0, 0, c) respectively.

The equations of the side BC of the Δ ABC are given by

$$\frac{x-0}{0}=\frac{y-b}{b}=\frac{z-0}{-c} \qquad \textbf{(Note)} \qquad ...(i)$$

The equation of the plane through OX *i.e.,* y = 0 = z is $y + \lambda z = 0$

If this plane is perpendicular to the line (i), then the normal to this plane must be parallel to (i), the condition for the same is

$$\frac{0}{0}=\frac{b}{1}=\frac{-c}{\lambda} \text{ or } \lambda=-\frac{c}{b}.$$

∴ The equation of the plane through OX and perpendicular to the side BC of the ΔABC is y – (c/b) z = 0 or by = cz ...(iii)

Similarly the equations of the other planes are

cz = ax ...(iv) and ax = by ...(v)

Evidently the line of intersection of the planes (iii), (iv), and (v) is

$$ax = by = cz \text{ or } \frac{x}{1/a}=\frac{y}{1/b}=\frac{z}{1/c}=r \text{ (say)} \qquad ...(vi)$$

Also the orthocentre of the triangle ABC lies on the three planes given by (iii), (iv) and (v) and it also lies on the plane of Δ ABC. **(Note)**

i.e., the orthocentre of Δ ABC lies on the line (vi) and the plane of Δ ABC viz.

x/a + y/b + z/c = 1.

i.e., the orthocentre of Δ ABC is the point of intersection of the line (vi) and the plane x/a + y/b + z/c = 1.

Now from (vi), any point on the line (vi) is (r/a, r/b, r/c).

If this point lies on the plane x/a + y/b + z/c = 1 we have

$$\frac{r}{a^2}+\frac{r}{b^2}+\frac{r}{c^2}=1 \text{ or } r = 1/(a^{-2} + b^{-2} + c^{-2}).$$

∴ The required coordinates of the orthocentre of Δ ABC are

$$\left(\frac{r}{a},\frac{r}{b},\frac{r}{c}\right), \text{ where } r=\frac{1}{a^{-2}+b^{-2}+c^{-2}}. \qquad \textbf{Ans.}$$

Example 153:

Show that the planes x + ay + (b + c) z + d = 0, x + by + (c + a) z + d = 0, x + cy + (a + b) z + d = 0 pass through one line.

Solution :

The rectangular array is given as

$$\left\|\begin{matrix} 1 & a & b-c & d \\ 1 & b & c+a & d \\ 1 & c & a+d & d \end{matrix}\right\|$$

$$\begin{vmatrix} 1 & a & b+c \\ 1 & b & c+a \\ 1 & c & a+b \end{vmatrix} = \begin{vmatrix} 1 & a & a+b+c \\ 1 & b & b+c+a \\ 1 & c & c+a+b \end{vmatrix}$$

$$\begin{vmatrix} 1 & a & 1 \\ 1 & b & 1 \\ 1 & c & 1 \end{vmatrix},$$

= 0 (c_1 and c_2 are identical)

$$\begin{vmatrix} 1 & b+c & d \\ 1 & c+a & d \\ 1 & a+b & d \end{vmatrix} \quad \begin{vmatrix} 1 & b+c & 1 \\ 1 & c+a & 1 \\ 1 & a+b & 1 \end{vmatrix}$$

Since $\Delta_4 = 0$, $\Delta_2 = 0$, therefore the given lines intersect in a line.

Example 154:

Prove that the three planes $x + y + z + 6 = 0$, $x + 2y + 2z + 6 = 0$, $x + 3y + 3z + 6 = 0$ intersect in a common line.

Solution :

The rectangular array is given by

$$\left\|\begin{matrix} 1 & 1 & 1 & 6 \\ 1 & 2 & 2 & 6 \\ 1 & 3 & 3 & 6 \end{matrix}\right\|$$

$$\begin{vmatrix} 1 & 1 & 1 \\ 1 & 2 & 2 \\ 1 & 3 & 3 \end{vmatrix}$$

$$\begin{vmatrix} 1 & 1 & 6 \\ 2 & 2 & 6 \\ 3 & 3 & 6 \end{vmatrix}$$

∴ As $\Delta_4 = 0$ and $\Delta_1 = 0$, so the given planes meet in a line.

Example 155:

For what values of λ do the planes

$$x - y + z + 1 = 0,\ \lambda x + 3y + 2z - 3 = 0,\ 3x + \lambda y + z - 2 = 0$$

(i) intersect in a point;

(ii) intersect along a line;

(iii) form a triangular prism?

Solution :

The equations of the given planes can be written as

$$x - y + z + 1 = 0$$

$$\lambda x + 3y + 2z - 3 = 0$$

$$-3x + \lambda y + z - 2 = 0$$

$\therefore$ The rectangular array is given by $\left\|\begin{matrix} 1 & -1 & 1 & 1 \\ \lambda & 3 & 2 & -3 \\ 3 & \lambda & 1 & -2 \end{matrix}\right\|$

$$\therefore \Delta_4 = \begin{vmatrix} 1 & -1 & 1 \\ \lambda & 3 & 2 \\ 3 & \lambda & 1 \end{vmatrix} \quad \begin{vmatrix} 0 & 0 & 1 \\ \lambda-2 & 5 & 2 \\ 2 & \lambda+1 & 1 \end{vmatrix}, \text{ replacing } C_1, C_2 \text{ by } C_1-C_3\ C_2+C_3$$

$= \begin{vmatrix} \lambda-2 & 5 \\ 2 & \lambda+1 \end{vmatrix}$, expanding with respect to first row. We have

$= (\lambda - 2)(\lambda + 1) - (5 \times 2) = \lambda^2 - \lambda - 12$

$= (\lambda - 4)(\lambda + 3)$ ...(i)

$$\text{Also } \Delta_1 = \begin{vmatrix} -1 & 1 & 1 \\ 3 & 2 & -3 \\ \lambda & 1 & -2 \end{vmatrix} = \begin{vmatrix} 0 & 1 & 1 \\ 0 & 2 & -3 \\ \lambda-2 & 1 & -2 \end{vmatrix}$$

$= (\lambda - 2)\begin{vmatrix} 1 & 1 \\ 2 & -3 \end{vmatrix}$, expanding with respect to first column. Then we have

$= (\lambda - 2)[-3 - 2] = -5(\lambda - 2);$ ...(ii)

$$\Delta_2 = \begin{vmatrix} 1 & 1 & 1 \\ \lambda & 2 & -3 \\ 3 & 1 & -2 \end{vmatrix} = \begin{vmatrix} 0 & 1 & 0 \\ \lambda-2 & 2 & -5 \\ 2 & 1 & -3 \end{vmatrix}, \text{ replacing } C_1, C_3 \text{ by } C_1-C_2\ C_3-C_2$$

$$= -\begin{vmatrix} \lambda-2 & -5 \\ 1 & -3 \end{vmatrix} = -[3(\lambda-2)+10] = 3\lambda - 16 \qquad \text{...(iii)}$$

Also $\Delta_3 = \begin{vmatrix} 1 & -1 & 1 \\ 1 & 3 & -3 \\ 3 & \lambda & -2 \end{vmatrix} = \begin{vmatrix} 0 & -1 & 0 \\ \lambda+3 & 3 & 0 \\ 3+\lambda & \lambda & \lambda-2 \end{vmatrix}$, replacing C_1, C_3 by $C_1 + C_2$ and $C_3 + C_2$

$$= (\lambda - 2)(\lambda + 3) \qquad \text{...(iv)}$$

(i) If the given planes intersect in a point, then $\Delta_4 \neq 0$ and so from (i) we must have $\lambda \neq 4$, $\lambda \neq -3$.

(ii) If the given planes intersect in a line, then we know that $\Delta_4 = 0$ and any one of Δ_1, Δ_2 and Δ_3 is zero.

Here from (i), (ii), (iii) and (iv) we find that if $\lambda = -3$, then $\Delta_4 = 0$ and $\Delta_3 = 0$.

Consequently for $\lambda = -3$, the given planes intersect in a line.

(iii) If the given planes form a triangular prism, then we know that $\Delta_4 = 0$ and none of Δ_1, Δ_2, Δ_3 is zero.

Here from (i), (ii), (iii) and (iv) we find that if $\lambda = 4$, then $\Delta_4 = 0$ and none of Δ_1, Δ_2, Δ_3 is zero

Consequently for $\lambda = 4$, the given planes form a triangular prism.

Example 156:

Prove that the planes $x = y \sin\psi + z \sin\phi$, $y = z \sin\theta + x \sin\psi$ and $z = x \sin\phi + y \sin\theta$ intersect in the line $\frac{x}{\cos\theta} = \frac{y}{\cos\phi} = \frac{z}{\cos\psi}$ if $\theta + \phi + \psi = \frac{1}{2}\pi$.

Solution :

The equation of the given planes can be written as follows

$$x - y\sin\psi - z\sin\phi = 0 \qquad \text{...(i)}$$

$$x\sin\psi - y + z\sin\theta = 0 \qquad \text{...(ii)}$$

and

$$x\sin\phi + y\sin\phi - z = 0 \qquad \text{...(iii)}$$

Let l, m, n be the d.c.'s of the line of intersection of the planes (i) and (ii), then as this line is perpendicular to the normal to the planes (i) and (ii), we have $l.1 - m.\sin\psi - n.\sin\phi = 0$, $l\sin\psi - m.1 + n\sin\theta = 0$

Solving these equations we obtain

$$\frac{l}{-\sin\theta\sin\psi-\sin\phi}=\frac{m}{-\sin\theta\sin\psi-\sin\phi}=\frac{n}{-1+\sin^2\psi} \quad ...(iv)$$

Now if $\theta+\phi+\psi=\frac{1}{2}\pi$, then $\phi=\frac{1}{2}\pi-(\phi+\psi)$

$$\therefore \sin\phi=\sin\left[\frac{1}{2}\pi-(\theta+\psi)\right]=\cos(\theta+\psi)$$

$$\Rightarrow \quad \sin\phi-\sin\phi\sin\psi$$

$$=\cos\theta\cos\psi-\sin\phi\sin\psi$$

$$\sin\phi+\sin\theta\sin\psi=\cos\theta\cos\psi$$

Similarly $\sin\theta+\sin\phi\sin\psi=\cos\phi\cos\psi$

Substituting these values in (iv) we have

$$\frac{l}{\cos\theta\cos\psi}=\frac{m}{\cos\phi\cos\psi}=\frac{b}{\cos^2\psi} \text{ or } \frac{l}{\cos\theta}=\frac{m}{\cos\phi}=\frac{n}{\cos\psi} \quad ...(v)$$

Also the planes (i) and (ii) pass through the origin, so the equations of the line of intersection of the planes (i) and (ii) is

$$\frac{x}{l}=\frac{y}{m}=\frac{z}{n} \text{ or } \frac{x}{\cos\theta}=\frac{y}{\cos\phi}=\frac{z}{\cos\psi} \quad ...(vi)$$

If this line (vi) lies on the plane (iii), then the point (0, 0, 0) on this line must be on the plane (iii) which is true, also the normal to the plane (iii) must be at right angles to the line (vi), the condition for the same is

$$\sin\phi\cos\theta+\sin\theta\cos\phi-1.\cos\psi=0$$

$$\Rightarrow \sin(\theta+\phi)-\cos\psi=0 \text{ or } \sin\left[\frac{1}{2}-\psi\right]-\cos\psi=0$$

$$\therefore \quad \theta+\phi+\psi=\frac{1}{2}\pi$$

$\Rightarrow \cos\psi-\cos\psi=0$, which being true the line (vi) lies on the plane (iii).

Example 157:

Show that the planes $cy-bz=l$, $az-cx=m$, $bx-ay=n$ intersect in a line if $al+bm+cn=0$, and the direction ratios of the line are a, b, c.

Solution :

The equations of given planes can be written follows

$$0.x+c.y-b.z-l=0$$

$$c.x+0.y-a.z+m=0$$

$b.x - a.y + 0.z - n = 0$.

$\therefore$ The rectangular array is $\begin{vmatrix} 0 & c & -b & -l \\ c & 0 & -a & m \\ b & -a & 0 & -n \end{vmatrix}$

$$\therefore \Delta_4 = \begin{vmatrix} 0 & c & -b \\ c & 0 & -a \\ b & -a & 0 \end{vmatrix} = -c\begin{vmatrix} c & -a \\ b & 0 \end{vmatrix} - b\begin{vmatrix} c & 0 \\ b & -a \end{vmatrix} = 0$$

and $$\Delta_1 = \begin{vmatrix} c & -b & -l \\ 0 & -a & m \\ a & 0 & -n \end{vmatrix} = c\begin{vmatrix} -a & m \\ 0 & -n \end{vmatrix} - a\begin{vmatrix} -b & -l \\ -a & m \end{vmatrix}$$

$$= acn - (-bm - al) = a(al + bm + cn)$$

If the given planes meet in a line then we must have $\Delta_4 = 0$ and $\Delta_1 = 0$

$\therefore$ If $\Delta_1 = 0$ we get $a(al + bm + cn) = 0$

$\Rightarrow \quad al + bm + cn = 0, \quad \therefore \quad a \neq 0.$ **Hence proved.**

Also if l_1, m_1, n_1 be the d.c.'s of the line of intersection of the given planes then as this line is perpendicular to the normal to the given planes, so we have $0.l_1 + c.m_1 - b.n_1 = 0$, $c.l_1 + 0.m_1 - a.n_1 = 0$

Solving these taking two at a time, we have

$$\frac{l_1}{-ac} = \frac{m_1}{-bc} = \frac{n_1}{-c^2}, \frac{l}{-a^2} = \frac{m_1}{-ab} = \frac{n_1}{-ac} \text{ and } \frac{l_1}{-ab} = \frac{m_1}{-b^2} = \frac{n_1}{-bc}.$$

All of these reduce to $\frac{l_1}{a} = \frac{m_1}{b} = \frac{n_1}{c}$, which shows that the direction ratios of the line of intersection of the given planes are a, b, c.

Example 158:

Prove that the planes $x = cy + bz$, $y = az + cx$, $z = bx + ay$ pass through one line if $a^2 + b^2 + c^2 + 2abc = 1$, and find its equations.

Solution :

The given planes are $x - cy - bz = 0$, $cx - y + az = 0$, $bx + ay - z = 0$

$\therefore$ The rectangular array is $\begin{vmatrix} 1 & -c & -b & 0 \\ c & -1 & a & 0 \\ b & a & -1 & 0 \end{vmatrix}$ **(Note)**

$$\therefore \Delta_4 = \begin{vmatrix} 1 & -c & -b \\ c & -1 & a \\ b & a & -1 \end{vmatrix} = \begin{vmatrix} 1 & -c & -b \\ 0 & c^2 - 1 & bc + a \\ 0 & a + bc & b^2 - 1 \end{vmatrix}$$

substracting c times 1st row from 2nd and b times from third.

$$= \begin{vmatrix} c^2-1 & bc+a \\ a+bc & b^2-1 \end{vmatrix}$$, expanding with respect to first column we have

$= (c^2 - 1)(b^2 - 1) - (bc + a)(bc + a)$

$= (c^2b^2 - c^2 - b^2 + 1) - (b^2c^2 + a^2 + 2abc)$

$= 1 - a^2 - b^2 - c^2 - 2abc$

and $\Delta_1 = \begin{vmatrix} -c & -b & 0 \\ -1 & a & 0 \\ a & -a & 0 \end{vmatrix} = 0.$

Now if the given planes intersect in a line then Δ_4 must be zero

i.e., $1 - a^2 - b^2 - c^2 - 2abc = 0$ *i.e.*, $a^2 + b^2 + c^2 + 2abc = 1.$

Hence proved.

If l, m, n be the d.c.'s of this line then this line being perpendicular to normals to the given three planes we have $1.l - c.m - b.n = 0$, $c.l - 1.m + a.n = 0$

and $b.l + a.m - 1.n = 0$

Solving the above three equations in pairs, we obtain

$$\frac{l}{-ac-b} = \frac{m}{-bc-a} = \frac{n}{-1+c^2}$$

$$\frac{l}{(1-a^2)} = \frac{n}{ab+c} = \frac{n}{ca+b} \qquad ...(ii)$$

and $$\frac{l}{c+ab} = \frac{n}{-b^2+1} = \frac{b}{a+bc} \qquad ...(iii)$$

From (ii) and (iii) we obtain

$$\frac{l^2}{(ab+c)(1-a^2)} = \frac{m^2}{(1-b^2)(ab+c)} \qquad \textbf{(Note)}$$

$\Rightarrow \quad l^2/(1 - a^2) = m^2/(1 - b^2)$

Similarly from (i) and (ii) we find that $l^2/(1 - a^2) = n^2/(1 - c^2)$

$\therefore$ We have $l^2/(1 - a^2) = m^2/(1 - b^2) = n^2/(1 - c^2)$

$$\Rightarrow \quad \frac{l}{\sqrt{(1-a^2)}} = \frac{n}{\sqrt{(1-b^2)}} = \frac{n}{\sqrt{(1-c^2)}}$$

Also the three given planes through the origin, so the equations of their line of intersection are given by

$$\frac{x}{\sqrt{(1-a^2)}} = \frac{y}{\sqrt{(1-b^2)}} = \frac{z}{\sqrt{(1-c^2)}}$$ **Ans.**

EXERCISES

1. Show that the lines $(x - 4) = -\frac{1}{4}(y + 3) = \frac{1}{7}(z + 1)$ and $\frac{1}{2}(x - 1) = -\frac{1}{3}(y + 1) = \frac{1}{8}(z + 10)$ intersect and find the coordinates of the point of intersection.
2. Show that the lines $x + 2y - 5z + 9 = 0 = 3x - y + 2z - 5$ and $2x + 3y - z - 3 = 0 = 4x - 5y + z + 3$ are coplanar.
3. Show that the two straight lines $x = mx + a$, $y = nz + b$ and $x = m'z + a', y = n'z + b'$ would intersect only if

   $$(a - a')(n - n') = (b - b')(m - m')$$
4. Find the equations of the perpendicular from the origin to the line $x + 4y + 4z - 27 = 0$, $2x + 2y + 3z - 21 = 0$. Also find the coordinates of the foot of the perpendicular.
5. Find the coordinates of the foot of the perpendicular drawn from the point P (5, 9, 3) to the line $\frac{1}{2}(x - 1) = \frac{1}{3}(y - 2) = \frac{1}{4}(z - 3)$.
6. Find the equations of the perpendicular from the point (1, 6, 3) to the line $x = \frac{1}{2}(y - 1) = \frac{1}{3}(z - 1)$.
7. Prove that the image of the line

   $(x - 1) = -9(y - 2) = -3(z + 3)$ in the plane $3x - 3y + 10z = 26$ is the line

   $(1/9) . (x - 4) = +(y + 1) = -(1/3)(z - 7)$.
8. Find the equation of the orthogonal projection of the line $\frac{1}{2}(x - 1) = \frac{1}{3}(y - 2) = \frac{1}{4}(z - 4)$ on the plane $x + 3y + z + 5 = 0$.
9. Find the image of the line $\frac{1}{2}(x + 1) = \frac{1}{3}(y + 2) = \frac{1}{4}(z + 3)$ in the plane $x - 2y + 3z - 4 = 0$.
10. Show that the line $\frac{1}{2}(x - 1) = \frac{1}{3}(y - 1) = \frac{1}{4}(z - 1)$ and $\frac{1}{3}(x - 5) = \frac{1}{2}(y - 7) = (z - 9)$ are coplanar and find the equation of the plane containing then

11. Prove that the two lines $2(x-1) = \frac{1}{2}(y-1) = \frac{1}{3}(z-1)$ and $\frac{1}{2}(x-2) = \frac{1}{3}(y-5) = \frac{1}{3}(z-7)$ are coplanar and find their point of intersection.

12. Prove that the lines $\frac{1}{2}(x+1) = \frac{1}{3}(y-2) = \frac{1}{4}(z-3)$ and $\frac{1}{3}(x+2) = \frac{1}{4}(y-3) = \frac{1}{5}(z-4)$ are coplanar.

13. Determine the values of k such that the following system of planes may (i) intersection in a line, (iii) form a triangular prism :

$3x + y + kz = 2,\ 2x + 3y + 4z = 3,\ x - 2y - 3z = -k.$

14. Find the condition that three planes

$$a_r x + b_r y + c_r z + d_r = 0,\ r = 1, 2, 3.$$

may intersect in a common line.

15. Examine the nature of intersection of the planes

$2x + 3y - z = 2,\ 3x + 3y + z = 4$ and $x - y + 2z = 5.$

16. Find the equations of the line which can be drawn from the point (2, – 1, 3) to intersect the lines

$$\frac{1}{2}(x+1) = \frac{1}{3}(y-2) = (z-3)$$

and $\frac{1}{4}(x-4) = \frac{1}{2}y = \frac{1}{3}(z+3).$

17. Find the equations of the line through the point (3, 1, 2) intersecting the line $x + 3 = y + 1 = 2(z-2)$ and parallel to the plane $4x + y + 5z = 0.$

18. Find the equations of the line through the point (1, 2, 3) and intersecting the lines.

$x + 2y - 3z = 0 = 3x - 2y - z$ and $x - y - 5 = 0 = z - x + 5.$

19. Find the equations to the line in intersecting the lines $x - 1 = y = z - 1$, $2x + 2 = 2y = z + 1$ and parallel to the line

$\frac{1}{2}(x-1) = (y-1) = \frac{1}{3}(z-2).$

20. Find the equations to the straight line drawn through the origin which will intersect both the lines.

$$(x-1) = \frac{1}{4}(y+3) = \frac{1}{3}(z-5);\ \frac{1}{2}(x-4) = \frac{1}{2}(y+3) = \frac{1}{4}(z-14).$$

21. Find the co-ordinates of the point where the joining the points (2, –3, 1) and (3, – 4, – 5) cuts the plane $2x + y + z = 7$.
22. Find the distance of the points (3, – 4, 5) from the plane $2x + 3y + 6z = 19$ measured along a straight line with direction cosines proportional to 2, 1, – 2.
23. Find the distance of the point (1, 2, 3) from the plane $x + y + z = 11$ measured parallel to the line $x + 1 = -\frac{1}{2}(y - 12) = \frac{1}{2}(z - 7)$.
24. Find the distance of the point (1, 2, 3) from the line
$$(x - 2) = \frac{1}{2}(y - 3) = \frac{1}{3}(z - 4).$$
25. Find the equations of the perpendicular drawn from the point (5, 9, 3) to the line $\frac{1}{2}(x - 1) = \frac{1}{3}(y - 2) = \frac{1}{4}(z - 3)$.
26. Find the distance of the point (– 1, – 5, – 10) from the point of intersection of the line $\frac{1}{3}(x - 2) = \frac{1}{4}(y + 1) = (1/12)(z - 2)$ and the plane $x - y + z = 5$.
27. Find the equation of the plane through (1, 2, 3) perpendicular to the line of intersection of the planes $x + 2y + 3z = 2$, $3x + 2y + 4z = 0$.
28. Find the equation of the plane through (3, 1, – 1) perpendicular to the line of intersection of the planes $3x + 4y + 7z + 4 = 0$ and $x - y + 2z + 3 = 0$.
29. Find the equation of the plane determined by the parallel lines $\frac{1}{3}(x + 1) = \frac{1}{2}(y - 2) = z$ and $\frac{1}{3}(x - 3) = \frac{1}{2}(y + 4) = z - 1$.
30. Find the equation of the plane containing the line $\frac{1}{3}(x + 3) = \frac{1}{4}(y - 1) = -\frac{1}{2}(z - 2)$ and the point (8, 2, 4).
31. Find the equation to the plane containing the line $-\frac{1}{3}(x + 1) = \frac{1}{2}(y - 3) = (z + 7)$ and the point (0, 7, – 7) and show that the line $x = -\frac{1}{3}(y - 7) = (z + 7)$ also lies in the same plane.+ z =0
32. Show that the line $\frac{1}{2}(x + 1) = -\frac{1}{3}(y - 1) = \frac{1}{4}(z + 5)$ lies on the plane $x + 2y + z + 4 = 0$.

33. Obtain the equation of the plane through the line $3x - 4y + 5z = 10$, $2x + 2y - 2z = 4$ and parallel to the line $x = 2y = 3z$.

34. Find the equation of the plane passing through the intersection of the planes $6x + 2y + z = 1$ and $2x + y + 3z = 4$ and which is parallel to the line $6x = 3y = 2z$.

35. Find the coordinates of the point of intersection of the line $x + 1 = \frac{1}{3}(y\ 3) = -\frac{1}{2}(z - 2)$ with the plane $3x + 4y + 5z = 20$

36. Find the coordinates of the point of intersection of the line $x + 1 = \frac{1}{3}(y + 3) = \frac{1}{2}(z - 2)$ with the plane $3x + 4y + 5z = 20$.

37. Find the equations to the line through the point (1, 2, 3) parallel to the line $x - y + 2z = 5$, $3x + y + z = 6$.

38. Find the equations of a straight line through the point (3, 1, – 6) and parallel to each of the planes.
$x + y + 2z - 4 = 0$ and $2x - 3y + z + 5 = 0$.

39. Prove that the line of intersection of the planes $4x + 4y - 5z = 12$, $8x + 12y - 13z = 32$ can be written as $\frac{1}{2}(x - 1) = \frac{1}{3}(y - 2) = \frac{1}{4}z$.

40. Find the distance of the point (– 1, 2, 3) from the line through (3, 4, 5) whose direction cosines are proportional to 2, – 3, 6.

41. Find the equations of the line drawn through the point (3, – 4, 1) parallel to the plane $2x + y - z = 5$ so as to cut the line $\frac{x-3}{2} = \frac{y+1}{-5} = \frac{z-2}{-1}$. Find also the co-ordinates of the point of intersection and the equation of the plane through given line and the required line.

42. Find the locus of the line which moves parallel to the yz-plane and meets the curves $x^2 + y^2 = a^2$, $z = 0$; $x^2 = az$, $y = 0$.

43. Find the locus of a point which is equidistant from two given lines $y = mx$, $z = c$ and $y = -mx$, $z = -c$.

44. Prove that the lines, which meet the lines $y = mx$, $z = c$; $y = -mx$, $z = -c$ and the hyperbola $xy = c^2$, $z = 0$ generates the surface
$$(cmx - yz)(mxz - cy) + m(c^2 - z^2) = 0.$$

45. Prove that by proper choice of axes, the equations of two skew lines can be put in the form $y = mx$, $z = c$ and $y = -mx$, $z = -c$

$$\Rightarrow \quad \frac{x}{1} = \frac{y}{m} = \frac{z-c}{0} \text{ and } \frac{x}{1} = \frac{y}{-m} = \frac{z+c}{0}.$$

46. Find the locus of the line which moves parallel to the zx-plane and meets the curves $xy = c^2$, $z = 0$; $y^2 = 4cz$. $x = 0$.

47. Find the length of S.D. length the lines

$$\frac{x-3}{1} = \frac{y-5}{-2} = \frac{z-7}{7}, \frac{x+1}{7} = \frac{y-1}{-6} = \frac{z+1}{1}$$

Also find the equations and the points, where it intersects the lines.

48. Prove that the locus of a line which intersects the three lines $y - z = 1$, $x = 0$; $z - x = 1$, $y = 0$; $x - y = 1$, $z = 0$ is

$$x^2 + y^2 + z^2 - 2yz - 2zx - 8xy = 1.$$

49. Find the length and equation of the S.D. between $x - 2y + z = 0 = x + y + z$ and $6x + 8y + 3z - 13 = 0 = x + 2y + z - 3$.

50. Find the equation, foot and length of the perpendicular from (2, 3, 5) to the line $(x - 15)/3 = (y - 29)(/8 = (z - 5)/(-3)$.

51. Find the distance of (– 2, 1, 5) from the line through (5, 7, 3) whose direction cosines are proportional to 2, – 3, 6.

52. A line through the origin makes angles α, β, γ with its projection on the co-ordinate planes, which are rectangular. The distance of any point (x, y, z) from the line and its projections are d, a, b, c respectively. Prove that

$$d^2 = (a^2 - x^2) \cos^2 a + (b^2 - y^2) \cos^2 b + (c^2 - z^2) \cos^2 \gamma.$$

53. Find the image of the straight line $x = 1 + t$, $y = 2 + 3t$, $z = 3 + 4t$ in the planes $z = 0$.

54. Find the image in the plane $x - y + 3z + 2 = 0$ of the line $x = 2 + t$, $y = 3 - 4t$, $z = 4 - 2t$.

55. Show that the lines $3x - 2y + z + 5 = 0 = 2x + 3y + 4z - 4$ and $\frac{x+4}{3} = \frac{y+6}{5} = \frac{z-1}{-2}$ are coplanar. Find also the coordinates of their point of intersection and the equation of the plane in which they lie.

56. Find the distance of the point (1, – 2, 3) from the plane $x - y + z = 5$ measured parallel to the line $x/2 = y/3 = -(z - 3)/4$. Find also the coordinates of the foot of the perpendicular.

57. Find the equation of a right circular cone whose vertex is P (2, –3, 5) and axis the line PQ, which is equally inclined to the axis and which passes through the point A (1, – 2, 5).

58. Find the perpendicular distance of the point (2, 4, – 1) from the line $x + 5 = \frac{1}{4}(y + 3) = -\frac{1}{9}(z - 6)$. Also find its equations.

59. Find the foot of the perpendicular from the origin to the plane 2x + 3y – 4z + 1 = 0. Also find the image of the origin in the plane.

60. Show that the distance of the point (α, β, γ) from the line $\frac{x - x_1}{l} = \frac{y - y_1}{m} = \frac{z - z_1}{n}$ measured parallel to the plane ax + by + cz + d = 0 is given by

$$d^2 = \frac{(a^2 + b^2 + c^2)\,\Sigma\{m(z_1 - \gamma) - n(y_1 - \beta)\}^2 - [\Sigma (x - a)(bn - cm)]^2}{(al + bm + cn)^2}$$

61. Find the equations of the line through (α, β, γ) at right angles to the lines

$$\frac{1}{2}(x - 1) = \frac{1}{3}(y - 2) = \frac{1}{4}(z - 4)$$

and $\frac{1}{3}(x - 2) = \frac{1}{4}(y - 4) = \frac{1}{5}(z - 5)$.

62. Find the coordinates of the image of the point (1, 3, 4) in the plane 2x – y + z + 3 = 0.

63. Find the image of the point (1, 3, 4) in the plane 2x – y + z = 0

64. Show that the lines $\frac{1}{2}(x + 3) = \frac{1}{3}(y + 5) = -\frac{1}{3}(z - 7)$ and $\frac{1}{4}(x + 1) = \frac{1}{5}(y + 1) = -(z + 1)$ are coplanar. Find the equation of the plane containing them.

65. Show that the lines $\frac{1}{4}(x - y) = \frac{1}{7}(y - 7) = -\frac{1}{5}(z + 3)$ and $\frac{1}{7}(x - 8) = (y - 4) = \frac{1}{3}(z - 5)$ are coplanar. Find their common point and equation of the plane in which they lie.

66. Prove that the lines $x = \frac{1}{2}(y - 2) = \frac{1}{3}(z + 3)$ and $\frac{1}{2}(x - 2) = \frac{1}{3}(y - 6) = \frac{1}{4}(z - 3)$ are coplanar and find the equation of the plane in which they lie and the point of intersection.

67. Prove that the lines $\frac{1}{3}(x-2) = \frac{1}{4}(y-3) = \frac{1}{5}(z-4)$ and $2x - 3x + z = x + y + 2z + 4$ are coplanar. Also find their point of intersection.

68. Show that the lines $\frac{1}{3}(x+5) = (y+4) = -\frac{1}{2}(z-7)$ and $3x + 2y + z - 2 = 0 = x - 3y + 2z - 13$ are coplanar and find the equation to the plane in which they lie.

69. Prove that the line $\frac{1}{2}(x+1) = \frac{1}{3}(y+2) = \frac{1}{4}(z+3)$ and $\frac{1}{3}(x-3) = \frac{1}{4}(y-2) = \frac{1}{5}(z-1)$ are coplanar and determine the plane containing them.

70. Find the equation, foot and length of the perpendicular from (2, 3, 5) to the line $(x - 15)/3 = (y - 29)(/8 = (z - 5)/(-3)$.

71. Find the angle between the lines $x - 2y + z = 0$, $x + y - z = 3$ and $x + 2y + z = 5$, $8x + 12y + 5z = 0$.

72. Let λ be the shortest distance between the lines $by + cz = 1$, $x = 0$ and $ax - cz = 1$, $y = 0$ and let μ be the length of a diagonal of the cuboid whose three concurrent edges are of lengths a, b, c respectively, then show that $\lambda\mu = 2$.

73. Prove that the equation of the plane through the line $x = x_0 + lt$,

74. Find the equations of the line $3x - 4y + 5 = 0$, $2x + 3y - 5z - 8 = 0$ in the symmetric form.

75. Find the equations of the line through (2, 3, 5) and parallel to the line given by $x + 2y - 2z = 7$, $6x + 8y - 9z = 1$.

76. Prove that the lines $2x + 3y - 4z = 0$, $3x - 4y + z = 7$ and $5x - y - 3z + 12 = 0$, $x - 2y + 5z - 6 = 0$ are parallel.

77. Find the angle between the lines $3x + 2y + z - 5 = 0 = x + y - 2z - 3$; $2x - y - z - 16 = 0 = 7z + 10y - 8z - 15$.

78. Find the condition that three planes

$$a_r x + b_r y + c_r z + d_r = 0,\ r = 1, 2, 3.$$

may intersect in a common line.

79. Examine the nature of intersection of the planes

$$2x + 3y - z = 2,\ 3x + 3y + z = 4 \text{ and } x - y + 2z = 5.$$

80. Find the equations of the line which can be drawn from the point (2, – 1, 3) to intersect the lines

$\frac{1}{2}(x+1) = \frac{1}{3}(y-2) = (z-3)$ and $\frac{1}{4}(x-4) = \frac{1}{2}y = \frac{1}{3}(z+3)$.

81. Find the equations of the line through the point (3, 1, 2) intersecting the line $x + 3 = y + 1 = 2(z - 2)$ and parallel to the plane $4x + y + 5z = 0$.

82. Find the equations of the line through the point (1, 2, 3) and intersecting the lines.

$x + 2y - 3z = 0 = 3x - 2y - z$ and $x - y - 5 = 0 = z - x + 5$.

83. Find the coordinates of the image of the point (a, b, c) with respect to the coordinate planes.

84. Find the distance from the point (3, 4, 5) to the point where the line $(x-3) = \frac{1}{2}(y-4) = \frac{1}{2}(z-5)$ meets the planes $x + y + z = 2$.

85. Find the equation of the plane containing the line $\frac{1}{2}(x+2) = \frac{1}{3}(y+3) = -\frac{1}{2}(z-4)$ and passing through the point (0, 6, 0).

86. Find the co-ordinates of the point of intersection of the line $(x-1) = \frac{1}{3}(y+3) = -\frac{1}{2}(z-4)$ with plane $3x + 4y + 5z - 6 = 0$.

87. Find the co-ordinates of the point where the straight line $\frac{1}{2}(x-1) = -(y+1) = \frac{1}{2}z$ intercepts the plane $3x + 2y - z = 5$.

88. Show that the following lines are parallel :

$$2x + 3y - 4z = 0 = 3x - 4y + z + 1$$

$$5x - y - 3z + 12 = 0 = x - 7y + 5z - 6$$

89. Find the equations of the line through (1, 2, –1) perpendicular to $3x - 5y + 4z = 5$ and deduce the length of the perpendicular from (1, 2, – 1) upon the plane and also the coordinates of the foot of the perpendicular.

90. Are the two lines

$$\frac{x-2}{3} = \frac{y-3}{2} = \frac{z+4}{4} \text{ and } \frac{x+1}{5} = \frac{y-2}{-6} = \frac{z+3}{2}$$

perpendicular to each other?

91. Find the angle between the lines $x - 2y + z = 0$, $x + y - z = 3$ and $x + 2y + z = 5$, $8x + 12y + 5z = 0$.

92. Find the equations of the line through (2, 3, 5) and parallel to the line given by $x + 2y - 2z = 7$, $6x + 8y - 9z = 1$.

93. Show that the condition for the lines $x = az + b$, $y = cz + d$, and $x = a_1z + b_1$, $y = c_1z + d_1$ to be perpendicular is $aa_1 + cc_1 = -1$.

94. Write the equations of the line parallel to x-axis in symmetrical form.

95. Find the equation of the plane through the points (1, 0, – 1), (3, 2, 2) and parallel to the line $(x - 1) = \frac{1}{2}(1 - y) = (z - 2)$.

96. Find the image of the point (2, – 1, 3) in the plane $3x - 2y + z = 9$.

97. Find the image of the point (2, 1, 3) in the plane $x + y - z + 2 = 0$.

98. Find the image of the point (– 2, 1, 3) in the plane

    $x + y - 2z + 1 = 0$.

99. Find the equation of the plane through the point (1, 2, 3) and perpendicular to the line $x + 2y + z = 2x - y + z - 1$.

100. Prove that the two lines in which the plane $3x - 7y - 5z = 1$ and $5x - 13y + 3z + 2 = 0$ cut the plane $8x - 11y + 2z = 0$ include a right angle.

101. Find the equations of the projection of the line

     $\frac{1}{2}(x - 1) = -(y + 1) = \frac{1}{4}(z - 3)$ on the plane $x + 2y + z = 6$.

102. A line with direction cosines proportional to (2, 1, 1) meet each of the lines given by the equations $x = y + a = z$; $x + a = 2y = 2z$.

     Find the coordinates of each of the points of intersection.

103. Examine the nature of intersection of the planes

     $x + 2y + 3z = 6$, $3x + 4y + 5z = 2$ and $5x + 4y + 3z + 18 = 0$.

104. Show that the planes $5x + 2y - 4z + 2 = 0$, $4x - 2y - 5z - 2 = 0$ and $2x + 8y - 2z + 1 = 0$ form a prism.

105. Interpret the following types of solutions of the set of simultaneous equations $a_1x + b_1y + c_1z + d_1 = 0$, $a_2x + b_2y + c_2z + d_2 = 0$, $a_3x + b_3y + c_3z + d_3 = 0$.

     (i) unique solution,

     (ii) no solution, that is the equations are inconsistent,

     (iii) infinitely many solutions.

106. If the lines $\frac{1}{4}(x - 5) = (1/6)(y - 3) = \frac{1}{4}(z - 7)$ and $\frac{1}{2}(x + 1) = (1/5)(y + 4) = (1/10)(z - 9)$ are in a plane then find the equation

of the plane containing them. If the lines are not in a plane then find the shortest distance between them.

107. Find the length of the S.D. between the lines

$$\frac{x-3}{1}=\frac{y-5}{-2}=\frac{z-7}{1}, \frac{y+1}{-6}=\frac{z+1}{1}$$

Find also its equations.

108. Find the points on the lines

$$\frac{1}{3}(x-6)=-(y-7)=z-4$$

and $\quad -\frac{1}{3}x=\frac{1}{2}x=\frac{1}{2}(y+9)=\frac{1}{4}(z-2)$

which are nearest to each other. Hence find the S.D. between the lines and also its equations.

109. If OA, OB, OC are three mutually perpendicular lines each of length a, show that the S.D. between OA and BC is $\frac{1}{3}a\sqrt{2}$ and its bisects BC.

110. Find the length and the equations of S.D between the lines $5x-y-z=x-2y+z+3$ and $7x-4y-2z=0=x-y+z-3$.

111. Find the magnitude and the equations of the line of S.D. between the lines $\frac{x-8}{3}=\frac{y+9}{-16}=\frac{z-10}{7}$ and $\frac{x-6}{3}=\frac{y-5}{8}=\frac{z-20}{-5}$

112. Find the point of intersection to the lines

$$\frac{1}{3}(x+1)=-\frac{1}{2}(y-3)=-(z+2); (x-3)=\frac{1}{2}(y-3)=\frac{1}{3}z.$$

Also find the equation of the plane passing through two lines.

113. Prove that the lines $\frac{1}{3}(x+1)=\frac{1}{5}(y+3)=-\frac{1}{7}(z-5)$ and $x-2=\frac{1}{3}(y-4)=\frac{1}{5}(z-6)$ intersect. Find the point of intersection and the plane in which they lie.

114. Find the equation of the plane which is perpendicular to the plane $5x+3y+6z+8=0$ and which contains the line of intersection of the planes $x+2y+3z-4=0$ and $2x+y-z+5=0$.

115. Find the S.D. between the lines

$$\frac{x-3}{1}=\frac{y-5}{-2}=\frac{z-7}{1} \text{ and } \frac{x+1}{7}=\frac{y+6}{-1}=\frac{z+1}{1}.$$

116. Show that if the equation $x + y + z = 0$ and $ayz + bzx + cxy = 0$ represent two lines and $abc = 0$, $a + b + c = 0$, then the lines are perpendicular to each other.

117. Find the equations of the line which intersects the line $x = 18 + 4t$, $y = 3t$, $z = 10 + t$ and $x = -11 + 4t$, $y = -10 + 3t$, $z = -7t$ perpendicularly.

118. Prove that non-parallel lines $x = \alpha + tl$, $y = \beta + tm$, $z = \gamma + tm$ and $x = \alpha' + t'l'$, $y = \beta' + t'm'$, $z' = \gamma' + l'n'$ are coplanar if

$$\begin{vmatrix} \alpha-\alpha' & \beta-\beta' & \gamma-\gamma' \\ l' & m' & n' \\ l & m & n \end{vmatrix} = 0.$$

119. A straight line with direction numbers $< 1, -1, 1 >$ is drawn to intersect the lines $x = 5 + 3t$, $y = 1 + t$, $z = -3 - 5t$ and $x = 4 + 2t$, $z = 7 - t$ $y = 1 - 5t$. Find the point of intersection and the length intercepted on it. Also find its equations.

120. Find the length of the perpendicular from the point (1, 2, 3) on the straight line $x = 7 + 3t$, $y = 6 + 2t$, $z = 7 - 2t$.

121. Show that the planes $x + y - z = 2$, $2x - y - z + 2 = 0$, $x - 5y + z + 4 = 0$ form a triangular prism and calculate the breadth of each face of the prism.